Sensory Analysis and Consumer Research in New Product Development

Sensory Analysis and Consumer Research in New Product Development

Editors

Claudia Ruiz-Capillas
Ana Herrero Herranz

MDPI • Basel • Beijing • Wuhan • Barcelona • Belgrade • Manchester • Tokyo • Cluj • Tianjin

Editors
Claudia Ruiz-Capillas
Instituto de Ciencia y Tecnología
de los Alimentos y Nutrición
(ICTAN-CSIC)
Consejo Superior de
Investigaciones Científicas
(CSIC)
Madrid
Spain

Ana Herrero Herranz
Instituto de Ciencia y Tecnología
de los Alimentos y Nutrición
(ICTAN-CSIC)
Consejo Superior de
Investigaciones Científicas
(CSIC)
Madrid
Spain

Editorial Office
MDPI
St. Alban-Anlage 66
4052 Basel, Switzerland

This is a reprint of articles from the Special Issue published online in the open access journal *Foods* (ISSN 2304-8158) (available at: www.mdpi.com/journal/foods/special_issues/sensory_consumer).

For citation purposes, cite each article independently as indicated on the article page online and as indicated below:

LastName, A.A.; LastName, B.B.; LastName, C.C. Article Title. *Journal Name* **Year**, *Volume Number*, Page Range.

ISBN 978-3-0365-1426-0 (Hbk)
ISBN 978-3-0365-1425-3 (PDF)

© 2021 by the authors. Articles in this book are Open Access and distributed under the Creative Commons Attribution (CC BY) license, which allows users to download, copy and build upon published articles, as long as the author and publisher are properly credited, which ensures maximum dissemination and a wider impact of our publications.

The book as a whole is distributed by MDPI under the terms and conditions of the Creative Commons license CC BY-NC-ND.

Contents

About the Editors vii

Preface to "Sensory Analysis and Consumer Research in New Product Development" ix

Claudia Ruiz-Capillas and Ana M. Herrero
Sensory Analysis and Consumer Research in New Product Development
Reprinted from: *Foods* **2021**, *10*, 582, doi:10.3390/foods10030582 1

Claudia Ruiz-Capillas, Ana M. Herrero, Tatiana Pintado and Gonzalo Delgado-Pando
Sensory Analysis and Consumer Research in New Meat Products Development
Reprinted from: *Foods* **2021**, *10*, 429, doi:10.3390/foods10020429 5

Artur Głuchowski, Ewa Czarniecka-Skubina, Eliza Kostyra, Grażyna Wasiak-Zys and Kacper Bylinka
Sensory Features, Liking and Emotions of Consumers towards Classical, Molecular and Note by Note Foods
Reprinted from: *Foods* **2021**, *10*, 133, doi:10.3390/foods10010133 21

Rajesh Kumar, Edgar Chambers, Delores H. Chambers and Jeehyun Lee
Generating New Snack Food Texture Ideas Using Sensory and Consumer Research Tools: A Case Study of the Japanese and South Korean Snack Food Markets
Reprinted from: *Foods* **2021**, *10*, 474, doi:10.3390/foods10020474 43

Katarzyna Świader and Magdalena Marczewska
Trends of Using Sensory Evaluation in New Product Development in the Food Industry in Countries That Belong to the EIT Regional Innovation Scheme
Reprinted from: *Foods* **2021**, *10*, 446, doi:10.3390/foods10020446 67

Simona Grasso and Sylvia Jaworska
Part Meat and Part Plant: Are Hybrid Meat Products Fad or Future?
Reprinted from: *Foods* **2020**, *9*, 1888, doi:10.3390/foods9121888 85

Dominika Guzek, Dominika Głąbska, Marta Sajdakowska and Krystyna Gutkowska
Analysis of Association between the Consumer Food Quality Perception and Acceptance of Enhanced Meat Products and Novel Packaging in a Population-Based Sample of Polish Consumers
Reprinted from: *Foods* **2020**, *9*, 1526, doi:10.3390/foods9111526 99

Krystyna Szymandera-Buszka, Katarzyna Waszkowiak, Anna Jedrusek-Golińska and Marzanna Heś
Sensory Analysis in Assessing the Possibility of Using Ethanol Extracts of Spices to Develop New Meat Products
Reprinted from: *Foods* **2020**, *9*, 209, doi:10.3390/foods9020209 111

Frederica Silva, Ana M. Duarte, Susana Mendes, Patrícia Borges, Elisabete Magalhães, Filipa R. Pinto, Sónia Barroso, Ana Neves, Vera Sequeira, Ana Rita Vieira, Maria Filomena Magalhães, Rui Rebelo, Carlos Assis, Leonel Serrano Gordo and Maria Manuel Gil
Adding Value to Bycatch Fish Species Captured in the Portuguese Coast—Development of New Food Products
Reprinted from: *Foods* **2020**, *10*, 68, doi:10.3390/foods10010068 127

Barbara Biró, Mária Anna Sipos, Anikó Kovács, Katalin Badak-Kerti, Klára Pásztor-Huszár and Attila Gere
Cricket-Enriched Oat Biscuit: Technological Analysis and Sensory Evaluation
Reprinted from: *Foods* **2020**, *9*, 1561, doi:10.3390/foods9111561 . . . . **149**

Katarzyna Świader, Anna Florowska, Zuzanna Konisiewicz and Yen-Po Chen
Functional Tea-Infused Set Yoghurt Development by Evaluation of Sensory Quality and Textural Properties
Reprinted from: *Foods* **2020**, *9*, 1848, doi:10.3390/foods9121848 . . . . **167**

Ran Tao and Sungeun Cho
Consumer-Based Sensory Characterization of Steviol Glycosides (Rebaudioside A, D, and M)
Reprinted from: *Foods* **2020**, *9*, 1026, doi:10.3390/foods9081026 . . . . **187**

Martin Kalumbi, Limbikani Matumba, Beatrice Mtimuni, Agnes Mwangwela and Aggrey P. Gama
Hydrothermally Treated Soybeans Can Enrich Maize Stiff Porridge (Africa's Main Staple) without Negating Sensory Acceptability
Reprinted from: *Foods* **2019**, *8*, 650, doi:10.3390/foods8120650 . . . . **203**

About the Editors

Claudia Ruiz-Capillas

Claudia Ruiz-Capillas has a degree in Veterinary Medicine, UCM (Spain). She is currently a senior research scientist at the Institute of Food Science Technology and Nutrition (ICTAN-CSIC). She addresses various lines of research: development of healthier meat products (fat analogues, reduction in salt and additives, etc.); meat analogues; sensory characteristics and consumer studies; chemical, physical and structural changes in components and properties of meat products; quality and safety characteristics, with special reference to biogenic amines; nitrates and nitrites; application of storage technologies (HP, protective atmospheres, etc.). She has participated in and led a number of national and international research projects. She have published over 150 peer-reviewed research articles, as well as book chapters (18), and scientific outreach articles (h-index: 36). She has also been a guest co-editor of several Special Issues of journals and the co-editor of several books.

Ana Herrero Herranz

Ana M. Herrero has a degree in Chemistry, UCM (Spain). She is currently a scientific researcher at the Institute of Food Science, Technology, and Nutrition (ICTAN-CSIC). Her research focuses on development of healthier meat products with reference to lipid contents and certain bioactive compounds (minerals, fibre, etc.). This research is based on using various reformulation strategies, studies of the relationship between physicochemical and textural properties of these products, and structural changes (using vibrational spectroscopy in situ). She has taken part in national and international projects and projects with private enterprises, acting as principal research. She has published articles in SCI journals (80), scientific outreach journals (19), and several book chapters (14). She has also given courses and seminars and taught at university level (degree and doctorate) and taken part in the direction of doctoral theses and training of students, technical personnel, etc.

Preface to "Sensory Analysis and Consumer Research in New Product Development"

Sensory analysis and consumer research are relevant tools in innovation and new product development, from design to commercialization. Sensory analysis techniques have evolved a lot in recent decades, from the more traditional techniques (discriminator and descriptive analysis or preference and hedonic tests) towards the more novel and rapid techniques (check all that apply, napping, flash profile, temporal dominance of sensations, etc.), with outstanding application at the different stages of food innovation. In addition to the application of these techniques, knowing how the consumer interacts with a food product in the different stages (purchase, ingestion, etc.) is essential to better understand and be able to measure consumer attitudes and behaviours. All this is crucial for the development of new products to be successful.

This Special Issue has collected 13 valuable scientific contributions, including 1 review, 12 original research articles and an editorial. The SI provides an interesting outlook and better understanding of sensorial analysis with different techniques and consumer research on new product development. Important practical applications have been reported for the development of different novel, functional and enhanced products (meat, fish, biscuits, yogurt, porridge, hybrid meat, molecular products, etc.), which helps increase knowledge in this field.

This SI is very useful for both present and future uses for the different players involved in this kind of product development (industry, companies, researchers, scientists, marketing, merchandising, consumers, etc.).

We would like to acknowledge the efforts of the authors of the publications in this Special Issue.

Claudia Ruiz-Capillas, Ana Herrero Herranz
Editors

Editorial

Sensory Analysis and Consumer Research in New Product Development

Claudia Ruiz-Capillas * and Ana M. Herrero

Institute of Food Science, Technology and Nutrition (ICTAN-CSIC), José Antonio Novais 10, 28040 Madrid, Spain; ana.herrero@ictan.csic.es
* Correspondence: claudia@ictan.csic.es

Abstract: Sensory analysis and consumer research are important tools in innovation and new product development (NPD), from design to commercialization. Innovation is necessary in companies to maintain their market position and attract new consumers. Sensory analysis techniques have evolved considerably in recent decades, from more traditional techniques (discriminator and descriptive analysis or preference and hedonic tests) towards more novel and faster techniques (check-all-that-apply, napping, flash profile, temporal dominance of sensations, etc.), with outstanding application at the different stages of NPD. In addition to the application of these techniques, knowing how the consumer interacts with a food product in the different stages (purchase, ingestion, etc.) is essential to better understand and be able to measure consumer attitudes, emotions and behavior. All this is crucial for the development of new products to be successful. This Special Issue has collected 11 original research articles and one review, which provide an interesting outlook and better understanding of sensorial analysis with the different techniques and consumer research on NPD. Important practical applications have been reported on the development of different novel, functional and enhanced products (meat, fish, biscuits, yogurt, porridge, hybrid meat, molecular products, etc.), which helps to increase knowledge in this field.

Keywords: sensory analysis; sensory properties; consumer research; new product development; healthier products; food quality and safety; food design; quantitative descriptive analysis (QDA); check-all-that-apply (CATA); napping; flash profile (FP); temporal dominance of sensations (TDS)

Citation: Ruiz-Capillas, C.; Herrero, A.M. Sensory Analysis and Consumer Research in New Product Development. *Foods* **2021**, *10*, 582. https://doi.org/10.3390/foods10030582

Received: 23 February 2021
Accepted: 9 March 2021
Published: 10 March 2021

Publisher's Note: MDPI stays neutral with regard to jurisdictional claims in published maps and institutional affiliations.

Copyright: © 2021 by the authors. Licensee MDPI, Basel, Switzerland. This article is an open access article distributed under the terms and conditions of the Creative Commons Attribution (CC BY) license (https://creativecommons.org/licenses/by/4.0/).

Sensory analysis examines the properties (texture, flavor, taste, appearance, smell, etc.) of a product or food through the senses (sight, smell, taste, touch and hearing) of the panelists. This type of analysis has been used for centuries for the purpose of accepting or rejecting food products. Historically, it was considered a methodology that complements technological and microbiological safety when assessing the quality of food. However, its important evolution and impact in recent decades has placed it as one of the most important methodologies for innovation and application to ensure final product acceptance by consumers. Traditional sensory techniques, such as discriminatory, descriptive evaluations, preference and hedonic tests, which are still widely used today, have evolved into newer, faster and more complete techniques: check-all-that-apply (CATA), napping (N), flash profile (FP), temporal dominance of sensations (TDS), etc., together with an important and adequate statistical analysis. All of these techniques, with their advantages and disadvantages, are very useful in the development of new foods. However, it is not only sensory characteristics that determine the acceptance or success of a new product. Factors such as social aspects, the environment, nutritional knowledge, specific diets, emotions, health, the nature of the products, packaging, etc., also have a very important influence. New food product developers should take into account the attitudes and expectations of potential consumers. Consumers describe a product's benefits by perceived intrinsic and extrinsic characteristics. For example, focus groups could be planned to identify different consumer expectations for new products. Properly measuring these factors and emotions will also have a very decisive influence on the success or failure of new product developments. A better understanding of the sensory experiences of potential consumers opens up a space and the inspiration for innovation. For all these reasons, sensory analysis, together with consumer research, is currently considered by the industry and researchers to be one of the most useful tools at the different stages of new product development, from design to commercialization, to improve the quality of products and to guarantee the success of innovation in market uptake among consumers. All these aspects are collected by Ruiz-Capillas et al. [1] in their review, which helps to better elucidate these techniques and enhance knowledge in this field, in order to facilitate the choice of the most appropriate aspect at the time of its application in the different stages of new product development, particularly regarding meat products.

Furthermore, this Special Issue (SI) also includes different sensory and consumer studies used for their application in the development of very different new products. Kalumbi et al. [2] evaluated consumer acceptability of new enriching maize-based stiff porridge with flour made from hydrothermally treated soybeans. This development could significantly contribute towards reducing the burden of energy–protein under-nutrition in populations in sub-Saharan Africa. On the other hand, Szymandera-Buszka et al. [3] employed consumer tests and sensory profiling to assess the impact of ethanol extracts of spices (lovage, marjoram, thyme, oregano, rosemary and basil) on the sensory quality of new pork meatballs and hamburgers. This work noted the usefulness of these techniques in the development of products with clean labeling, replacing synthetic preservatives with natural plant extracts. Tao and Cho [4] evaluated the sensory characteristics of Rebaudioside (Reb) A, D and M compared with sucrose, using a consumer panel, and explored the relationship between 6-n-Propylthiouracil (PROP) taster status (i.e., non-tasters, medium tasters, supertasters) and the perceived intensity of sweet and bitter tastes of the three steviol glycosides. Consumers were instructed to rate the sweetness and bitterness intensities of different solutions and a check-all-that-apply (CATA) question was used to evaluate the taste, which are considered to be more flexible and less time-consuming methodologies than traditional ones.

Other authors have analyzed the association between consumer perceptions of food quality and their acceptance of enhanced meat products and novel packaging [5]. This study was conducted using the Computer-Assisted Personal Interview (CAPI) method in a random group of 1009 respondents. The results suggested that educating consumers may improve their acceptance of product enhancement, as concerns about the addition of food preservatives may lead them to reject enhanced products. Biro et al. [6] also study sensory evaluation (CATA) together with technological parameters (color, hardness, etc.) to assess consumer acceptance of insects as food. These authors valued the acceptance or rejection of oat biscuits enriched with insect powders [*Acheta domesticus* (house cricket)]. An important part of this study was to discover how the insect content of the products affects the overall liking (OAL) and which attributes are the drivers of liking.

The potential to design natural tea-infused set yoghurt was also investigated using quantitative descriptive profile analysis and a consumer hedonic test together with technological properties (texture analysis, yield stress, physical stability and color) [7]. Three types of tea (*Camellia sinensis*)—black, green and oolong tea—as well as lemon balm (*Melissa officinalis* L.), were used to produce set yoghurt and compare this with plain yoghurt. Both types of yoghurt were also characterized by a high consumer willingness to buy. Principal component analysis (PCA) was used to analyze differences between samples and the correlation of selected variables.

Studies of different consumers from different countries have also been presented in this SI. Grasso and Jaworska [8] studied the presence of hybrid meat products in the UK market, extracted UK online consumer reviews on hybrid meat products and gathered preliminary consumer insights, utilizing the tools and techniques of corpus linguistics. These studies are of great importance since hybrid meat products could open up new business opportunities for the food industry and a greater diversity of products for consumers. Silva et al. [9] also carried out a preliminary inquiry with 155 consumers from Região de Lisboa and

Vale do Tejo (Center of Portugal) to assess fish consumption, the applicability of fish product innovation and the importance of valorizing discarded fish (blue jack mackerel, black seabream, piper gurnard, etc.). Five products (black seabream ceviche, smoked blue jack mackerel pâté, dehydrated piper gurnard, fried boarfish and comber pastries) were developed and investigated for their sensory characteristics and consumer liking by hedonic tests by 90 consumers. The knowledge of consumers' interests acquired with the data from the initial survey allowed for the development of new fish products, with the addition or substitution of the fish species under study through the reformulation of existing products that are familiar to the consumer. Głuchowski et al. [10] also used descriptive quantitative analyses and consumer tests to explore sensory characteristics, consumer liking of key attributes, their declared sensations and emotions, as well as consumers' facial expressions when responding to the six dishes prepared using lemon or tomatoes and made in the traditional (classical), molecular and Note by Note (NbN) versions. Tests included a nine-point hedonic scale for degree of liking a dish, check-all-that-apply (CATA) for declared sensations and FaceReader for facial expressions. The influence of factors associated with consumer attitudes toward new food and willingness to try the dishes in the future were also determined. Such an approach was valuable in modifying features such as the taste, flavor and texture of dishes according to consumers' points of view. The goal of the Kumar et al. [11] study was to highlight one strategic framework to find white spaces in the marketplace and then develop new snack texture concepts to fit the sensory concepts identified as white spaces. This paper shows one method of how new product concepts can be developed using such sensory science tools as product categorization, projective mapping and descriptive profiling. This research approach for novel and distinctive market opportunities displays an innovative, practical side of NPD research as a complement. The methodology produced in this study can be used by food product developers to explore new opportunities in the global marketplace.

Finally, Świąder and Marczewska [12] explored the current trends of using sensory evaluation in NPD in the food industry in countries that belong to the EIT Regional Innovation Scheme (RIS). A computer-assisted self-interviewing (CASI) technique for survey data collection was used. The research results showed that almost 70% of companies apply sensory evaluation methods in NPD and more so the bigger the company. Here, it was also noted that most companies prefer consumer (affective) tests to expert tests.

Conclusions and Future Outlook

To summarize, the works collected in this Special Issue serve to offer an interesting contribution to the field and a better understanding of sensorial techniques and consumer research as a tool for innovation in new product development in order to satisfy the demands of consumers who increasingly seek new flavors, pleasure and fun and for companies to gain a better market position. Therefore, this SI is very useful for both present and future use for the different players involved in this kind of product development (industry, companies, researchers, scientists, marketing, merchandising, consumers, etc.). However, although these techniques have evolved substantially in recent years, the potential of using sensory evaluation methods is not yet fully exploited. A great future is predicted for them and a significant development is expected in the coming years. Finding new opportunities in food product development is a challenging assignment.

Author Contributions: C.R.-C.; A.M.H. writing—original draft preparation; C.R.-C.; A.M.H. writing—review and editing; C.R.-C.; A.M.H. funding acquisition. All authors have read and agreed to the published version of the manuscript.

Funding: This research was funded by the Spanish Ministry of Science and Innovation (PID2019-107542RB-C21), by the CSIC Intramural projects (grant numbers 201470E073 and 202070E242), CYTED (Reference 119RT0568; Healthy Meat Network).

Acknowledgments: We thank to all the authors who have collaborated and have made this special issue possible.

Conflicts of Interest: The authors declare no conflict of interest.

References

1. Ruiz-Capillas, C.; Herrero, A.; Pintado, T.; Delgado-Pando, G. Sensory Analysis and Consumer Research in New Meat Products Development. *Foods* **2021**, *10*, 429. [CrossRef] [PubMed]
2. Kalumbi, M.; Matumba, L.; Mtimuni, B.; Mwangwela, A.; Gama, A.P. Hydrothermally Treated Soybeans Can Enrich Maize Stiff Porridge (Africa's Main Staple) without Negating Sensory Acceptability. *Foods* **2019**, *8*, 650. [CrossRef] [PubMed]
3. Szymandera-Buszka, K.; Waszkowiak, K.; Jędrusek-Golińska, A.; Hęś, M. Sensory Analysis in Assessing the Possibility of Using Ethanol Extracts of Spices to Develop New Meat Products. *Foods* **2020**, *9*, 209. [CrossRef] [PubMed]
4. Tao, R.; Cho, S. Consumer-Based Sensory Characterization of Steviol Glycosides (Rebaudioside A, D, and M). *Foods* **2020**, *9*, 1026. [CrossRef] [PubMed]
5. Guzek, D.; Głąbska, D.; Sajdakowska, M.; Gutkowska, K. Analysis of Association between the Consumer Food Quality Perception and Acceptance of Enhanced Meat Products and Novel Packaging in a Population-Based Sample of Polish Consumers. *Foods* **2020**, *9*, 1526. [CrossRef] [PubMed]
6. Biró, B.; Sipos, M.A.; Kovács, A.; Badak-Kerti, K.; Pásztor-Huszár, K.; Gere, A. Cricket-Enriched Oat Biscuit: Technological Analysis and Sensory Evaluation. *Foods* **2020**, *9*, 1561. [CrossRef] [PubMed]
7. Świąder, K.; Florowska, A.; Konisiewicz, Z.; Chen, Y.-P. Functional Tea-Infused Set Yoghurt Development by Evaluation of Sensory Quality and Textural Properties. *Foods* **2020**, *9*, 1848. [CrossRef] [PubMed]
8. Grasso, S.; Jaworska, S. Part Meat and Part Plant: Are Hybrid Meat Products Fad or Future? *Foods* **2020**, *9*, 1888. [CrossRef] [PubMed]
9. Silva, F.; Duarte, A.M.; Mendes, S.; Borges, P.; Magalhães, E.; Pinto, F.R.; Barroso, S.; Neves, A.; Sequeira, V.; Vieira, A.R.; et al. Adding Value to Bycatch Fish Species Captured in the Portuguese Coast—Development of New Food Products. *Foods* **2020**, *10*, 68. [CrossRef] [PubMed]
10. Głuchowski, A.; Czarniecka-Skubina, E.; Kostyra, E.; Wasiak-Zys, G.; Bylinka, K. Sensory Features, Liking and Emotions of Consumers towards Classical, Molecular and Note by Note Foods. *Foods* **2021**, *10*, 133. [CrossRef] [PubMed]
11. Kumar, R.; Chambers, E.; Chambers, D.; Lee, J. Generating New Snack Food Texture Ideas Using Sensory and Consumer Research Tools: A Case Study of the Japanese and South Korean Snack Food Markets. *Foods* **2021**, *10*, 474. [CrossRef] [PubMed]
12. Świąder, K.; Marczewska, M. Trends of Using Sensory Evaluation in New Product Development in the Food Industry in Countries That Belong to the EIT Regional Innovation Scheme. *Foods* **2021**, *10*, 446. [CrossRef] [PubMed]

Review

Sensory Analysis and Consumer Research in New Meat Products Development

Claudia Ruiz-Capillas *, Ana M. Herrero, Tatiana Pintado and Gonzalo Delgado-Pando

Institute of Food Science, Technology and Nutrition (ICTAN-CSIC), José Antonio Novais 10, 28040 Madrid, Spain; ana.herrero@ictan.csic.es (A.M.H.); tatianap@ictan.csic.es (T.P.); g.delgado@ictan.csic.es (G.D.-P.)
* Correspondence: claudia@ictan.csic.es

Abstract: This review summarises the main sensory methods (traditional techniques and the most recent ones) together with consumer research as a key part in the development of new products, particularly meat products. Different types of sensory analyses (analytical and affective), from conventional methods (Quantitative Descriptive Analysis) to new rapid sensory techniques (Check All That Apply, Napping, Flash Profile, Temporal Dominance of Sensations, etc.) have been used as crucial techniques in new product development to assess the quality and marketable feasibility of the novel products. Moreover, an important part of these new developments is analysing consumer attitudes, behaviours, and emotions, in order to understand the complex consumer–product interaction. In addition to implicit and explicit methodologies to measure consumers' emotions, the analysis of physiological responses can also provide information of the emotional state a food product can generate. Virtual reality is being used as an instrument to take sensory analysis out of traditional booths and configure conditions that are more realistic. This review will help to better understand these techniques and to facilitate the choice of the most appropriate at the time of its application at the different stages of the new product development, particularly on meat products.

Keywords: sensory analysis; food quality; sensory attributes; new meat product development; healthier meat products; Quantitative Descriptive Analysis (QDA); Check All That Apply (CATA); Napping; Flash Profile; Temporal Dominance of Sensations (TDS); consumer research

Citation: Ruiz-Capillas, C.; Herrero, A.M.; Pintado, T.; Delgado-Pando, G. Sensory Analysis and Consumer Research in New Meat Products Development. *Foods* **2021**, *10*, 429. https://doi.org/10.3390/foods10020429

Academic Editor: Sandra Sofia Quinteiro Rodrigues

Received: 31 December 2020
Accepted: 12 February 2021
Published: 16 February 2021

Publisher's Note: MDPI stays neutral with regard to jurisdictional claims in published maps and institutional affiliations.

Copyright: © 2021 by the authors. Licensee MDPI, Basel, Switzerland. This article is an open access article distributed under the terms and conditions of the Creative Commons Attribution (CC BY) license (https://creativecommons.org/licenses/by/4.0/).

1. Introduction

Sensory evaluation has been used since ancient times with the purpose of accepting or rejecting food products. However, it started developing as a hard science in the last century, when sensory analysis grew rapidly together with the growth of industry and processed food. It boomed during the second world war when the food industry began to prepare food rations for soldiers and there was a need for them to be palatable. This promoted the development of different sensory techniques, and progress was made on the knowledge of human perception [1,2].

Sensory analysis is a scientific specialty used to assess, study, and explain the response of the particularities of food that are observed and interpreted by the panellists using their senses of sight, smell, taste, touch, and hearing [3,4]. This human-panellist reply is quantitatively assessed. Sensory analysis has a subjective connotation due to human involvement. In general, data collected from human perception shows great variability among the participants (cultural, educational, environmental, habits, weaknesses, variability in sensory capacities and predilection, etc.). A lot of the answers from individuals cannot be mastered in this type of analysis. Therefore, in order to limit the subjectivity of the test, the circumstances during its development have to be attentively carried out. In this way, the sensory evaluation results will be more objective [5]. Many factors have to be taken into account to address these variations and increase the accuracy of the analysis: Adequate selection of personnel, training, preparation, and information to the panel, the place where the sensorial analysis will be carried out (tasting room with individual test booths), preparation and serving of samples, labelling the samples with random numbers, etc. [6,7]. Moreover, and due to the potential variability, proper data analysis and interpretation is a key part of the sensory techniques. Therefore, evaluation of the results and statistical analysis are a critical part of sensory testing. This requires advanced and diverse statistical skills both from the quantitative and qualitative fields [8,9].

On the other hand, sensory analysis is a very useful tool for the elaboration of new products. Apart from technological and safety analysis, foods stand out for their organoleptic properties (taste, smell, texture, etc.), and they must be taken into account when innovating, since they are the properties that will determine if the consumer will purchase the product and if it will choose the same product again. More studies focused on the stakeholder requirements in the final products' demands, such as analysis of sensory analysis and the consumers' research, can significantly improve the quality of products and their success in the market. All these sensory studies involve human participants. Therefore, they should be performed according to the indications of the Declaration of Helsinki of 1975, checked in 2013 [10].

Based on the importance of these sensorial techniques and their great potential at the different stages of new product development, from design to commercialisation, this manuscript aims to give an overview of the sensory and consumer techniques. From the traditional sensorial techniques to the most recent ones that have been used in sensory analysis, together with studies on consumers and their fundamental importance as an analysis stage in the development of new products, particularly meat products. These include a classification, their bases, importance, and advantages and disadvantages at the different stages of new product development. The review aims to consolidate the knowledge in order to help both industry and sensory scientists.

2. Traditional Sensory Analysis

Initially, the quality control of industrial productions was carried out by one person or a small number of people. They would assess the goodness or not of a production process and its resulting product quality through precarious sensory tests. The conducted tests were changed progressively by others more disciplined and directed, which were more quantifiable and exact, more reliable, less risky, and with eliminated segmentation [1,3].

In general, traditional sensory analysis can be divided in two: Analytical and affective. Analytical tests, which include discriminatory and descriptive evaluations, try to describe and differentiate the products. On the other hand, affective tests try to evaluate the acceptance of the product and are divided into preference and hedonic tests [7,11] (Table 1).

Table 1. Different traditional and novel sensory tests used to evaluate food.

Sensory Test	Types	Subtypes	Panellists	Question?
Analytic	Discrimination	Duo-trio Triangle PM Napping	Trained/Consumers	Are the new products different?
	Descriptive	Flavour Profile Texture profile QDA Free choice profiling Flash Descriptive Analysis Spectrum CATA FP RATA	Greater training/ Semi-trained/Consumers	How are the new products different?
Affective	Preferences or choice	Pair-comparative PSP	Naive	Which sample do you prefer?
	Hedonic		Naive	How do you like the sample?

QDA: Quantitative Descriptive Analysis; CATA: Check-all-that-apply; FP: Flash profile; RATA: Rate-all-that-apply; PM: Projective mapping; PSP: Polarized sensory positioning.

2.1. Analytical Tests

Analytical tests can address analysis such as discrimination or differentiation between new products (are the new products different?) or product description (how different are the new products?). This will provide information that can be employed with different purposes in the optimisation of technological developments.

Discrimination (difference tests) are the simplest sensory analysis that try to dilute if the panellists are able to detect any difference between two samples, as well as the magnitude of the perceived difference between two confounding stimuli. Attributes are not valued. It is important to eliminate the component due to chance in the analysis, so an important number of evaluators must appreciate the differences between the products for them to be significant. The panellists require a certain degree of training. The most commonly used discrimination techniques are: The paired-comparison method, duo-trio, and triangular test (Table 1). For example, the duo-trio presents a selection between 2 samples (A and B) establishing similarity or difference of a known pattern (R). In the triangular, the panellist must identify between 3 samples, (A, B, R), which are the same and which one is different [1].

Descriptive tests consist of a full sensory description of the products and need a trained sensory panel; the results can be quantified (Table 1). For these analyses, it is necessary to establish and find descriptors that could provide maximum information about the sensory properties of the product [1]. The panellists have to evaluate their perception with quantitative values proportional to an intensity. To obtain a significant and meaningful result, the panellists must have gone through thorough training. Some of these techniques, mostly novel sensory techniques, can also be carried out by semi-trained panellists [5,8].

Different descriptive methods, such as flavour profile method, or the texture profile method, use trained judges [1,12]. For example, texture profile has been used to identify particular intensities in a product using control products. An improvement of these methods that can be applied not only to taste and texture was achieved with the Quantitative Descriptive Analysis (QDA) [3]. Free choice profiling, flash descriptive, and spectrum method are other descriptive procedures [6].

Structured and equidistant scales are usually used for descriptive analysis, where the panellists through these scales assess his/her perception assigned to a particular attribute with a determined intensity. The strength of the attribute is indicated on the horizontal scale with a generally vertical mark, so that its numerical assignment is easier to assess. These scales can be of a single attribute or multiple attributes or descriptors, which represent the descriptive profile of the products as in the QDA. In these scales, the descriptors are arranged according to a logical order of perception: sight, smell and sensation in mouth. Descriptors are a critical point in these analyses and must be accurately chosen to describe the impulse. They must be specific and clear about the sensation they describe and they must have certain relevance and discrimination power in the products to be analysed [13]. In general, these scales benefit from the use of fewer tasting samples and fewer trained tasters, although fatigue errors can also occur [14]. The excess of parameters that are subjected to evaluation is one of the main problems when using semi-trained tasters, and this fact can negatively affect the final results, since differences between very similar parameters are a difficulty for them losing interest in the analysis [8].

In general, descriptive analysis are presented as one of the most adequate sensory tests, they provide the greatest amount of information and are easily interpreted in the elaboration of new products [5].

2.2. Affective Tests

Affective tests assess the preference or choice of a product (preferences analysis and consumers' willingness to pay) and the level of acceptance (hedonic evaluation) using the subjective criteria of the tasters. In most cases, the panellists correspond to naïve consumers not trained in the description of preferences, where their evaluation is based on taste and focused on the purchase decision and general acceptance [3,5]. There are two types of affective techniques: Preference and hedonic (Table 1).

The preference or choice tests allow us to ascertain the preference (or not) for a new product based on the majoritarian response of a panel. Traditionally, they are applied to different products in pairs [3]. It is also recommended to include the "no preference" option, as it will provide more information to facilitate the interpretation of the results. These preference techniques are very useful and are usually employed for market research of new products. They allow us to obtain important information regarding different population targets. However, the main drawback is that this methodology does not give any information about the magnitude of the liking or disliking from the respondents, as panellists only choose whether they like a product or not. To obtain more information about it, hedonic tests can be utilised:

The hedonic method offers an assessment of the liking of the product being tested, using hedonic scales (9-pt hedonic) [15] (Figure 1). In this scale, the panellists have to choose the expression more in relation to their perception and acceptance of the product. The use of this type of scale allows us to transform this answer into a numerical value, for example, 1 = dislike extremely to 9 = like extremely. This type of evaluation provides quick information on the capacity and potential for success of the new developed product. Hedonic tests can also provide information of the various cluster of consumers for different products, different textures, different composition, etc. These results would help to better understand the justification for liking or disliking a product [5]. However, this technique also has some limitations, such as: The number of necessary panellists (representative consumers), and the atmosphere and circumstances, that should be similar to the real situations in which consumers would find themselves. Usually, more than 60 representative consumers are used. It should be taken into account that the result of this type of test is not indicative of the consumer purchase intention, as other types of factors, apart from the linking, influence it. Assessing the purchase intention requires a greater number of participants (usually more than 100).

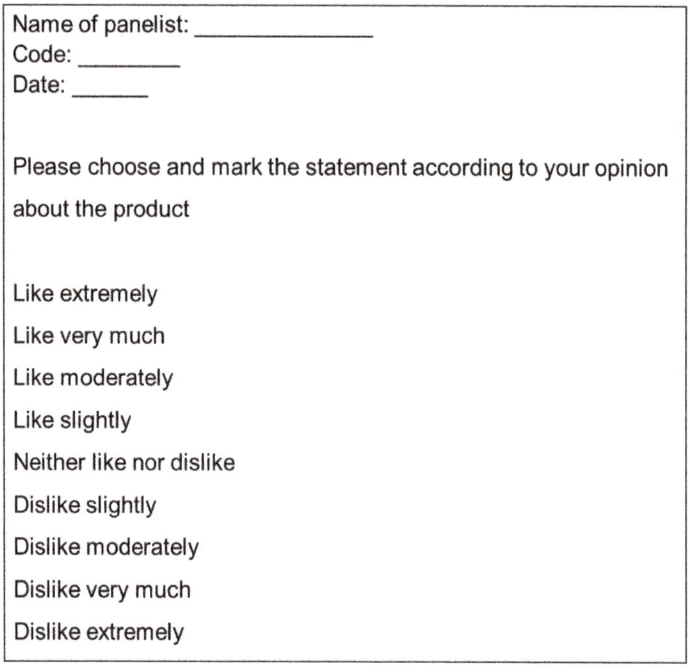

```
Name of panelist: _____
Code: _____
Date: _____


Please choose and mark the statement according to your opinion
about the product


Like extremely

Like very much

Like moderately

Like slightly

Neither like nor dislike

Dislike slightly

Dislike moderately

Dislike very much

Dislike extremely
```

Figure 1. An example of a 9-point hedonic scale useful for evaluating the acceptance of a new products [16].

Currently, a combination of affective and descriptive sensory technologies is applied during the processing and elaboration of new products. This allows us to take advantage of each technique's convenience limiting the disadvantages and helps in understanding, through acceptance or consumer preferences (affective), what qualities should be improved, maintained (descriptive), or formulated during the development of new products. However, some of these sensory analyses have shown their limitations. Some aspects in relation to the whole complexity of the consumer-product interactions are often forgotten in traditional sensory techniques. These interactions go further than the conscious response stamped on a liking scale, as external stimuli are also affecting the decision and the degree of acceptance of a food product. To understand the consumers' preferences for a product, it is also necessary to understand their needs and restrictions, purchasing power, prices of fresh or processed products, product quality, the connotation of healthiness (fat content, salt additives, etc.), the environment of its consumption, etc. In order to solve some of these limitations, new sensory and consumer research techniques have been developed.

3. New Sensory Methods

During the past decades, efforts have been put into developing new methodologies for sensory characterization of food with the aim of gaining speed and simplicity in relation to the traditional ones (Table 1). These new techniques try to provide complete information in innovation and product development and in proper approach of their marketing campaigns, to ensure success. These new alternatives have been categorized into three types depending on the nature of the evaluation task assigned to the panellists [17]:

3.1. Methods Based on Written Descriptions of the Products

Check-All-That-Apply (CATA) is a method that traditionally has been used with trained assessors, however, its use has recently become popular for food products' sensory analysis with consumers. CATA is a versatile multiple-choice questionnaire where different options of words or sentences are shown for the panellists to give their free opinion of without any type of limitation [18]. Consumers could use terms related with sensory attributes, hedonic responses, or other non-sensory properties such as: When are the products consumed? In which situation and atmosphere? What are the emotions or feelings while consuming? etc. One important thing to consider in this analysis is that the attributes are chosen by the consumer.

Flash Profiling (FP) is a method that in the first step develops the descriptive terms together with the participants and on a second step uses these descriptive terms to rank (e.g., from low to high, or least to most, etc.) the tasted products. Panellists are forced to generate discriminative attributes of the whole sample set, which is more important than the individual attributes of the products. This test allows combining free-choice profiling with a comparative evaluation of the set of products [19]. The number of needed panellists will depend on the objective and dissimilarities among the products. Even though panellists could be untrained, there is a need for at least familiarisation with the products. That is why semi-trained panellists are recommended. Moreover, FP can be more discriminating than conventional profiling for similar product categories [19]. Some limitations of FP are the need of presenting all the products at the same time and the difficulties when trying to compare results from this methodology and more traditional ones. FP is considered one of the more agile and malleable sensory methodologies to characterise food products.

Rate-All-That-Apply (RATA) is a type of CATA that is based querying consumers to classify the level of strength of descriptors that are applicable for defining/labelling samples [20]. This test has an increased ability to differentiate between samples which have a similar sensory response in terms of attributes, and is able to differentiate them based on the intensity of that response [21]. Although RATA has been tested on a different range of products, methodological studies on their reliability are still limited [20].

3.2. Methods Based on the Measurements of the Similarity or the Differences between Products

Napping is an evolved version of projective mapping a methodology developed to solve the limitations showed by the traditional techniques [22]. Untrained participants evaluate the samples taking into account their similarities (close to each other) and differences (further apart). The test allows for a comparison between all the samples presented at the same time, but it is not suitable if the samples have to be previously prepared [23]. Napping is usually combined with other sensory tests, for example, with Ultra Flash Profile, where participants can write down the properties that they consider best describe the samples, in this way, extra qualitative information is provided to the analysis [24].

3.3. Methods Based on the Comparison of Individual Products with a Reference

Polarized Sensory Positioning (PSP) uses reference products (poles) to determine the similarities or differences between samples to be evaluated. The reference poles must be different from the products to be evaluated, but they must represent the main characteristics in the products they represent [25].

3.4. Dynamic Sensory Methods

The aforementioned sensory techniques assess the perception of attributes as a "static" phenomenon. However, sensory perception is a dynamic practice, so its assessment, intensity, etc., changes with time while consuming a food product. In that sense, dynamic sensory techniques allow us to describe these changes in sensory perception during the test. Some examples are:

The Time-Intensity (TI), first to be developed, and temporal dominance of sensations (TDS) are the main dynamic sensory evaluation techniques currently used [26]. TI presented the modification of strength the one single appreciation over time; however, TDS assesses multiple attributes, trying to elucidate the sequence of dominant attributes throughout the test. The choice of one or the other method mainly depends on the objective of the analysis: Qualitative, quantitative, evolution of the quality and perception along testing, etc.

Temporal Check-All-That-Apply (TCATA) is a temporal addition of CATA. Currently, evaluating the multidimensional sensory characteristics in food products as they evolve over time during consumption has gained a lot of attention. For this technique, trained panellists must select sensory attributes (less than 10) freely and continuously, resulting in a temporal classification of the products. However, TCATA does not offer data on the dominant impressions, and none of them calculate consumers' hedonic insights of the products. Combining TCATA and TDS has shown good results [18].

4. Complementary Measures for Consumer Research

It has been studied that there is more to eating behaviour than sensory liking: External context, social factors, nutritional status, emotional state, etc., all have an impact on how a consumer interacts with a food product [27]. For this reason, in consumer research, more tools than affective testing are needed to understand and measure the attitudes, emotions, and behaviours for the successful development of new products. Some authors indicated [28] that non-verbal emotion punctuation enhanced food choice prediction when employed in conjunction with hedonic scales. Measuring emotions after food ingestion or food purchase seems to be an important step to take when developing new products. However, emotions are usually disregarded by food companies when launching new products.

Several tools have been developed to assess the consumers' emotions based on both explicit and implicit methods. Explicit means that the methods are based on self-reporting, and thus implies a direct and conscious measurement of the emotions, whereas in the implicit methods, there is no self-reporting and the emotions are measured indirectly.

A verbal self-reporting question sheet is the greatest employed tool for emotion measurements due to their rapidness, discrimination power, and ease of application [29]. These questionnaires consist of an emotional lexicon the consumers select while consuming the products. Some examples of these are already predefined, like EsSense Profile®, EsSense25, PANAS, Food Experience' Scale, etc., but some others are defined by the consumers during different sessions. As pointed out by Kaneko et al. [30], these verbal self-reporting questionnaires have some associated shortcomings: a) Difficulties to verbalise emotions, b) language dependence of the lexicon, c) interference with food experience, and d) only capturing conscious emotions. In an attempt to improve and facilitate the capture of emotions with little impact on the food testing, a self-reported questionnaire called PrEmo was developed based only on images and animations.

On the other hand, implicit methods are based either on physiological and/or visual measurements, or on behavioural tasks measurements. The latter are based on psychological tools such as the Implicit Association Test (IAT) and the Affective Priming Paradigm (APP). IAT consists of measuring the speed at which words are associated with one of two pairs of concepts. For example, we could have four categories (two products and pleasant and unpleasant words) the consumer has to recognise by clicking a certain key. Monnery-Patris et al. [31] have used an IAT to assess children's food choices. In APP, consumers undertake a categorisation task with target words preceded by food primes. The APP has been confirmed as a robust indirect measure of food enjoyment, although there is not enough evidence of its utility to measure eating behaviours [32].

Measuring involuntary physiological responses governed by the Autonomic Nervous System (ANS) and other physiological characteristics, such as face recognition, heart rate, eye-movement, body temperature, skin temperature and conductivity, etc., can provide information of the emotional state a food product can generate. Gunaratne et al. [33] used measurements of skin temperature, facial manifestations, and heart speed to analyse the relationships between short and unconscious answers to different chocolate testing. The authors found that sweet chocolate was contrariwise related with displeased emotion and salted chocolate was positively connected with sadness. Another interesting application has been proposed by Fuentes et al. [34], where the authors were able to derive models from heart speed, blood pressure, facial manifestations, and skin-temperature modifications to predict the liking of insect-based foods with the help of machine learning.

There has been significant interest in an enhanced comprehension of the position of the context in consumer sensory testing as it is widely accepted that context participates in how emotional and hedonic responses are shaped. Hathaway and Simons [35] found that the distinguishable and consistency of consumer acceptance information increases with the level of immersion the consumer experiences. The use of VR to increase the immersion level has proved to be successful on a few food products such as cookies, vegetables, and coffee. Another recent application of this technology in sensory science has been the possibility of transporting the consumer to virtual stores. The more realistic the setting was—e.g., consumers able to walk in a virtual supermarket—the better the evaluation of the purchase decision [36].

In addition to emotions and context, there are other factors that have a significant effect on food choice. These factors are product and person-dependent, as they deviate from the intrinsic quality attributes to be more external ones. Some examples of these are: Healthiness, price, familiarity, pleasure, convenience, ethical issues (e.g., vegans), cultural disgust (e.g., entomophagy), etc. In 1995, Steptoe et al. [37] developed the Food Choice Question sheet (FCQ) as an instrument to assess the reasons for accepting a food. This questionnaire was later improved on its ethical dimension with the addition of animal welfare, environmental defence, and political and religious principles [38]. The original questionnaire comprises 36 four-point matters (e.g., "It is significant to me that the food I eat on a usual day maintains me healthy", where 1= not at all important, and 4= very important), and has been used extensively. Another extensively used test has been the Food Neophobia Scale (FNS). Food neophobia is the unwillingness to eat unusual foods, such as insects in occidental culture. Pliner and Hobden [39] developed the FNS, consisting on a 10-item test, and it was validated through confirmatory factor analysis. The FNS is a completely balanced neophobia analysis and has been frequently exposed to predict real replies to novel food.

Qualitative investigation is widely employed to study consumer behaviour and extract ideas for the development of new products. One of the qualitative techniques more used in consumer research is focus groups. Focus groups were traditionally used in social sciences with the aim to help the researcher to find questions for future questionnaires. Focus groups are one of the most appropriate methods to obtain qualitative data while boosting the participants' interaction to interchange ideas, establishing a non-aggressive environment to encourage dialogue among them [40,41]. Focus groups are formed with a small number of individuals and set in a closed environment, although online meetings are now also used, where participants have an informal discussion about a specific issue or several established topics. The advances in specialised software to analyse the results as well as the possibility of combining them with other exploratory and projective techniques has made focus groups an interesting tool for consumer research. Ethnography is another qualitative tool that has gained popularity in consumer research. It aims to provide a cultural comprehension of consumers through sharing events, moving from the lab to their homes as a method to make more useful communication procedures [42]. The scientist must become a member of the community, but should also maintain distance and objectivity while observing.

5. Sensory Analysis as Tool for the Development of New Meat Products

Sensory analyses are important tools used by sensory scientists and food companies to achieve data applicable to technology, quality assessment, consumer insights, marketing, and the development of new products. Sensory analysis involves consumers, offering a relationship with technology and the market strategies [43–45].

Sensory analysis methods can be used at many stages of the process to assess the quality of the new product, but also the consumers' expectations and reactions to the product. However, traditionally, the development of new products appears to be disconnected between the understandings of consumers and the different stages (research, design, process, packaging, labelling, etc.) in the productions and commercialisation of these new products. These phases are critical, and it has been demonstrated that more studies and more participation of sensory panels and consumers in the products' design and development processes affect products success in their commercialisation.

In general, the growth of the global market for food, meat, and meat products especially, is a good opportunity for the development of new products that satisfy the demands of consumers around the world.

The development of novel products passes strict quality controls (physico-chemical, microbiological, and sensory) to guarantee their safety and their success among consumers. Sensory analysis to assess a product's quality are a significant part of a quality control program, since the consumer is the final evaluator of the quality of a new product [46].

Although we can find different meanings of quality in the scientific literature, we can say that one of the most used describes quality as the entirety of features and characteristics of a product that bear on its capability to please a given need. Some of them included also some quality properties such as safety, nutritional quality, availability, convenience and integrity, and freshness quality. Other definitions incorporate an extensive variety of other features such as value for money, legal value, technological importance, socio-ecological value, and even psychological, political, and ecological abilities according to their specific expertise and interests [47]. With this regard, the perception of food quality should be based on the manufacturer's, the consumer's, and the surveillance and legislative bodies' diverse requests. Then, there is both an objective and a subjective understanding of the quality [48,49]. The objective understanding is connected to the material characteristics that can be explained and objectively calculated. The subjective definitions depend on the consumer's view and assessment, being criteria implicated in consumer approval, mostly sensorial parameters such as colour, odour, flavour, etc. (Figure 2) [50].

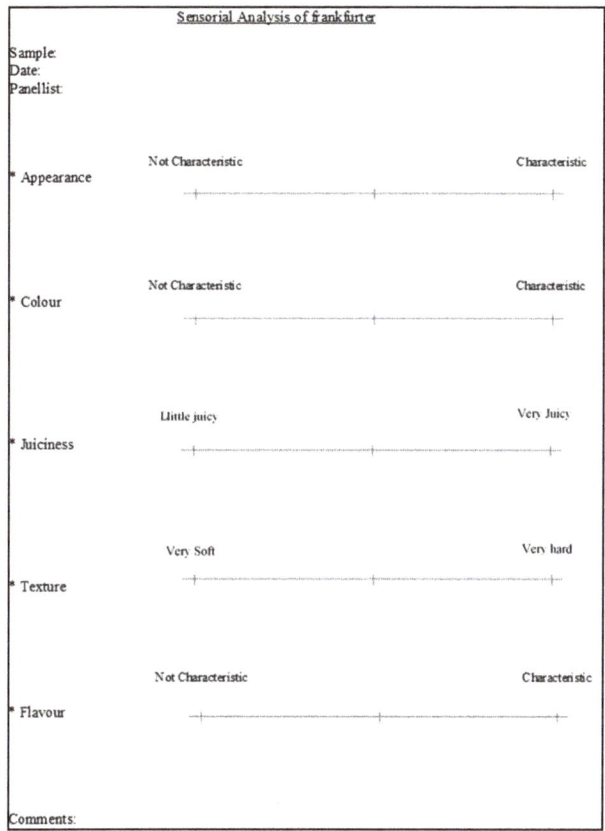

Figure 2. An example of a scale of sensorial analysis applied for development of new meat products [50].

The sensory analysis plays a very significant role in the successful elaboration of new meat products in the whole production process, from research and development to quality control and marketing. These sensory analyses bring important information to the different sectors involved in the production and commercialisation chain (industry, commerce, R+D, consumers, consumer agencies, etc.). The success or failure of the new meat products in the market will depend to some extent on these analyses, their correct application, and the adequate interpretation of the results. One of the key points is to choose the most adequate analysis depending on the type of product and target population. An optimised design of the analyses at the different stages of the processing and commercialisation steps can entail great savings of both time and money. To this regard, it is noteworthy that the meat product sector encompasses a huge variety of products with different manufacturers, processing conditions, packaging, flavours, composition, etc., and thus, sensory analysis should consider these specificities as well as the appropriate panellist selection (purchase capacity, eating routines, special requirements, etc.).

On the other hand, an important part of the development of meat products is the one addressing the design and development of new healthier meat products. The elaboration of these healthier products involves changed composition and/or processing settings to reduce the presence of specific possibly harmful compounds, and/or the option of incorporating specific appropriate substances, either naturally or by incorporation, with the consequent additional benefits to health status. The aim of this development is to enhance the nutritional profile and the health characteristics of the product, while maintaining acceptable taste and flavour [51].

Healthier meat products are a response to the increasing demand from the consumers of safer and healthier products. One of the most studied healthier meat products has been the optimisation of the lipid content [51,52], mainly due to the relationship between the animal fat in the meat products and the risk of certain diseases.

Traditional sensory tests, mainly discrimination tests, are the most used for the evaluation of the organoleptic properties of new healthier meat products at the industrial level. However, in research studies, it is the hedonic or descriptive tests that are used the most. Tenderness and juiciness have been the most sensory analysed attributes in meat product research. However, it was observed that using only these two parameters limited the overall assessment of the products, and extra relevant attributes were considered: Appearance, colour, tenderness, juiciness, aroma, and flavour [50,53] (Figure 2). This allowed for a more objective and accurate judgment, which can give a better indication of consumer acceptance [12].

Different healthier meat products with improved lipid profile (frankfurters, fresh sausages, dry fermented sausages, burger patties, etc.) have been developed with the support of sensory analysis results [54–60]. In dry fermented sausages, such as chorizo, reformulated with healthier lipid content, a hedonic scale rating test was performed where panellists evaluated appearance, flavour, firmness, juiciness, and overall acceptability, which refers to a general point of view of the product [59]. Although, the panel considered that the organoleptic properties of the new healthier dry fermented sausages in general were acceptable, the greatest sensorial limitation was the firmness score, which was considered as mainly responsible for the reduction in the general acceptability of the new products (Figure 3a). In other types of meat products with enhanced fat content, such as frankfurters and fresh sausages, a sensory panel were instructed to evaluate some parameters such as texture, colour, flavour, and general acceptability [54,56,58,60]. Generally, the panellists considered that all products were acceptable at moderately high scores (Figure 3b,c).

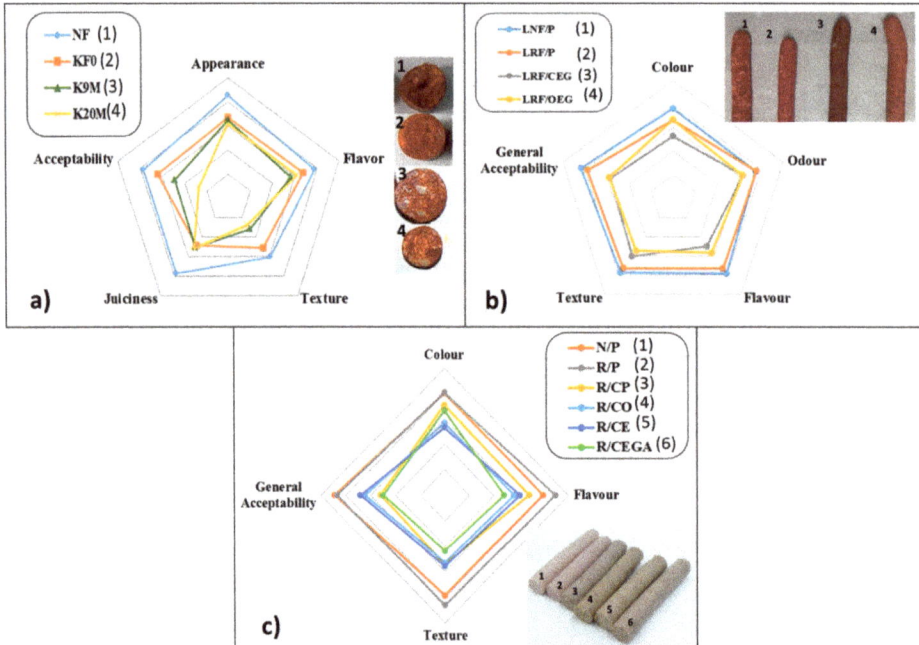

Figure 3. Example of sensory analysis results obtained in studies based on the improvement of the lipid content in meat products: (**a**) Dry fermented sausages (adapted from Jiménez-Colmenero et al. [59]; (**b**) fresh sausages (adapted from Pintado et al. [58]; (**c**) cooked sausages, frankfurter type (adapted from Pintado et al. [54]).

In the formulation of other healthy meat products based on minimising the presence of deleterious compounds, such as sodium or nitrites, sensory analysis has also been utilised. In this context, non-structured descriptive scales with fixed extremes have been employed in the elaboration of low-fat sodium reduced fresh merguez sausage, observing that the reduction of salt did not undesirably affected the sensory evaluation [57]. Moreover, sensory analysis has also been incorporated in the formulation of healthier meat products such as hot dog without nitrites, and the panellists considered all products acceptable [61,62].

On the other hand, an important part in the formulation of new products is the correlation of these sensorial results with the instrumental measures for the different attributes by means of statistical methods such as regression and correlation, thus achieving greater objectivity in sensory analyses [50,63]. However, the main problem is the lack of homogeneity in the attributes and descriptors, as well as establishing which attribute is the main one in an analysis. Since for each taster it may be different parameters (juiciness, hardness, favour, etc.) the ones that determine their acceptance or rejection of a product.

Despite this, the correlation results are an important measurement of new products quality. Colour is an important attribute of acceptance or rejection and constitutes a direct and efficient measure of the commercial acceptance of meat. Different studies have correlated instrumental measures of L^*, a^*, and b^* (colour parameters—CIELAB) with the results of descriptive sensory analysis [50,54,57]. Moreover, lower juiciness values from a sensorial analysis were correlated with greater weight loss during processing in dry fermented sausages [50]. Similar studies have carried out the correlation between instrumental and sensory hardness [64].

Spectroscopic techniques combined with chemometric analysis in the sensory analysis of meat and meat products and the elaboration of healthy meat products have been a recent novel approach. Near infrared spectroscopy (NIR) has been used as a method to quickly determine some organoleptic characteristics of meat such as appearance (colour, marbling, etc.), odour, flavour, juiciness, tenderness, or firmness [65–67]. On the other hand, Raman spectra from cooked beef samples has been correlated with organoleptic properties (juiciness and texture) using PLSR [68]. Attenuated total reflectance–Fourier transform infrared spectroscopy (ATR-FTIR) has been used in the development of healthier meat products for the evaluation of both their technological and sensory properties. Results showed that these healthier products involved more lipid–protein interactions, but their sensory properties were not affected and the new products were judged acceptable [54,60].

Novel sensory techniques have also been employed in the sensory characterisation and development of traditional and healthier meat products. In this sense, flash profiling has also been applied for the sensory analysis of meat products such as hams or hot dogs. The results derived from Flash Profile were comparable to those obtained applying quantitative descriptive analysis (QDA) [24]. Flash Profile demonstrated an efficient discriminant ability between a traditional Madagascar meat product elaborated with pork and beef and a traditional Portuguese sausage [69]. Lorido et al. [70] applied Flash Profile to differentiate between dry-cured loins made with various quantities of NaCl. These works also combined Flash Profile with other sensory analysis such as napping or dynamic sensory techniques [69,70]. Alves et al. [71] through a CATA analysis chose the expressions to bologna-type sausages from a previous dialogue with a team of 15 consumers, and consumers were requested to conclude the CATA questionnaire with 19 descriptors connected to the organoleptic characteristics of the Bologna-type sausage. In another study, a total of 32 sensory descriptors were developed on the adapted "Kelly Repertory Grid Method". These terms were clustered (appearance, colour, favour/taste, texture, and odour) and were used to determinate the organoleptic properties of healthier bolognas (enriched with ω3 fatty acids) by CATA [72]. Both studies concluded that the employment of CATA showed some significant considerations in the formulation of healthier bolognas since it was capable to explain relevant characteristics. Other authors have compared CATA analysis with trained panellists' results [73], Descriptive Analysis (DA) and their relationship with overall liking (OL) [74], acceptance testing [75], etc. According to these authors, the CATA questions successfully distinguished between the meat products regarding their organoleptic properties. In addition, these attributes were connected to chemical and instrumental quality parameters.

The use of CATA has been applied to commercial and healthier reformulated meat products to analyse the acceptance and the impact that some modifications (partial protein replacement, lipid content improvement, etc.) have on consumers [24]. This method was able to indicate some relevant considerations in the elaboration of meat products and was able to describe important characteristics.

Many recent works have indicated the application of napping-UFP in assorted meat products. The method allowed for a good discrimination among pork tested samples in relation to different cooking methods and conditions [76]. Napping-UFP successfully characterised bacon samples smoked with different woods, discerning the woods employed for smoking. The samples characterisation of samples and the results were correlated with volatile compounds [74]. Moreover, the great discrimination ability of Napping-UFP in healthier reformulated products has been proven: With bioactive components (e.g., fibres, prebiotics), with different fat or salt levels [24].

With respect to dynamic sensory analysis on meat products, TI was applied to determine the temporal opinion of tenderness in cooked pork and beef [77,78]. TDS was performed by Paulsen et al. [79], who considered the influence of NaCl replacement on the temporal perception of flavour and texture on sausages. TDS indicated unidentified sensory descriptions of NaCl replacement in meat products when it was compared with the results obtained from the classic QDA. TI and TDS have been applied to dry-cured hams elaborated from pigs with diverse feeding backgrounds and varying in NaCl content. TDS allowed a more effective discernment between different types of ham [80]. TI and TDS were found to offer complementary results, and thus using both temporal methods are recommended when a thorough sensory evaluation of the samples is expected. Paglarini et al. [81] evaluated the effect of salt and fat reduction on Bologna sausage with incorporation of emulsion gel in the dynamic sensorial perception by using TDS and TCATA methods contemplating overall enjoyment. The TDS and TCATA curves indicated that texture attributes were relevant at the beginning of the estimation for all samples, and TCATA also exhibited that juiciness was prevailing in the first 15 s of the eating period.

6. Conclusions

Sensory analysis and consumer research are a relevant tool in the development of health-enhanced meat products. Although sensory analysis techniques have evolved greatly in the last decades, these advances must continue. It should be noted that sensory analysis is a science that determines, analyses, and interprets the replies of people to products as perceived by the human senses, which implicate many factors and variability. The different sensory techniques that are applied in the development of new products must reduce and control the variability due to human involvement for these new developments to be successful. This is in line with the objective 9 (industries, innovation, and infrastructure) of the UN Agenda 2030, as it will modernize and innovate the sector while increasing efficiency. In order to do this, novel technologies and methodologies have to be further explored and implemented in a more holistic way, not only taking into account likeness and affectivity, but also emotions, context, and preference factors.

Author Contributions: Conceptualization C.R.-C., and A.M.H.; formal analysis C.R.-C., A.M.H., T.P., G.D.-P.; investigation C.R.-C., A.M.H., T.P., G.D.-P., project administration, C.R.-C. and A.M.H.; resources, C.R.-C. and A.M.H.; writing—original draft, C.R.-C., A.M.H., T.P., G.D.-P.; writing—review and editing, C.R.-C., A.M.H., T.P., G.D.-P. All authors have read and agreed to the published version of the manuscript.

Funding: This research was funded by the Spanish Ministry of Science and Innovation (PID2019-107542RB-C21), by the CSIC Intramural projects (grant number 201470E073 and 202070E242), CYTED (grant number reference 119RT0568; HealthyMeat network) and the EIT Food Project 20206.

Conflicts of Interest: There is no conflict of interest.

References

1. Stone, H.; Bleibaum, R.N.; Thomas, H.A. *Sensory Evaluation Practices*, 4th ed.; Food Science and Technology International Series; Academic Press: London, UK, 2012.
2. Barda, N. Análisis Sensorial de los Alimentos. Available online: https://inta.gob.ar/sites/default/files/script-tmp-inta-_anlisis_sensorial_de_los_alimentos_fruticultura.pdf (accessed on 25 October 2020).
3. Stone, H.; Sidel, J. *Sensory Evaluation Practices*, 3rd ed.; Elservier Academic Press: California, CA, USA, 2004.
4. UNE-EN ISO 5492:2010 A1:2017. Sensory Analysis-Vocabulary-Amendment 1 (ISO 5492:2008/Amd 1:2016). Available online: https://www.une.org/encuentra-tu-norma/busca-tu-norma/norma?c=N0058898 (accessed on 25 October 2020).
5. Lawless, H.T.; Heymann, H. *Sensory Evaluation of Food*; Food Science Text Series; Springer: New York, NY, USA, 2010.
6. Meilgaard, M.; Civille, G.V.; Carr, B.T. *Sensory Evaluation Techniques*, 5th ed.; CRC Press: Boca Raton, FL, USA, 2016.
7. Delarue, J.; Ben Lawlor, J.; Rogeaux, M. *Rapid Sensory Profiling Techniques and Related Methods*; Woodhead Publishing, Elsevier: Cambridge, UK, 2015.
8. Ruiz-Capillas, C.; Moral, A.; Villagarcia, T. Use of semi-trained panel members in the sensory evaluation of hake (*Merluccius merluccius*, L.) analyzed statistically. *J. Food Qual.* **2003**, *26*, 181–195. [CrossRef]
9. Luciano, G.; Tormod, N. Interpreting sensory data by combining principal component analysis and analysis of variance. *Food Qual. Prefer.* **2009**, *20*, 167–175. [CrossRef]
10. Declaration of Helsinki of 1975–2013. Available online: https://www.wma.net/what-we-do/medical-ethics/declaration-of-helsinki/ (accessed on 25 October 2020).
11. Sánchez, I.C.; Albarracín, W. Sensory analysis of meat. *Rev. Colomb. Cienc. Pecu.* **2010**, *23*, 227–239.
12. Murray, J.M.; Delahunty, C.M.; Baxter, I.A. Descriptive sensory analysis: Past, present and future. *Food Res. Int.* **2001**, *34*, 461–471. [CrossRef]
13. Urdapilleta, I.; Tuus, C.A.; Nicklaus, S. Sensory evaluation based on verbal judgments. *J. Sens. Stud.* **1999**, *14*, 79–95. [CrossRef]
14. Cartie, R.; Rytz, A.; Lecomte, A.; Poblete, F.; Krystlik, J.; Belin, E.; Martin, N. Sorting procedure as an alternative to quantitative descriptive analysis to obtain a product sensory map. *Food Qual. Prefer.* **2006**, *17*, 562–571. [CrossRef]
15. Lim, J. Hedonic scaling: A review of methods and theory. *Food Qual. Prefer.* **2011**, *22*, 733–747. [CrossRef]
16. Peryam, D.R.; Pilgrim, F.J. Hedonic scale method of measuring food preferences. *Food Technol.* **1957**, *December*, 9–14.
17. Varela, P.; Ares, G. Sensory Profiling, the blurred line between sensory and consumer science. A review of novel methods for product characterization. *Food Res. Int.* **2012**, *48*, 893–908. [CrossRef]
18. Ares, G.; Jaeger, S.R.; Antúnez, L.; Vidal, L.; Giménez, A.; Coste, B.; Picalloc, A.; Casturad, J.S.R. Comparison of TCATA and TDS for dynamic sensory characterization of food products. *Food Res. Int.* **2015**, *78*, 148–158. [CrossRef]
19. Delarue, J. Flash Profile, its evolution and uses in sensory and consumer science. In *Rapid Sensory Profiling Techniques. Applications in New Product Development and Consumer Research*; Delarue, J., Lawlor, J.B., Rogeaux, M., Eds.; Woodhead Publishing: Cambridge, UK, 2015; pp. 121–151.
20. Ares, G.; Picallo, A.; Coste, B.; Antúnez, L.; Vidal, L.; Giménez, A.; Jaeger, S.R. A comparison of RATA questions with descriptive analysis: Insights from three studies with complex/similar products. *J. Sens. Stud.* **2018**, *33*, e12458. [CrossRef]
21. Vidal, L.; Ares, G.; Hedderley, D.L.; Meyners, M.; Jaeger, S.R. Comparison of rate-all-that-apply (RATA) and check-all-that-apply (CATA) questions across seven consumer studies. *Food Qual Prefer.* **2016**, *67*, 49–58. [CrossRef]
22. Risvik, E.; McEwan, J.A.; Colwill, J.S.; Rogers, R.; Lyon, D.H. Projective mapping: A tool for sensory analysis and consumer research. *Food Qual. Prefer.* **1994**, *5*, 263–269. [CrossRef]
23. Lê, S.; Lê, T.M.; Cadoret, M. Napping and sorted Napping as a sensory profiling technique. In *Rapid Sensory Profiling Techniques. Applications in New Product Development and Consumer Research*; Delarue, J., Lawlor, J.B., Rogeaux, M., Eds.; Woodhead Publishing: Cambridge, UK, 2015; pp. 197–213.
24. Ventanas, S.; Gonzalez-Mohino, A.; Estevez, M.; Carvalho, L. Innovation in sensory assessment of meat and meat products. In *Meat Quality Analysis: Advanced Evaluation Methods, Techniques, and Technologies*; Biswas, A.K., Mandal, P.K., Eds.; Academic Press Ltd.-Elsevier Science Ltd.: London, UK, 2020; pp. 393–418.
25. Ares, G.; Antúnez, L.; de Saldamando, L.; Giménez, A. *Polarized Sensory Positioning Chapter 16.*, In *Descriptive Analysis in Sensory Evaluation*; Sarah, E., Kemp, S.E., Hort, J., Hollowood, T., Eds.; John Wiley & Sons Ltd.: Chichester, UK, 2018; pp. 561–577.
26. Pineau, N.; Schilch, P. Temporal dominance of sensations (TDS) as a sensory profiling technique. In *Meat Quality Analysis: Advanced Evaluation Methods, Techniques, and Technologies*; Biswas, A.K., Mandal, P.K., Eds.; Academic Press Ltd-Elsevier Science Ltd.: London, UK, 2020; pp. 269–306.
27. He, W.; Boesveldt, S.; de Graaf, C.; de Wijk, R.A. The relation between continuous and discrete emotional responses to food odors with facial expressions and non-verbal reports. *Food Qual. Prefer.* **2016**, *48*, 130–137. [CrossRef]
28. Dalenberg, J.R.; Gutjar, S.; Ter Horst, G.J.; De Graaf, K.; Renken, R.J.; Jager, G. Evoked emotions predict food choice. *PLoS ONE* **2014**, *9*, e115388. [CrossRef] [PubMed]
29. Dorado, R.; Chaya, C.; Tarrega, A.; Hort, J. The impact of using a written scenario when measuring emotional response to beer. *Food Qual. Prefer.* **2016**, *50*, 38–47. [CrossRef]

30. Kaneko, D.; Toet, A.; Brouwer, A.M.; Kallen, V.; van Erp, J.B.F. Methods for evaluating emotions evoked by food experiences: A literature review. *Front. Psychol.* **2018**, *9*, 911. [CrossRef]
31. Monnery-Patris, S.; Marty, L.; Bayer, F.; Nicklaus, S.; Chambaron, S. Explicit and implicit tasks for assessing hedonic-versus nutrition-based attitudes towards food in French children. *Appetite* **2016**, *96*, 580–587. [CrossRef]
32. Tzavella, L.; Maizey, L.; Lawrence, A.D.; Chambers, C.D. The affective priming paradigm as an indirect measure of food attitudes and related choice behaviour. *Psychon. Bull. Rev.* **2020**, *27*, 1397–1415. [CrossRef]
33. Gunaratne, T.M.; Sigfredo Fuentes, S.; Nadeesha, M.; Gunaratne, N.M.; Damir, D.; Torrico, D.D.; Gonzalez Viejo, C.; Frank, R.; Dunshea, F.R. Physiological Responses to Basic Tastes for Sensory Evaluation of Chocolate Using Biometric Techniques. *Foods* **2019**, *8*, 243. [CrossRef]
34. Fuentes, S.; Wong, Y.Y.; Gonzalez Viejo, C. Non-Invasive Biometrics and Machine Learning Modeling to Obtain Sensory and Emotional Responses from Panellists during Entomophagy. *Foods* **2020**, *9*, 903. [CrossRef]
35. Hathaway, D.; Simons, C.T. The impact of multiple immersion levels on data quality and panelist engagement for the evaluation of cookies under a preparation-based scenario. *Food Qual. Prefer.* **2017**, *57*, 114–125. [CrossRef]
36. Siegrist, M.; Ung, C.Y.; Zank, M.; Marinello, M.; Kunz, A.; Hartmann, C.; Menozzi, M. Consumers' food selection behaviors in three-dimensional (3D) virtual reality. *Food Res. Int.* **2019**, *117*, 50–59. [CrossRef]
37. Steptoe, A.; Pollard, T.M.; Wardle, J. Development of a measure of the motives underlying the selection of food: The food choice questionnaire. *Appetite* **1995**, *25*, 267–284. [CrossRef]
38. Lindeman, M.; Väänänen, M. Measurement of ethical food choice motives. *Appetite* **2000**, *34*, 55–59. [CrossRef] [PubMed]
39. Pliner, P.; Hobden, K. Development of a scale to measure the trait of food neophobia in humans. *Appetite* **1992**, *19*, 105–120. [CrossRef]
40. Wong, L.P. Focus group discussion: A tool for health and medical research. *Singapore Med. J.* **2008**, *49*, 256–260. [PubMed]
41. Guerrero, L.; Xicola, J. New approaches to focus groups, Chapter 3. In *Methods in Consumer Research, Volume 1*; Ares, G., Varela, P., Eds.; Woodhead Publishing: Cambridge, UK, 2018; pp. 49–77.
42. Valentin, D.; Gomez-Corona, C. Using Ethnography in Consumer Research, Chapter 5. In *Methods in Consumer Research, Volume 1*; Ares, G., Varela, P., Eds.; Woodhead Publishing: Cambridge, UK, 2018; pp. 103–122.
43. Botta, J.R. Sensory evaluation: Freshness quality grading. In *Evaluation of Seafood Freshness Quality*; Bota, J.R., Ed.; VCH Publisher Ltd.: Cambridge, UK, 1995.
44. Hough, G.; Garitta, L. Methodology for sensory shelf-life estimation: A review. *J. Sens. Stud.* **2012**, *27*, 137–147. [CrossRef]
45. Yang, J.; Lee, J. Application of sensory descriptive analysis and consumer studies to investigate traditional and authentic foods: A review. *Foods* **2019**, *8*, 54. [CrossRef]
46. O'Sullivan, M.G.A. *Handbook for Sensory and Consumer-Driven New Product Development. Innovative Technologies for the Food and Beverage Industry*; Elsevier: Cambridge, UK, 2017.
47. Bremner, H.A. Toward practical definitions of quality for food science. *Crit. Rev. Food Sci. Nutr.* **2000**, *40*, 83–90. [CrossRef]
48. Grunert, K.G.; Bredahl, L.; Bunsø, K. Consumer perception of meat quality and implications for product development in the meat sector—A review. *Meat Sci.* **2004**, *66*, 259–272. [CrossRef]
49. Röhr, A.; Lüddecke, K.; Drusch, S.; Müller, M.J.; Alvensleben, R.V. Food quality and safety—Consumer perception and public health concern. *Food Control.* **2005**, *16*, 649–655.
50. Ruiz-Capillas, C.; Triki, M.; Herrero, A.M.; Rodriguez-Salas, L.; Jiménez-Colmenero, F. Konjac gel as pork backfat replacer in dry fermented sausages: Processing and quality characteristics. *Meat Sci.* **2012**, *92*, 144–150. [CrossRef] [PubMed]
51. Jiménez-Colmenero, F.; Salcedo-Sandoval, L.; Bou, R.; Cofrades, S.; Herrero, A.M.; Ruiz-Capillas, C. Novel applications of oil structuring methods as a strategy to improve the fat content of meat products. *Trends Food Sci. Technol.* **2015**, *44*, 177–188. [CrossRef]
52. Herrero, A.M.; Ruiz-Capillas, C. Novel lipid material based on gelling procedures as fat analogues in the development of healthier meat products. *Curr. Opin. Food Sci.* **2021**, *39*, 1–6. [CrossRef]
53. O'Sullivan, M.G.; Kerry, J.P.; Byrne, D.V. Use of sensory science as a practical commercial tool in the development of consumer-led processed meat products. In *Processed Meats*; Kerry, J.P., Kerry, J.F., Eds.; Woodhead Publishing: Cambridge, UK, 2011.
54. Pintado, T.; Herrero, A.M.; Ruiz-Capillas, C.; Triki, M.; Carmona, P.; Jiménez-Colmenero, F. Effects of emulsion gels containing bioactive compounds on sensorial, technological and structural properties of frankfurters. *Food Sci. Tech. Int.* **2016**, *22*, 132–145. [CrossRef] [PubMed]
55. Alejandre, M.; Poyato, C.; Ansorena, D.; Astiasarán, I. Linseed oil gelled emulsion: A successful fat replacer in dry fermented sausages. *Meat Sci.* **2016**, *121*, 107–113. [CrossRef] [PubMed]
56. Triki, M.; Herrero, A.M.; Jiménez-Colmenero, F.; Ruiz-Capillas, C. Effect of preformed konjac gels, with and without olive oil, on the technological attributes and storage stability of merguez sausage. *Meat Sci.* **2013**, *93*, 351–360. [CrossRef]
57. Triki, M.; Herrero, A.M.; Jiménez-Colmenero, F.; Ruiz-Capillas, C. Storage stability of low-fat sodium reduced fresh merguez sausage prepared with olive oil in konjac gel matrix. *Meat Sci.* **2013**, *94*, 438–446. [CrossRef] [PubMed]
58. Pintado, T.; Herrero, A.M.; Jiménez-Colmenero, F.; Pasqualin Cavalheiro, C.; Ruiz-Capillas, C. Chia and oat emulsion gels as new animal fat replacers and healthy bioactive sources in fresh sausage formulation. *Meat Sci.* **2018**, *135*, 6–13. [CrossRef]

59. Jiménez-Colmenero, F.; Triki, M.; Herrero, A.M.; Rodríguez-Salas, L.; Ruiz-Capillas, C. Healthy oil combination stabilized in a konjac matrix as pork fat replacement in low-fat, PUFA-enriched, dry fermented sausages. *LWT—Food Sci. Technol.* **2013**, *51*, 158–163. [CrossRef]
60. Pintado, T.; Muñoz-González, I.; Salvador, M.; Ruiz-Capillas, C.; Herrero, A.M. Phenolic compounds in emulsion gel-based delivery systems applied as animal fat replacers in frankfurters: Physico-chemical, structural and microbiological approach. *Food Chem.* **2021**, *340*, 128095. [CrossRef]
61. Ruiz-Capillas, C.; Tahmouzi, S.; Triki, M.; Rodríguez-Salas, L.; Jiménez-Colmenero, F.; Herrero, A.M. Nitrite-free Asian hot dog sausages reformulated with nitrite replacers. *J. Food Sci. Technol.* **2015**, *52*, 4333–4341. [CrossRef]
62. Ruiz-Capillas, C.; Herrero, A.M.; Tahmouzi, S.; Razavi, S.H.; Triki, M.; Rodríguez-Salas, L.; Samcova, K.; Jiménez-Colmenero, F. Properties of reformulated hot dog sausage without added nitrites during chilled storage. *Food Sci. Technol. Int.* **2016**, *22*, 21–30. [CrossRef] [PubMed]
63. Giménez, A.; Ares, F.; Ares, G. Sensory shelf-life estimation: A review of current methodological approaches. *Food Res. Int.* **2012**, *49*, 311–325.
64. Combes, S.; González, I.; Déjean, S.; Baccini, A.; Jehl, N.; Juin, H.; Cauquil, L.; Gabinaud, B.; Lebas, F.O.; Larzul, C. Relationships between sensory and physicochemical measurements in meat of rabbit from three different breeding systems using canonical correlation analysis. *Meat Sci.* **2008**, *80*, 835–841. [CrossRef] [PubMed]
65. Venel, C.; Mullen, A.; Downey, G.; Troy, D. Prediction of tenderness and other quality attributes of beef by near infrared reflectance spectroscopy between 750 and 1100 nm; further studies. *J. Near Infrared Spectrosc.* **2001**, *9*, 185–198. [CrossRef]
66. Liu, Y.; Lyon, B.G.; Windham, W.R.; Lyon, C.E.; Savage, E.M. Prediction of physical, color, and sensory characteristics of broiler breasts by visible/near infrared reflectance spectroscopy. *Poultry Sci.* **2004**, *83*, 1467–1474. [CrossRef] [PubMed]
67. Andres, S.; Murray, I.; Navajas, E.A.; Fisher, A.V.; Lambe, N.R.; Bünger, L. Prediction of sensory characteristics of lamb meat samples by near infrared reflectance spectroscopy. *Meat Sci.* **2007**, *76*, 509–516. [CrossRef]
68. Beattie, R.J.; Bell, S.J.; Farmer, L.J.; Moss, B.W.; Patterson, D. Preliminary investigation of the application of Raman spectroscopy to the prediction of the sensory quality of beef silverside. *Meat Sci.* **2004**, *66*, 903–913. [CrossRef] [PubMed]
69. Pintado, A.I.E.; Monteiro, M.J.P.; Régine Talon, R.; Sabine Leroy, S.; Valérie Scislowski, V.; Geneviève Fliedel, G.; Rakoto, D.; Maraval, I.; Costa, A.I.A.; Silva, A.P.; et al. Consumer acceptance and sensory profiling of reengineered kitoza products. *Food Chem.* **2016**, *198*, 75–84. [CrossRef]
70. Lorido, L.; Estévez, M.; Ventanas, S. Fast and dynamic descriptive techniques (Flash Profile, Time-intensity and Temporal Dominance of Sensations) for sensory characterization of dry-cured loins. *Meat Sci.* **2018**, *145*, 154–162. [CrossRef]
71. dos Santos Alves, L.A.A.; Lorenzo, J.M.; Gonçalves, C.A.A.; dos Santos, B.A.; Heck, R.T.; Cichoski, A.J.; Campagnol, P.C.B. Impact of lysine and liquid smoke as favor enhancers on the quality of low-fat Bologna-type sausages with 50% replacement of NaCl by KCl. *Meat Sci.* **2017**, *123*, 50–56. [CrossRef]
72. Pires, M.A.; Rodrigues, I.; Barros, J.C.; Trindade, M.A. Kelly's repertory grid method applied to develop sensory terms for consumer characterization (check-all-that-apply) of omega-3 enriched bologna sausages with reduced sodium content. *Eur. Food Res. Technol.* **2021**, *247*, 285–293. [CrossRef]
73. Dos Santos, B.A.; Bastianello Campagnol, P.C.; da Cruz, A.G.; Galvão, M.T.E.L.; Monteiro, R.A.; Wagner, R.; Pollonio, M.A.R. Check all that apply and free listing to describe the sensory characteristics of low sodium dry fermented sausages: Comparison with trained panel. *Food Res. Int.* **2015**, *76*, 725–734. [CrossRef] [PubMed]
74. Saldaña, E.; Oliveira Garcia, A.; Selani, M.M.; Haguiwara, M.H.; Almeida, M.A.; Siche, R.; Contreras-Castillo, C.J. A sensometric approach to the development of mortadella with healthier fats. *Meat Sci.* **2018**, *137*, 176–190. [CrossRef]
75. Conceição, E.; JorgeAndressa, J.; Gaione Mendes, C.; Auriema, B.E.; Pereira Cazedey, H.; Rogério Fontes, P.; Souza Ramos, A.L.; MendesRamos, E. Application of a check-all-that-apply question for evaluating and characterizing meat products. *Meat Sci.* **2015**, *100*, 124–133.
76. Gonzalez-Mohino, A.; Antequera, T.; Pérez-Palacios, T.; Ventanas, S. Napping combined with ultra-flash profile (UFP) methodology for sensory assessment of cod and pork subjected to different cooking methods and conditions. *Eur. Food Res. Technol.* **2019**, *245*, 2221–2231. [CrossRef]
77. Duizer, L.M.; Gullett, E.A.; Findlay, C.J. Time-Intensity methodology for beef tenderness perception. *J. Food Sci.* **1993**, *58*, 943–947. [CrossRef]
78. Brown, W.E.; Gérault, S.; Wakeling, I. Diversity of perceptions of meat tenderness and juiciness by consumers: A Time-Intensity study. *J. Text. Stud.* **1996**, *27*, 475–492. [CrossRef]
79. Paulsen, M.T.; Nys, A.; Kvarberg, R.; Hersleth, M. Effects of NaCl substitution on the sensory properties of sausages: Temporal aspects. *Meat Sci.* **2014**, *98*, 164–170. [CrossRef] [PubMed]
80. Lorido, L.; Hort, J.; Estévez, M.; Ventanas, S. Reporting the sensory properties of dry-cured ham using a new language: Time intensity (TI) and temporal dominance of sensations (TDS). *Meat Sci.* **2016**, *121*, 166–174. [CrossRef] [PubMed]
81. Paglarini, C.D.; Vidal, V.A.S.; dos Santos, M.; Coimbra, L.O.; Esmerino, E.A.; Pollonio, M.A.R. Using dynamic sensory techniques to determine drivers of liking in sodium and fat-reduced Bologna sausage containing functional emulsion gels. *Food Res. Int.* **2020**, *132*, 109066. [CrossRef] [PubMed]

Article

Sensory Features, Liking and Emotions of Consumers towards Classical, Molecular and Note by Note Foods

Artur Głuchowski [1], Ewa Czarniecka-Skubina [1,*], Eliza Kostyra [2], Grażyna Wasiak-Zys [2] and Kacper Bylinka [1]

[1] Department of Food Gastronomy and Food Hygiene, Institute of Human Nutrition Sciences, Warsaw University of Life Sciences (WULS), 02-787 Warsaw, Poland; artur_gluchowski@sggw.edu.pl (A.G.); kacperbylinka@gmail.com (K.B.)

[2] Department of Functional and Organic Food, Institute of Human Nutrition Sciences, Warsaw University of Life Sciences (WULS), 02-787 Warsaw, Poland; eliza_kostyra@sggw.edu.pl (E.K.); grazyna_wasiak_zys@sggw.edu.pl (G.W.-Z.)

* Correspondence: ewa_czarniecka_skubina@sggw.edu.pl; Tel.: +48-22-593-7063

Citation: Głuchowski, A.; Czarniecka-Skubina, E.; Kostyra, E.; Wasiak-Zys, G.; Bylinka, K. Sensory Features, Liking and Emotions of Consumers towards Classical, Molecular and Note by Note Foods. *Foods* **2021**, *10*, 133. https://doi.org/10.3390/foods10010133

Received: 29 November 2020
Accepted: 8 January 2021
Published: 10 January 2021

Publisher's Note: MDPI stays neutral with regard to jurisdictional claims in published maps and institutional affiliations.

Copyright: © 2021 by the authors. Licensee MDPI, Basel, Switzerland. This article is an open access article distributed under the terms and conditions of the Creative Commons Attribution (CC BY) license (https://creativecommons.org/licenses/by/4.0/).

Abstract: Modern cuisine served at top-end restaurants attempts to attract customers, who increasingly demand new flavor, pleasure and fun. The materials were six dishes prepared using lemon or tomatoes and made in the traditional (classical), molecular and Note by Note (NbN) versions. The study explores sensory characteristics, consumer liking of key attributes, their declared sensations and emotions, as well as consumers' facial expressions responding to the dishes. These objectives were investigated by descriptive quantitative analysis and consumer tests. Tests included a 9-point hedonic scale for degree of liking a dish, Check-All-That-Apply (CATA) for declared sensations and FaceReader for facial expressions. The influence of factors associated with consumer attitudes toward new food and willingness to try the dishes in the future were also determined. It was stated that the product profiles represent different sensory characteristics due to the technology of food production and the ingredients used. The food neophobia and consumer innovativeness had a significant ($p \leq 0.05$) effect on liking. The odor-, flavor-, texture- and overall-liking of the NbN dishes were lower than that of traditional versions but did not vary from scores for molecular samples. The expected liking of NbN dishes was higher than experienced-liking. Traditional and modern products differed in CATA terms. Classical dishes were perceived by consumers as more tasty, traditional and typical while modern cuisine dishes were perceived as more surprising, intriguing, innovative and trendy. Mimic expressions assessment by FaceReader showed similar trends in some emotions in both classical dishes and separate temporal patterns in modern products.

Keywords: new product development; molecular and Note by Note products; classical dishes; sensory profiling; emotions; CATA; consumer research

1. Introduction

Choice of food is driven by numerous biological, economic, physical and sociopsychological determinants. The changes in global dietary patterns are influenced by various trends and factors. For example, when eating out, consumers do not just want to meet their physiological needs but are in search of food that brings emotional benefits and gives the pleasure of new flavors and tastes [1–3]. There are many factors that contribute to food satisfaction for consumers' eating out experiences. The most common factors include satiation, sensory experience, food variety and quality (nutritional value, origin, healthiness, visual attractiveness, freshness), financial aspects (price/value, promotion), company of friends or family [4–21], restaurant service, food presentation, taste of the food and the physical details of the restaurant (e.g., interior colors and design, music and mood) [6,8,16,18–21].

Molecular cuisine and Note by Note cooking (NbN) are novelties that have appeared in recent years. Molecular cooking is typically defined as producing food in professional kitchens using "new" tools (siphons, evaporators, etc.), ingredients (food additives, surprising texturizers) and methods (sous-vide cooking, flash freezing, etc.) [22]. Modern cuisine uses the "food deconstruction" technique, which consists of preparing a classic dish, with its same ingredients and flavors, and presenting it in a very unconventional and nontraditional manner [23]. Note by Note cuisine is the use of pure compounds or their mixtures in the production of dishes. Compounds are used to design the shape, color, taste, odor, temperature, trigeminal stimulation, texture and nutritional aspects of the developed meal [24–27]. The concept of modern cuisines is to create avant-garde, unique dishes that surprise customers. Modern restaurants' dishes should provide a multisensory experience involving emotions, memory, culture [28] and sensory perception, especially by aesthetic visual appearance. It also typically provides rich contrasts of flavors (through, for example, surprising ingredient pairings), textures and temperatures [29,30]. This sensory incongruity and the associated surprise evoke various responses from different people that depend on a consumer's fear and willingness to try new foods [31].

Consumers still do not trust new food technologies, especially when they do not know their effect on human health and the environment [32]. Acceptance of the modification of traditional food depends on their character that results from the changes. Innovations like diversification of shapes and textures and unexpected combinations of ingredients to create new flavors were perceived as the least acceptable and as the most harmful to the traditional character of food [33].

Success of markets for new food technologies depends on consumers' behavioral responses and emotional states [34,35] and is primarily associated with sensory properties [36]. Disgust, food neophobia and related features have been recognized as the main barrier to accepting novel food [37], and negative and suspicious consumer attitudes towards food technologies may lead to product failure [34]. Additionally, the preferences of modern consumers for food prepared at catering establishments are contradictory. On the one hand, consumers appreciate the naturalness of food [38], which is perceived as a crucial feature [39]. On the other hand, consumers accept novel technologies like modern cuisine, which contains food additives [40].

The evaluation of molecular and Note by Note meals is not easy because the idea of creating them is based on the constant novelty and uniqueness of dishes. Our results previously revealed a positive attitude of consumers toward molecular dishes, their moderately high sensory-experience ratings, and openness towards new ideas [41].

According to the bidirectional effect, foods that people consume affect their emotions and vice versa. Most of the choices associated with eating behavior are unarticulated and occur without insight or awareness. Thus, facial expression measurements reflect the dynamic sequence of emotional responses, whereas non-verbal reports correspond to the emotional status at the end of assessment [42].

There is no comprehensive sensory and consumer research in the literature [2,41,43–49] on the molecular and Note by Note meals in both cognitive impression and hedonic terms, which requires a special methodological approach.

It is worth determining the relationship between the expected and the experienced-liking of food products. The sensory expectations are associated with expectations of sensory descriptors such as sweetness and creaminess (at certain intensity), while the hedonic expectations are related to the extent the consumer likes or dislikes the food products [50]. According to Piqueras-Fiszman and Spence [51], the consumer's expectations and perceptions may match or diverge; i.e., their expectations may be confirmed or disconfirmed. This phenomenon likely depends on the consumers and their attitudes, beliefs, personality and product familiarity. Many authors describe new, modernist dishes as a phenomenon in the restaurant business but have not studied sensory quality and emotions associated with it.

The aims of the study were (1) to determine the sensory characteristics of traditional, molecular, and Note by Note dishes; (2) to evaluate liking of key attributes (including expected- and experienced-liking) by consumers and their declared perceptibility of features and emotions in relation to dishes and (3) to observe the type and level of emotions and facial expressions of consumers in response to consumed dishes. As modernist cuisine is a relatively new alternative form of meals, food neophobia and innovativeness level of consumers have been determined. Willingness to try them in the future was also determined.

2. Materials and Methods

2.1. Material

A total of 6 dishes were prepared using lemon or tomatoes made in the traditional, molecular and Note by Note versions (Table 1). As a reference point, two traditional dishes were selected: tomato soup with rice and lemon butter cookies. The molecular version of the conventional dishes was prepared by modifying their texture, shape and/or temperature. The NbN version was designed by replacing conventional ingredients with as many technologically possible chemical compounds, distillates and NbN evocations (IQEMUSU SAS, France). A short ingredients list and description of the dishes' preparation are presented in Appendix A (Table A1).

Table 1. The versions of dishes used in the study.

Main Ingredients	Version of Dish		
	Traditional (Classical)	Molecular	Note by Note
Tomato-based dishes (tomatoes, rice)	creamy tomato soup with rice technique: boiling	puffed rice snack with heirloom tomato and roasted meat powders techniques: powdering; rice snack: deep frying	rice soufflé with the roasted meat essence and tomato gel technique: foaming
Lemon-based dishes (lemon, butter)	lemon cookie technique: baking	butter cookie with lemon peel sherbet with and lemon consommé techniques: baking, freezing, evaporating	cookie-flavored sphere with a lemon coating techniques: freezing, coating, centrifugation

The idea was to obtain completely new dishes that capture the quintessence of classic meals. The dishes (Figure 1) were designed and prepared by a professional chef who has worked in a Michelin-starred restaurant. The plan of the research was to evaluate the new version of foods that were visually different but in congruence with color and taste/flavor of classic products (lemon cakes) alongside dishes that were not in congruence in color (for example of tomato soup). Typical gastronomy equipment, and also Thermomix®VORVERK (Vorwerk Poland, Wrocław), Pacojet®2 (Zug, Switzerland) for ultra-comminution, and Centrifuge 5804 R (Eppendorf, Hamburg, Germany) for centrifugation were used for the preparation of molecular and NbN versions of dishes. The prepared products were complex and involved the presence of a sous-chef during all measurements. The dishes were not suitable for storage, so the chef prepared them immediately before the assessment.

2.2. Experiment Design and Methods

The study design involved two types of research:
1. sensory analytical tests,
2. consumer tests.

In the sensory analytical test, profiling was performed by trained panelists to provide objective data about products. The consumer tests included a survey aimed to characterize the tested group, as well as hedonic scaling and CATA to explore, respectively, liking of some attributes and declared perception of sensory features and emotions by consumers toward dishes. To investigate the mimic expressions assessment of consumers, the FaceReader was used.

Figure 1. Plating proposition of: (**A**) lemon cookie, (**B**) molecular "lemon cookie", (**C**) Note by Note "lemon cookie", (**D**) traditional tomato soup, (**E**) molecular "tomato soup', (**F**) Note by Note 'tomato soup".

2.2.1. Sensory Analytical Method

Sensory profiling. The assessment was carried out using a Quantitative Descriptive Analysis. Sensory descriptors were selected and defined in accordance with the procedure of ISO 13299:2016 [52]. As a result, 16 of them were chosen for lemon-based dishes and 18 for tomato ones. The intensity of sensory attributes was assessed on an unstructured linear graphic scale (100 mm long), with specific word anchors on the edges (none on the left to very intensive on the right). The evaluation was conducted in three repetitions by a group of 8 panelists, who held expert qualifications [53] and have extensive experience in sensory analysis of innovative products.

2.2.2. Consumer Study

Dishes testing included a survey and hedonic assessment performed by consumers (hedonic rating, CATA, FaceReader). Participants were informed that they would assess traditional dishes and their modernist versions.

Questionnaire was validated in a group of 4 people and consisted of 9 questions on their familiarity with the molecular and Note by Note cuisines' terms and the frequency of their consumption. To estimate food neophobia level, the Food Neophobia Scale (FNS) by Pilner and Hobden [54] was used, while consumer innovativeness level was measured using the Domain-Specific Innovativeness (DSI) by Goldsmith and Hofacker [55] with modifications of Huotilainen et al. [56]. Ten statements applied from FNS and six from DSI were assessed on a 7-point scale and ranged from 1 = "strongly disagree" to 7 = "strongly agree" (Appendix A, Table A2). Consumers were also tasked with determining their willingness to try similar dishes in the future.

Hedonic scaling. A 9-point hedonic scale with expressions ranging from "extremely dislike" to "extremely like" [57] was used to quantify the degree of appearance, odor, taste/flavor, consistency and overall-liking (experienced). Additionally, prior to sensory testing, consumers were asked to observe samples and indicate their expected overall-liking of dishes based on their associations related to taste, flavor, and texture of presented dishes. The answer alternatives ranged from "extremely dislike" to "extremely like".

Check-All-That-Apply (CATA). The multiple-choice method was used to evaluate sensory-hedonic sensations and emotions associated with the dish consumption. Consumers had a choice of 22 different emotional attributes (surprising, common, intriguing, typical, disappointing, traditional, dietetic, artificial, delicate, rich, unique, flat flavor, atypical, tasty, interesting, innovative, trendy, boring, fresh, natural, aesthetic and processed).

In our study, attributes were selected by the assessors with a panel leader (after the profiling sessions of the dishes) and on the basis of a literature review made by the authors that was related to for example the use of CATA questions for product evaluation [58,59]. Knowing the sensory characteristics of the dishes was important in proposing the attributes to be identified. The final list of terms was developed by taking into consideration the (1) specificity of the assessed dishes, (2) panel discussion basing on the literature review made by the authors, (3) clarity of meaning terms for consumers (after translation into national language). The coordinator of the consumer assessment explained the principle of CATA questions prior to the evaluation of dishes [60]. Consumers had time to familiarize themselves with the selected attributes and ask questions. No problems with understanding the attributes were reported by consumers.

Mimic expressions assessment. The FaceReader 6 analyzing software (Noldus Information Technology, Wageningen, The Netherlands), connected to a web camera facing participants (angle < 40°), was used to assess mimic expressions in relation to the presented and tested dishes. The program, based on 491 model points, allows the recognition and real-time recording of seven facial reaction patterns: happy, sad, angry, surprised, scared, disgusted and neutral. A more detailed description of the software was published in a previous article [61,62]. The intensity of measured emotion was presented on a numerical scale, wherein 0 means not presented at all and 1 means the maximum value of the fitted model. The procedure for evaluation of the dishes was explained to participants before measurement began. In this study, due to various product textures and moments of swallowing, the emotions of subjects were recorded after the sample was placed in the mouth and for the next 50 s.

2.2.3. Characteristics of Consumers

The consumer testing (CATA, hedonic scaling) was conducted in a group of 56 students and academics, who had reported any food allergies. Most were women (88.4%) and under the age of 25 (80.3%) Table 2. The FaceReader mimic expression was evaluated by 15 participants who declared a good mood and agreed to take part in the test.

FNS and DSI scales were used to sort respondents into appropriate groups. The range of possible FNS scores was from 10 to 70 c.u., wherein higher values corresponded to a greater neophobia level. The actual scores ranged from 12 to 50. The respondents were divided into three groups: the most neophilic (10.0–18.3), the most neutral (18.4–33.8) and the most neophobic subjects (33.9–70.0), in which the cutoff points were calculated by adding or subtracting one standard deviation (7.6) from the mean value (25.9). This classification has been applied in many studies, as reported by Vidigal et al. [32]. DSI scores could range between 6 and 42, but the actual scores were between 10 and 42. The mean value in the group was 26.3 (SD ± 5.6). Based on 33rd and 66th percentile points as cutoffs, consumers were classified into three groups: adapters (6–23), neutrals (24–28) and innovators (29–42).

More than half of the group (65.2%) was classified as the most neutral, while 20.5% of subjects were regarded as the most neophilic, that is, eager to try new products (Table 2). This is also supported by a higher percentage of innovators (34.8%) when compared with adapters (27.7%).

The majority (90.2%) of participants were familiar with the molecular cuisine term, but a smaller percentage (only 33.9%) had tasted it before (Table 2). The concept of the Note by Note cuisine is relatively recent, hence, not many consumers (26.8%) participating in the survey had encountered it before.

Table 2. Characteristic of participants (n = 56).

Characteristic		Percent of Consumers (%)
Gender	Woman	88.4
	Man	11.6
Age	<25	80.3
	25–40	12.5
	41–56	7.2
Food Neophobia Level *	The most neophilic (10.0–18.3)	20.5
	The most neutral (18.4–33.8)	65.2
	The most neophobic (33.9–70.0)	14.3
Domain-Specific Innovativeness level **	Adapters (6–23)	27.7
	Neutrals (24–28)	37.5
	Innovators (29–42)	34.8
Familiarity with the molecular cuisine term	Yes	90.2
	No	9.8
Possibility of earlier tasting of molecular cuisine	Yes	33.9
	No	66.1
Familiarity with the Note by Note cuisine term	Yes	26.8
	No	72.4
Application of ingredients based on synthetic substances at home (e.g., vanillin, bouillon cube)	Yes	84.8
	No	15.2

* Adapted Food Neophobia Scale (10–70 c.u.) [34]; ** Domain-Specific Innovativeness Scale (6–42 c.u.) (Goldsmith and Hofacker [55] with modifications of Huotilainen et al. [56]).

2.2.4. Sample Presentation

Samples of the food (Table 1) were prepared and served immediately. Due to their different textures and consistencies, each was served at the temperature at which it would normally be consumed. They were given as a whole dish for sensory evaluation. Tomato soup and rice soufflé were served at 65 °C; lemon sorbet at −10 °C; cookie sphere and lemon consommé at 5 °C; lemon cookie and puffy rice snack at room temperature. The samples in appropriate quantities (20 g, the only exception was puffy rice snack, which weighed 10 g), were placed in plastic, transparent containers (100 mL) and then covered with a lid, marked with a 3-digit code and given to the assessors in a random order using the sequential monadic test [63]. Natural water was provided as a taste neutralizer between products.

2.2.5. Testing Condition

Both sensory profiling and consumer study were carried out in an accredited sensory laboratory (contract No AB 564) that meets the requirements of ISO 8589:2007 [64]. The products were assessed in the same lighting and conditions to ensure focus on the assessor's perception of sensory characteristics, emotions and liking.

To perform sensory profiling of the products and collect data, the computerized system ANALSENS was used. Panelists that had been trained in sensory analysis were used for profiling in two sessions per day, with sufficient relaxation time (3 h) interval between them. The assessors taking part in the test had about 15 years of experience in profiling of various products, including dishes.

Consumers evaluated the samples during a session in the afternoon of an assessment day that lasted approximately 25 min.

Mimic expressions of participants in relation to each food were observed during two sessions that lasted 8 min each and were separated by relaxation time of about 12 min. Lemon-based meals were followed by tomato-based meals.

2.3. Data Analysis

Statistical analyses were performed using XLStat 2017 (Addinsoft, Paris, France). The Shapiro-Wilk test was used to verify the normality of data distribution. The results were considered to be statistically significant at the level of materiality equal to 0.05.

Analysis of variance (ANOVA) with Fisher's Least Significant Difference (LSD) post hoc test was performed to examine the differences in the intensity of attributes between tomato and lemon-based meals (traditional, molecular, NbN) considering products, assessors, and their interactions as fixed variables (the model two-way ANOVA with interactions).

Pearson correlation coefficients were calculated to relate the liking of given sensory characteristics and results of hedonic scaling of facial expression measurements in consumer tests.

The Kruskal–Wallis test was performed to analyze the differences in degree of liking of examined dishes. The relationships between degree of expected/experienced-liking and liking of particular sensory traits were analyzed by Spearman's rank correlation.

Frequency of use of each one of the terms of the CATA question was determined by counting the number of consumers using a particular term to describe each product. To compare the type of dishes on CATA, results of the correspondence analysis were used. Cochran's Q test was applied to CATA counts to determine whether there was significant difference in consumer perception for a given attribute among the examined dishes in terms of their preparation process. If significant differences were found among the variables, post hoc multiple pairwise comparisons were carried out using McNemar's test with Bonferroni alpha adjustment. Correspondence analysis, based on chi-square distance, was performed to visualize associations between the CATA attributes and the tested products.

To determine the intensity of mimic expressions (FaceReader) in consumption time, the facial expressions of subjects were recorded after the sample was placed in the mouth and for the next 50 s. Then, from 750 records (individual face shots) per person consisting of 7 numbers quantifying emotions, 11 records corresponding to every 5 s of measurement were extracted, and the average for the group was calculated. To determine the intensity of the expressions for swallowing-related time, the moment of swallowing for individuals was determined using video recordings; then, 6 records representing $-10, -5, 0, +5, +10, +15$ in relation to swallowing were extracted, and the average for the group was computed.

3. Results

3.1. Sensory Characteristics of Different Dishes

The modern dishes may differ in sensory properties from traditional counterparts despite similar general assumptions. According to the "food deconstruction" technique, a new dish should preserve the essence of classic one (in our case: tomato soup—tomatoes, rice, roasted meat; lemon cookie—lemon, cookie dough). It becomes essential to recognize the qualitative and quantitative sensory dimensions of the examined dishes.

Sensory characteristics depended mainly on the type of dishes (traditional, molecular or Note by Note) for both lemon cookies and tomato soups (Figure 2A,B). Traditional versions of the dishes, regardless of their type, highlighted the characteristic (typical) attributes. For example, the intense butter note and crispy texture distinguished traditional cookies from others. Among different versions of tomato soup, the highest intensity of typical natural tomato odor and flavor, vegetable note and sour taste were found in traditional products.

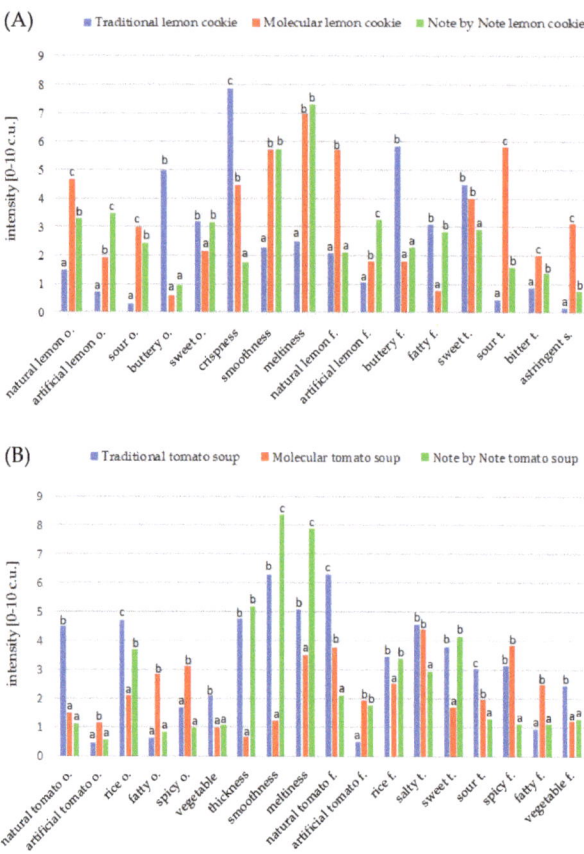

Figure 2. Sensory properties of (**A**) lemon cookie and (**B**) tomato soup that are prepared in traditional, molecular and Note by Note versions: a, b, c—mean values marked by different letters in versions of dishes differ significantly at $p \leq 0.05$.

In contrast, modern cuisine caused a statistical increase or decrease in the intensity of some key attributes that are related to odor, flavor, taste and texture. The molecular and NbN versions of cookies showed a greater level of smoothness and meltiness than the traditional ones. The intensity of many odors and taste/flavor attributes were also different. Molecular cookies were characterized by the highest level of natural lemon odor and flavor, sour taste, and astringency, while the Note by Note cookies revealed the most artificial lemon note (Figure 2A). In turn, the lowest sensation of rice odor and flavor and of sweet taste and the greatest fatty and spicy odor and flavor were observed in molecular tomato soup (Figure 2B). In terms of consistency, traditional and NbN soups revealed a very similar level of thickness, but the NbN sample had the greatest smoothness and meltiness. The molecular version had a completely different texture.

3.2. Consumer Liking of Innovative Food

The evaluation of dishes is a complex experience that includes sensory, emotional and ideational components. For this reason, the experienced and expected-liking and the influence of factors associated with consumer attitudes toward new food were explored.

Hedonic Image. Expected and experienced-liking of traditional and molecular versions of lemon- and tomato-based dishes were not varied (Figure 3). However, the expected-liking of NbN in both meals was significantly higher than the experienced-liking.

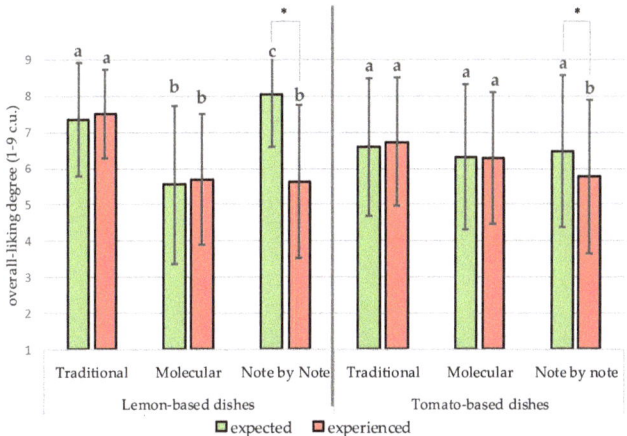

Figure 3. Expected and experienced overall liking in relation to dishes: a, b, c—values of expected-liking or experienced-liking that are marked by different lowercase letters between various versions of dishes with the same main ingredients differ significantly at $p \leq 0.05$; * asterisk brackets indicate a significant difference between expected and experienced overall liking of the given sample at $p \leq 0.05$.

The highest positive correlations were observed between the expected-liking and appearance-liking (rho = [0.33–0.73]; $p \leq 0.05$). In turn, the experienced-liking depended significantly on flavor-liking (rho = [0.82–0.88]; $p \leq 0.05$) and texture-liking (rho = [0.61–0.80]; $p \leq 0.05$). Generally, the liking based on appearance of NbN dishes had the lowest effect on experienced liking.

All the dishes received moderately high hedonic scores; however, their deeper examination showed differences in the degree of liking, depending on type of dish and their version. The hedonic scores for the experienced-liking of the Note by Note dishes were lower than for traditional versions ($p \leq 0.05$) but did not vary from scores of molecular courses (Figure 4). The odor, taste/flavor, and texture-liking for traditional meals were significantly higher ($p \leq 0.05$) than molecular and NbN versions of lemon-based meals and the NbN version of tomato meals (Table 3).

Table 3. Degree of liking of evaluated dishes.

Sensory Trait	Lemon-Based Dishes			Tomato-Based Dishes		
	Traditional	Molecular	NbN	Traditional	Molecular	NbN
	Degree of Liking, (1–9 c.u.) $\bar{x} \pm SE$; $n = 56$					
appearance	7.4 ± 0.2 [b]	5.2 ± 0.3 [a]	8.2 ± 0.2 [c]	6.5 ± 0.3 [a]	6.6 ± 0.3 [a]	6.0 ± 0.3 [a]
odor	8.0 ± 0.2 [c]	6.3 ± 0.2 [a]	6.9 ± 0.3 [b]	6.8 ± 0.2 [b]	5.4 ± 0.3 [a]	6.0 ± 0.3 [a]
taste/flavor	7.5 ± 0.2 [b]	5.5 ± 0.3 [a]	5.2 ± 0.3 [a]	6.8 ± 0.3 [b]	6.2 ± 0.3 [ab]	5.5 ± 0.3 [a]
texture	7.9 ± 0.2 [b]	5.5 ± 0.3 [a]	5.8 ± 0.3 [a]	7.0 ± 0.2 [b]	6.9 ± 0.3 [ab]	6.2 ± 0.3 [a]

[a,b,c]—mean values marked by different letters in verses, differ significantly at $p \leq 0.05$.

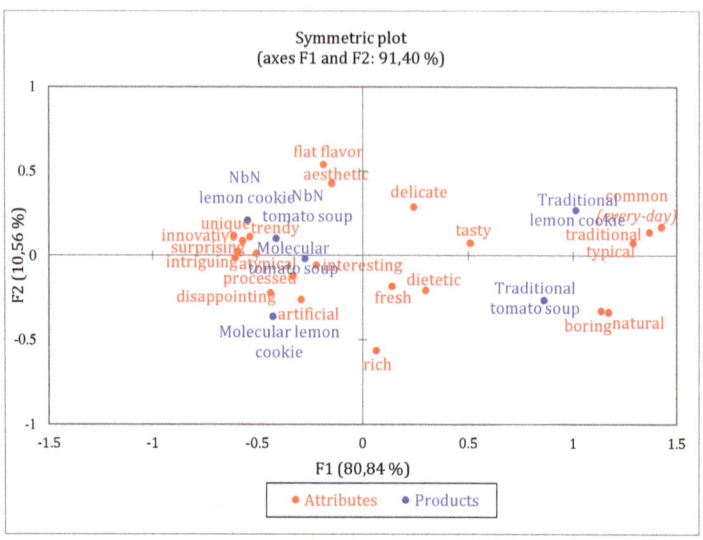

Figure 4. The dishes and term representation in the first and second coordinates of the Correspondence Analysis performed on the frequency of use of attributes by consumers (according to check-all-that-apply).

Factors influencing liking. The higher standard error of the mean hedonic scores exhibits varied affective response of consumers (Table 3, Figure 3). The effect of food neophobia and consumer innovativeness' levels were evaluated. The explanation of the results might be found in the Food Neophobia Level (Table 2). Higher levels of Food Neophobia (FNL) of evaluators resulted in less liking for texture (r = −0.27, $p \leq 0.05$) for the molecular cookie. Moreover, the higher FNL, the lower the experienced-liking for the NbN cookie (r = −37, $p \leq 0.05$). This phenomenon was especially evident in the group of the most neophobic participants. The difference between mean values of expected and experienced-liking was the highest (2.0–4.7 c.u.) in the group of the most neophobic subjects, while in the other groups (the most neutral and the most neophilic participants), it ranged between 0.4–2.2 c.u.

As consumer innovativeness level increased, the flavor-liking (r = −0.36, $p \leq 0.05$) and experienced-liking (r = −0.40, $p \leq 0.05$) of traditional lemon cookie decreased. The familiarity with novel cuisines did not show any significant effect on the liking.

Willingness to try by consumers innovative foods in the future. A higher percentage of consumers showed a willingness to try similar traditional dishes (71.9–86.0%) in the future, than the molecular (54.4–61.4%) or Note by Note (59.6–64.9%) ones (not shown in tables). However, the percentage of people who did not want to try molecular (17.5–31.6%) or NbN (17.5–19.3%) versions was similar to that of the undecided consumers (15.8–21.1% and 17.5–22.8%, respectively). The mean values of liking in a group of consumers who would not eat modernist dishes again were lower than 4.5 c.u. The mean score of undecided consumers (5.1–6.5 c.u.) was lower than those who declared a willingness to eat again (6.5–6.8 c.u.). This may indicate that the consumption of dishes prepared with these innovative techniques introduces a dissonance between unfulfilled sensory expectations and the desire to be surprised.

3.3. Emotions and Sensory-Hedonic Sensations of Meals

The sum of word expressions indicated by consumers used for the survey (n = 56) revealed (data not presented) that molecular and Note by Note dishes evoked more emotions than traditional versions. The NbN cuisine aroused the most emotions (366–405 indications in total), followed by molecular cuisine (300–330) and the least traditional cuisine (303–308).

Data analysis further showed that traditional cuisine compared to modernist ones were associated with word expressions such as tasty (43 indications per meal), traditional (35) and typical (31). The proportions of selections by consumers for each attribute of CATA question for all dishes was presented in a contingency table (Table 4).

Table 4. A contingency table of data set with 6 products and 22 emotional attributes of the CATA question.

Attributes	Lemon-Based Dishes			Tomato-Based Dishes			p-Value (Cochran's Q)
	Traditional	Molecular	NbN	Traditional	Molecular	NbN	
surprising	0.018 a	0.298 b	0.412 b	0.035 a	0.316 b	0.395 b	0.000
common	0.237 b	0.000 a	0.000 a	0.175 b	0.018 a	0.000 a	0.000
intriguing	0.053 a	0.281 b	0.368 b	0.053 a	0.263 b	0.333 b	0.000
typical	0.289 b	0.009 a	0.000 a	0.246 b	0.035 a	0.018 a	0.000
disappointing	0.009 a	0.140 b	0.149 b	0.053 ab	0.061 ab	0.079 ab	0.000
traditional	0.342 b	0.000 a	0.000 a	0.272 b	0.026 a	0.018 a	0.000
dietetic	0.026 a	0.009 a	0.000 a	0.070 a	0.070 a	0.053 a	0.011
artificial	0.018 a	0.132 a	0.114 a	0.079 a	0.105 a	0.096 a	0.052
delicate	0.298 b	0.061 a	0.167 ab	0.167 ab	0.158 ab	0.281 b	0.000
rich	0.070 a	0.281 b	0.088 a	0.263 b	0.167 ab	0.114 a	0.000
unique	0.009 a	0.096 ab	0.184 b	0.035 a	0.140 b	0.237 b	0.000
flat flavor	0.035 ab	0.000 a	0.123 b	0.026 ab	0.035 ab	0.026 ab	0.000
atypical	0.000 a	0.175 b	0.246 b	0.035 a	0.088 ab	0.211 b	0.000
tasty	0.430 b	0.114 a	0.114 a	0.325 b	0.272 ab	0.175 a	0.000
interesting	0.132 a	0.263 ab	0.263 ab	0.158 a	0.289 ab	0.333 b	0.003
innovative	0.009 a	0.149 b	0.298 c	0.018 a	0.211 bc	0.219 bc	0.000
trendy	0.018 ab	0.105 abc	0.175 c	0.009 a	0.123 bc	0.123 bc	0.000
boring	0.053 ab	0.009 ab	0.009 ab	0.096 b	0.009 ab	0.000 a	0.000
fresh	0.132 a	0.158 a	0.079 a	0.114 a	0.088 a	0.096 a	0.379
natural	0.158 b	0.026 a	0.000 a	0.272 b	0.026 a	0.018 a	0.000
aesthetic	0.149 ab	0.053 a	0.263 b	0.044 a	0.132 ab	0.149 ab	0.000
processed	0.035 a	0.123 ab	0.123 ab	0.053 a	0.193 b	0.096 ab	0.001

Cochran's Q test was carried out to determine whether the proportions of selection by consumers for each attribute of the CATA question differed, taking into account the assessed dishes. Post hoc multiple pairwise comparisons were carried out using McNemar's test with Bonferroni alpha adjustment. Different superscript letters within each row denote significant differences ($p \leq 0.05$).

A higher proportion, e.g., closer to 1.00, indicates that, among the six dishes, the attribute was more frequently chosen by consumers. Cochran's Q test revealed differences for 19 of the 22 terms of the CATA questions used to characterize the dishes. The dishes did not vary in attributes such as dietetic, artificial, and fresh. As the use of processed ingredients increased, the number of the following indications increased: surprising, intriguing, innovative, atypical, unique and trendy. These relationships can be clearly seen on a bi-plot of correspondence analysis (explaining 91.4% of total variance), which show associations between the type of dishes (culinary techniques) and emotional descriptors or hedonic sensations. Along with the level of food processing, e.g., for lemon-based dishes (Classic → Molecular, Note by Note cuisine), the disappointment increased when the naturalness sensation decreased (Figure 4). Interestingly, a larger share of consumers perceived NbN dishes as slightly more unique, trendy and innovative than molecular ones.

3.4. Mimic Expressions Assessment: Type and Level of Emotions in Relation to Dishes in Time

Consumption of traditional meals aroused fewer facial expressions among consumers than molecular or Note by Note ones. In the case of traditional courses, the intensity of neutral facial expression ranged between 0.38 and 0.58 c.u., while in molecular and NbN dishes were lower (respectively, 0.28–0.53 and 0.26–0.53 c.u.). The level of neutrality depended on the moment of consumption. It was at a stable level in traditional dishes, while in the case of modernist cuisines, it had different dynamics in time. The consumption of molecular and Note by Note dishes evoked fewer neutral facial expressions while chewing, which rose significantly after swallowing samples. Neutrality rose 4.9–9.2% in traditional versions, 16.6-24.5% in molecular cuisine and 29.1–33.4% in NbN.

Mean values from facial expression measurement (n = 15) and mean values of experienced-liking (n = 56) revealed a significant positive correlation ($p \leq 0.05$). Higher intensity of neutrality (only during the first 10s of measurement) and lower intensity of disgust in 15s of assessment (average deglutition time) resulted in higher scores for experienced-liking (respectively r= 0.82–0.92 and r = 0.96).

Analysis of facial expressions during the entire consumption time revealed that the dominant expression (excluding neutrality) associated with the eating of traditional lemon cookie and tomato soup was happiness (Figure 5a,c). Although kept at a relatively constant level, the emotion was more intense than other emotions from 5 to 20 s after placing the sample in the mouth and from 35 s until the end of measurement.

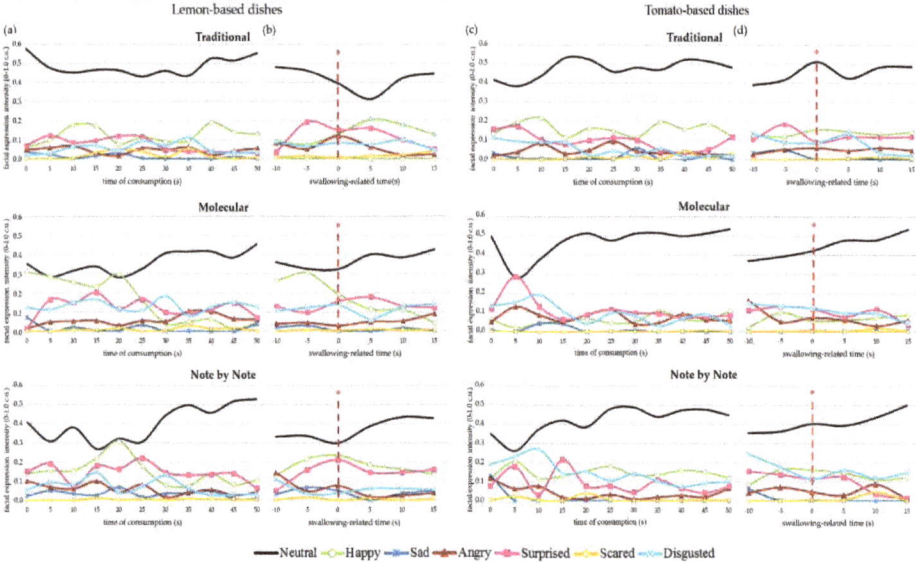

Figure 5. Real-time mimic expresions intensity towards dishes based on (**a**,**b**) lemon and (**c**,**d**) tomato: (**a**)—during entire consumption time of lemon-based dishes; (**b**)—for the swallowing-related time of lemon-based dishes; (**c**)—during entire consumption time of tomato-based dishes; (**d**)—for the swallowing-related time of tomato-based dishes.

In the case of molecular and NbN versions, quite different dynamic changes in expressions can be observed. Moreover, facial expression status during consumption depended on the type of dish. The common feature for both of the lemon-based products examined is that the highest peaks of happiness appear around 20 s from placing the sample in the mouth and then disappear (Figure 5a). Simultaneously, moderately high intensities of surprise mixed with disgust are exhibited during the entire consumption.

Consumers reacted differently for the tomato puffy rice snack. Its consumption did not evoke any happiness, and during the first 10 s of FaceReader measurement, a high level of surprise rose and then decreased (Figure 5c). After 15 s of measurement, the emotions decreased, probably as the result of snack texture. Mimic expressions assessment of the rice soufflé with tomato gel revealed that the dominant expression prior to 15 s of measurement was disgust, followed by happiness. Afterwards, notable peaks of surprise and happiness were recorded. Some facial mimic expression may be related to the sensory properties of dishes (Figure 3) and the feeling of aftertaste in time, both positive and negative.

The analysis of facial expressions dynamics for swallowing-related time (Figure 5b,d) revealed that prior to the swallowing (−5 s) of both traditional dishes, slight surprise peaks were noticed, followed by a neutrality decrease (+5 s). In the case of molecular and NbN lemon cookies, notable (especially in the molecular version) peaks of happiness were noticed 5 s before swallowing. At 5 s before the swallowing of molecular and NbN versions of tomato soup, a slight facial expression of disgust was noticed. In all molecular and NbN versions, a growing trend of neutrality was observed, which was opposite from classical ones.

4. Discussion

The detailed sensory profile is a very important tool for new product development. It allows one to explore the effects of ingredients and processing variables on the final sensory quality of foods and dishes. Moreover, such information is very useful for understanding consumer responses in relation to products' acceptance. The present study revealed that each version of the lemon and tomato dishes (traditional, molecular, NbN) had a separate sensory pattern (profile) related to technological aspects of their production and the recipe used. It is crucial for cooks and producers to maintain a balanced intensity of attributes in various meals that determine their identity and specificity regardless of the modernization process. Thus, too intense or weak intensity of, for example, key flavor descriptors and basic tastes like sourness or sweetness can result in a hedonic response in relation to a product. That, in turn, can potentially affect the degree of liking by consumers and their positive or negative experiences, including evoked emotions [65]. According to Spinelli and Jaeger [66], the sensory characteristics of a product may be altered in order to increase specific positive emotions or even to decrease negative emotions among consumers. This seems crucial in modern cuisine and it is also intriguing due to the atypicality of the dishes.

Visual appeal of the product is an important driver of sensory expectations, which was confirmed in the present study by the highest correlation of expected-liking and appearance-liking. The experienced-liking of NbN dishes was lower than that of traditional ones. The most important factor for consumers was the flavor/taste of the product. These results are in accordance with findings of Santagiuliana et al. [67], who stated that the visual appearance of novel heterogeneous foods can significantly affect expected-liking but not the actual-liking. The experienced-liking is affected by textural and flavor oral sensory perceptions. However, visual cues can significantly influence consumer texture perception. Moreover, literature [68] confirmed that consumers primarily pay attention to flavor-liking when evaluating the overall-liking of products. The examined meals, depending on the type, had a completely different appearance, which could determine the perception of taste and flavor. It is intriguing the extent to which the consonance of the color with the product (yellow lemon cookies) affected the sensory perception of consumers in contrast to the incongruent color of dishes in the case of tomato soup (e.g., white NbN version). Moreover, different textures of the dishes could affect the results of the assessment. Liking of evaluated dishes may be related to their sensory profile. For example, low intensity of buttery attributes as well as a higher level of lemon note in the molecular cookie and artificial lemon attributes (odor and flavor) in NbN compared to the traditional products could determine a lower hedonic response of consumers. Similarly, the significantly lower intensity of tomato flavor and aroma in modern cuisine (molecular, NbN) probably affected the lower results of liking compared to the traditional option. According to other studies [69,70], change of texture and appearance of familiar food in novel food development are only possible when similar retro-nasal sensation is ensured. Consumers are sensitive to the modification of the expected texture. Traynor et al. [48] emphasize that besides intrinsic sensory properties, extrinsic motivation like emotional reactions have a significant effect on the acceptance of novel foods.

Consumer expectations regarding odor, taste, and texture-liking of modernized versions of dishes were probably higher. Similar results were demonstrated in a study by Stolzenbach et al. [71], where the mean scores for liking were significantly higher for the traditional honey than for the novel kind. Moderately high sensory ratings of the molecular version of Portuguese custard tart 'Pastel de Nata' were stated by Oliveira et al. [72]. The lowest appearance-liking for the molecular version of lemon cookies may be the result of food looking more chaotic (cookie chunk stuck into sherbet, which is served in soup). On the other hand, a congruent color with a product representing an unusual shape influenced a significantly higher expected overall-liking when compared to the tomato version (white soufflé soup color, NbN version). This is in line with the research of other authors. Zellner et al. [73] and Zellner et al. [74] provided scientific evidence that foods presented in a less neat manner are less liked. Similar to our findings, results of studies with novel salad dressings [75,76] and with other novel foods [77] indicated a higher mean of expected acceptability in a group of neophiliacs than neophobics.

According to the psychological theories, there are differences in consumer's product expectations [51]. Piqueras-Fiszman and Spence [51] stated that disconfirmation (to any degree) between a consumer's product expectations and the experience can be explained by the two phenomena. The assimilation is one of them, and it happens when the consumer tries to minimize differences between the perception of the product and its expectation, while the contrast happens when the consumer tries to maximize these differences. In our study, consumer attitudes toward examined meals include these two aspects. The assimilation probably occurred for the first two methods (traditional and molecular meals), while the contrast appeared for NbN. This suggests that expectations of consumers were set too high, which resulted in a certain level of dissatisfaction. This also may indicate that the consumption of dishes prepared with these innovative techniques creates dissonance between unfulfilled sensory expectations and the desire to be surprised by novelties. It is worth mentioning that small deviations from the level of consumer adaptation can be seen as interesting and novel. Large deviations tend to result in disgust or neophobia instead [51,54]. It is worth emphasizing that the research results could be somewhat different if the meals were assessed in a different surrounding, for example, in a restaurant. According to many authors [31,44,78] consumers may be more willing to tolerate or expect a certain level of incongruency in some contexts (e.g., modernist/experimental restaurants).

CATA questions provided more profound insight and better explain consumers' perceptions and liking of modern cuisines (molecular and NbN versions of dishes). Traditional cuisine compared to modernist ones were associated with being tasty, traditional, and typical. In contrast, novel cuisines we more frequently perceived as surprising, intriguing and innovative. Our results appear to be well substantiated by the results of several authors, who compared traditional foods with their novel versions. Consumers perceived novel beers as "unusual", "complex", "intriguing" and for "special occasions" [79]; molecular dishes as "surprising", "arousing curiosity", "innovative" and "liked" [41]. Moreover, a familiar type of chocolate was associated with positive emotion terms (e.g., sweet—"happy"), but new chocolates (e.g., with salt) evoked various emotions, including "bored", "interested", or "a little naughty" [80]. Loss et al. [81] also concluded that the acceptance of very innovative foods go beyond neophobia. The results of previous studies [81,82] reveal that traditional versions of selected foods were more liked in comparison to their modernized versions. Additionally, the most unusual foods, arousing curiosity and surprise and challenging for senses, were least liked. Acceptance of unfamiliar food is mainly dependent on its degree of similarity to familiar food [44], which aligns to some extent with the results of our research. In this context, it should be emphasized that the unique, aesthetic and atypical appearance of new products should also correspond with the experienced and balanced taste and flavor that provide satisfaction, evoke positive emotions and affect the degree of liking. It seems that this will be met in particular when the expected-liking is relatively high, but equal to the experienced-liking or even lower. In a previous study [41], modernist cuisine also aroused more positive emotions (35.3%) in consumers than the traditional one.

Mimic expressions assessment by FaceReader was a measure to capture the type and level of emotions as temporal effects in relation to examined meals. According to Danner and Duerrschmid [83], the unfolding of emotions can provide deeper insight into the evaluation process itself and the formation of liking. Our study revealed that higher neutrality during 10 s and disgust in 15 s of consumption was related with higher experienced-liking. These findings align with the result of Danner et al. [84], who found a high positive correlation between "neutral" facial expression with liking, as well as a high negative correlation of "angry" and "disgusted" facial expression with liking. Neutral emotion is frequently used to describe hedonically liked stimuli [85]. In He et al. [42] study, pleasant odors were related to more intense neutral and surprised facial expressions as well as less intense disgust emotion. On the other hand, unpleasant odors evoked fewer neutral reactions and a more intensified disgust. Generally, a higher concentration of odors resulted in lower facial neutrality and more scared expressions. It should be noted that fewer consumers (n = 15) agreed and participated in the FaceReader test than in other tests performed with consumers. According to Crist et al. [86], a range of 10–50 participants would take part in the study using automated facial expression analysis software. They found that the number of participants would vary depending on flavors, flavors intensity and expected treatment acceptability. It was emphasized that products with smaller flavor differences require more participants. Our dishes varied significantly in sensory characteristics (profiling data). Nevertheless, the results of the research with FaceReader should be considered as preliminary.

More consumers would like to eat similar traditional dishes in the future than molecular or NbN dishes. Consumers were a little disappointed by NbN. This was a result of the unfamiliar appearance of dishes, especially the tomato-based dish. The lemon-based dish in NbN version was similar to a dessert, and it affected the evaluation. According to Burke et al. [87], making the NbN dish that looks familiar to the consumers may help to overcome their neophobic reaction. The inspiration from traditional foods may contribute to the success of NbN cooking and cuisine. In previous studies, consumers declared moderately high willingness to buy the molecular vinaigrette jellies (42%), a port wine faux caviar (58%) or powdered olive oil with flavors (72%) [2,45,46]. Most of the participants (88%) in our recent study [41] would like to taste different molecular courses in the future. In turn, consumers in the study by Mielby and Frøst [44] did not want to eat the most innovative and highly unusual molecular dishes again. This is supported by the research of Oliveira et al. [72], wherein purchase intention was moderately high. Tan et al. [88] have also reported that willingness to eat novel burgers with insect additions was lower than regular beef patties. Their findings suggest that consumers may be inclined to taste novel (unusual) foods to satisfy their curiosity, but not to consume them again, especially in the case of low sensory-liking or inappropriateness for consumption. The results of Barenna and Sánchez's [36] study suggest that decisions to consume novel foods have a more pronounced emotional component in a group of neophobic consumers. Hence, neophiliacs try novel foods in restaurants more often than the neophobes [89]. Neophiliacs had a higher awareness, willingness to try and rated unfamiliar food more favorably [75].

The detailed sensory profile is a very important tool for new product development. It allows us to explore the effects of ingredients and processing variables on the final sensory quality of foods and dishes. Moreover, such information is very useful for understanding consumer responses to modernist products such as molecular and NbN dishes. Higher results of expected-liking compared to Note by Note experienced-liking showed disappointment with this version which may be related to the flavor and taste of the dishes. Additionally, significantly lower overall liking of NbN dishes compared to classic ones suggests that the success of novel dishes might be related to the other elements of multisensory experience. This type of modern cuisine is not designed for a large audience. It seems to be intended for small groups of consumers, with low neophobia and who are very familiar with modern cuisine. This finding requires further research.

5. Conclusions

Our findings show that the expected-liking of dishes depended on their type. The results may be related to appearance and especially color–taste congruency with the dishes, which affect consumer associations and expectations regarding sensory characteristics.

Moderately high liking scores of molecular and Note by Note (NbN) cuisines combined with the structure of the emotional response suggests the new insight that consumption of such dishes is as result of temporary emotional arousal derived from an element of novelty. Consumption of modernist dishes evoked more mimic expressions among participants than traditional ones, especially during the first phase (chewing). This facial emotional arousal was confirmed by a number of declared emotions and hedonic sensations. The success of modernist dishes that are served at high-end restaurants is not only strongly associated with acceptable intrinsic sensory characteristics but also with enhanced emotional reactions.

On the other hand, it is worth emphasizing that food neophobia and consumer innovativeness had significant effects on the liking of given products and affected the attitude of consumers towards modern dishes and their willingness to try them again. A higher percentage of consumers showed a willingness to try similar traditional dishes in the future than molecular or NbN dishes, with more than half of the consumers indicating they would eat similar modernist dishes. This may indicate that the consumption of dishes prepared with these innovative techniques introduces a dissonance between unfulfilled sensory expectations and the desire to be surprised by novelties.

Traditional cuisine was associated with a greater number of consumers' word expressions such as tasty, traditional and typical compared to modern ones. A higher level of food processing (molecular cuisine → Note by Note cuisine), caused more participants to perceive meals as surprising, intriguing/interesting and innovative. The uniqueness of modern dishes was most related to innovativeness and being surprising and intriguing, while disappointment in modernist dishes may have been associated with a lower sensation of naturalness (e.g., lemon-based dishes) compared to classical ones.

New food product developers should take into account the attitudes and expectations of potential consumers. In the future, focus groups could be planned to identify different consumer expectations for new products. Such an approach would be valuable in modifying features like the taste, flavor and texture of dishes according to consumers' points of view. Culinary workshops for consumers combined with the preparation and evaluation of such kinds of dishes would be also interesting.

Learning about the perception of modernist cooking solutions enables a better understanding of the sensory experiences of potential consumers, which becomes the space and inspiration for innovation. According to these findings, and taking into account the fact that it is difficult to assess the type of customers trying the dishes (neophiliacs, neophobes, neutrals), we believe that more attention should be given to the appearance of new products. The acceptance of innovation is also linked to the risks of failure. If we change the appearance of the food too much, the exposure to failure will be increased.

6. Limitation

A limitation of our research is that assessments of hedonic and other tests were carried out under laboratory conditions, while the multisensory experience of modernist cuisine implies that it is associated with the restaurant atmosphere, including social contacts and service provided. Moreover, the evaluation of these innovative dishes was performed by comparison to traditional dishes, which could have had an impact on the results.

Application of the immersive approach would recognize contextual influences (e.g., external variables like colors, light, temperature) on perception of modernist dishes, simultaneously with consumer engagement. Such research would allow a deeper analysis and understanding of consumers behavior and expectations towards molecular cuisine and Note by Note cooking.

Another limitation of our research is the number of consumers taking part in the experiment. There were only 56 consumers that took part in the experiment. Amongst them, more than a half were classified as the most neutral, and about 20% were categorized as the most neophiliac, who are eager to try new products.

Author Contributions: Conceptualization, A.G. and E.C.-S.; methodology, A.G., E.C.-S. and E.K.; data curation, A.G., G.W.-Z. and K.B.; writing—original draft preparation, A.G., E.C.S. and E.K.; writing—review and editing, A.G., E.C.-S. and E.K.; visualization, A.G.; supervision, E.C.-S.; data analysis, A.G. and E.K. All authors have read and agreed to the published version of the manuscript.

Funding: Research financed by Polish Ministry of Science and Higher Education within funds of Institute of Human Nutrition Sciences, Warsaw University of Life Sciences (WULS) for scientific research.

Institutional Review Board Statement: The study was conducted according to the guidelines of the Declaration of Helsinki, and approved by the Ethics Committee of the Faculty of Human Nutrition and Consumer Sciences, Warsaw University of Life Sciences–SGGW (Poland) on 22nd January 2019 (Resolution No. 03/2019).

Informed Consent Statement: Informed consent was obtained from all subjects involved in the study.Acknowledgments: Thanks to Andrea Camastra, the Head Chef for enabling Kacper Bylinka to prepare Note by Note dishes in Senses Restaurant in Warsaw, Poland.

Conflicts of Interest: The authors declare no conflict of interest.

Appendix A

Table A1. The preparation procedures of dishes used in research.

Version of Dish	Name of Dish	Ingredients	The Preparation Procedures
		Lemon-Based Dishes	
Traditional (classical)	lemon cookie	flour, egg yolks, milk, butter, sugar, baking powder and lemon zest	Traditional lemon biscuit. The dough was prepared by mixing cold butter with the rest of the ingredients. Next, the dough was rolled, molded and baked.
Molecular	butter cookie with lemon peel sherbet with and lemon consommé	flour, egg yolks, butter, sugar, lemon, glucose, stabilizer, citric acid, xanthan gum and heavy cream	Lemon sherbet was made in PACOJET by comminution of the frozen purée. The puree consisted of cooked lemon zests, lemon juice, sugar sirup, stabilizer, xanthan gum and heavy cream. Lemon consommé was made by reducing a mixture of glucose syrup and lemon juice in a rotary evaporator.
Note by Note	cookie-flavored sphere with a lemon coating	clarified cookie water, xanthan gum, cocoa butter, milk powder, NbN evocation (biscuit and lemon flavors), yellow food colorant (Tartrazine)	In this version, the butter biscuits were mixed with water. After maceration, the liquid was separated using centrifuge. It was then mixed with xanthan gum, milk powder and NbN biscuit evocation followed by freezing in a silicone spherical form. Once frozen, the spheres were then immersed in a coating. The coating was made with cocoa butter, yellow colorant and NbN lemon evocation.
		Tomato-based dishes	
Traditional (classical)	creamy tomato soup with rice	chicken wings, milk powder, sunflower oil, carrot, celery, parsley, onion, fresh bay leaf, allspice, thyme, pelati tomatoes, onion, garlic, tomato paste, salt, pepper, basmati rice	The stock of roasted wings, vegetables and seasonings was pressure cooked and then reduced. Next tomato, onion, garlic, and tomato paste was added. All ingredients were subsequently blended in Thermomix®. Finally, the soup was served with cooked rice.

Table A1. Cont.

Version of Dish	Name of Dish	Ingredients	The Preparation Procedures
		Tomato-based dishes	
Molecular	puffed rice snack with heirloom tomato and roasted meat powders	basmati rice, sunflower oil, maltodextrin, dried heirloom tomatoes, pork sirloin, bay leaf, allspice, thyme, salt, pepper	A mix of heirloom tomato puree and maltodextrin and roasted pork meat (with seasoning) were dried, powdered and served on a puffed rice snack. The snack was made of overcooked rice blended with salt, spread thinly on the sheet pan, dehydrated and then deep-fried in sunflower oil.
Note by Note	rice soufflé with the roasted meat essence and tomato gel	basmati rice, whole milk, kuzu starch, xanthan gum, egg white, sucrose, dextrose, red colorant, fructose, NbN evocation (heirloom tomato flavor, roasted chicken flavor), salt	Heirloom tomato essence was mixed with red colorant, salt, fructose and xanthan gum. It was then frozen in a spherical mold. Rice soufflé was prepared from starch, milk, xanthan gum, sucrose, dextrose and egg whites. Finally, the tomato gel was added to the mixture, steamed in a silicone mold and drizzled with chicken essence.

Table A2. Statements from the Food Neophobia Scale and Domain-Specific Innovativeness used for research.

	To What Extent Do You Agree with the Following Statements (Please Mark X in the Appropriate Box)?	Strongly Disagree	Disagree	More or Less Disagree	Undecided	More or Less Agree	Agree	Strongly Agree
		1	2	3	4	5	6	7
1	I am constantly sampling new and different cuisines. (R)							
2	I do not trust new foods.							
3	If I do not know what is in a food, I will not try it.							
4	I like foods from different countries. (R)							
5	Modernist foods look too weird to eat.							
6	At dinner parties I will try a new food. (R)							
7	I am afraid to eat things I have never had before.							
8	I am very particular about the foods I will eat.							
9	I will eat almost anything. (R)							
10	I like to try restaurants with new cuisines. (R)							
11	I buy new foods before other people do							
12	In general, I am among the first in my circle of friends to buy new foods							
13	Compared to my friends I buy more new foods							
14	Even though new foods are available in the store, I do not buy them (R)							
15	In general, I am the last in my circle of friends to know the trademarks of new foods (R)							
16	I will not buy new foods, if I have not tasted them yet (R)							

R—reversed; Questions 1–10. Adapted Food Neophobia Scale by Pilner and Hobden (1992); Questions 11–16. Domain-Specific Innovativeness Scale by Goldsmith and Hofacker (1991) with modifications of Huotilainen et al. (2006).

References

1. Siro, I.; Kápolna, E.; Kápolna, B.; Lugasi, A. Functional food. Product development, marketing and consumer acceptance—A review. *Appetite* **2008**, *51*, 456–467. [CrossRef]
2. Guiné, R.P.; Barros, A.; Queirós, A.; Pina, A.; Vale, A.; Ramoa, H.; Folha, J.; Carneiro, R. Development of a solid vinaigrette and product testing. *J. Culin. Sci. Technol.* **2013**, *11*, 259–274. [CrossRef]
3. Mărkut, L.; Mărkut, A.; Mârza, B. Modern Tendencies in Changing the Consumers' Preference. *Procedia Econ. Financ.* **2014**, *16*, 535–539. [CrossRef]
4. Baldwin, C.; Wilberforce, N.; Kapur, A. Restaurant and food service life cycle assessment and development of a sustainability standard. *Int. J. Life Cycle Assess.* **2011**, *16*, 40–49. [CrossRef]
5. Haghighi, M.; Dorosti, A.; Rahnama, A.; Hoseinpour, A. Evaluation of factors affecting customer loyalty in the restaurant industry. *Afr. J. Bus. Manag.* **2012**, *6*, 5039–5046. [CrossRef]
6. Walter, U.; Edvardsson, B. The physical environment as a driver of customers' service experiences at restaurants. *Int. J. Qual. Serv. Sci.* **2012**, *4*, 104–119. [CrossRef]
7. Gneezy, A.; Gneezy, U.; Lauga, D.O. A reference-dependent model of the price–quality heuristic. *J. Mark. Res.* **2014**, *51*, 153–164. [CrossRef]
8. Pecotić, M.; Bazdan, V.; Samardžija, J. Interior Design in Restaurants as a Factor Influencing Customer Satisfaction. *RIThink* **2014**, *4*, 10–14.
9. Andersen, B.V.; Hyldig, G. Consumers' view on determinants to food satisfaction. A qualitative Approach. *Appetite* **2015**, *95*, 9–16. [CrossRef]
10. Pilelienė, L.; Almeida, N.; Grigaliunaite, V. Customer satisfaction in catering industry: Contrasts between Lithuania and Portugal. *Tour. Manag. Stud.* **2016**, *12*, 53–59. [CrossRef]
11. Almohaimmeed, B.M.A. Restaurant Quality and Customer Satisfaction. *Int. Rev. Manag. Mark.* **2017**, *7*, 42–49.
12. Konuk Faruk, A. The influence of perceived food quality, price fairness, perceived value and satisfaction on customers' revisit and word-of-mouth intentions towards organic food restaurants. *J. Retail. Consum. Serv.* **2019**, *50*, 103–110. [CrossRef]
13. Liu, Y.; Song, Y.; Sun, J.; Sun, C.; Liu, C.; Chen, X. Understanding the relationship between food experiential quality and customer dining satisfaction: A perspective on negative bias. *Int. J. Hosp. Manag.* **2020**. [CrossRef]
14. Harrington, R.J.; Ottenbacher, M.C.; Kendall, K.W. Fine-dining restaurant selection: Direct and moderating effects of customer attributes. *J. Foodserv. Bus. Res.* **2011**, *14*, 272–289. [CrossRef]
15. Namkung, Y.; Jang, S. Does food quality really matter in restaurants? Its impact on customer satisfaction and behavioral intentions. *J. Hosp. Tour. Res.* **2007**, *31*, 387–410. [CrossRef]
16. Ha, J.; Jang, S. The effects of dining atmospherics on behavioral intentions through quality perception. *J. Serv. Mark.* **2012**, *26*, 204–215. [CrossRef]
17. Ryu, K.; Han, H. Influence of the quality of food, service, and physical environment on customer satisfaction and behavioral intention in quick-casual restaurants: Moderating role of perceived price. *J. Hosp. Tour. Res.* **2010**, *34*, 310–329. [CrossRef]
18. Namkung, Y.; Jang, S. Are highly satisfied restaurant customers really different? A quality perception perspective. *Int. J. Contemp. Hosp. Manag.* **2008**, *20*, 142–155. [CrossRef]
19. Ryu, K.; Jang, S. The effect of environmental perceptions on behavioral intentions through emotions: The case of upscale restaurants. *J. Hosp. Tour. Res.* **2007**, *31*, 56–72. [CrossRef]
20. Ha, J.; Jang, S. Effects of service quality and food quality: The moderating role of atmospherics in and ethnic restaurant segment. *Int. J. Hosp. Manag.* **2010**, *29*, 520–529. [CrossRef]
21. Biswas, D.; Szocs, C.; Chacko, R.; Wansink, B. Shining light on atmospherics: How ambient light influences food choices. *J. Mark. Res.* **2017**, *54*, 111–123. [CrossRef]
22. Ivanovic, S.; Mikinac, K.; Perman, L. Molecular gastronomy in function of scientific implementation in practice. *UTMS J. Econ.* **2011**, *2*, 139–150.
23. Armesto, F.F. Modern Techniques in Kitchen Chemistry. In *Modern Garde Manger: A Global Perspective*; Garlough, R.B., Camp-bell, A., Eds.; Delmar Cengage Learning: New York, NY, USA, 2012; p. 60.
24. This, H. Molecular gastronomy is a scientific discipline, and note by note cuisine is the next culinary trend. *Flavour* **2013**, *2*, 1. [CrossRef]
25. This, H. *Note by Note Cooking—The Future of Food*; Columbia University Press: New York, NY, USA, 2014.
26. Burke, R.; This, H.; Kelly, A.L. Molecular Gastronomy. *Ref. Modul. Food Sci.* **2016**. [CrossRef]
27. Głuchowski, A.; Czarniecka-Skubina, E. Designing innovative food Note by Note (Projektowanie Innowacyjnej Żywności Note by note). *Zesz. Probl. Postępów Nauk Rol.* **2017**, *588*, 25–36. [CrossRef]
28. Torrico, D.D.; Fuentes, S.; Viejo, C.G.; Ashman, H.; Dunshea, F.R. Cross-cultural effects of food product familiarity on sensory acceptability and non-invasive physiological responses of consumers. *Food Res. Int.* **2019**, *115*, 439–450. [CrossRef]
29. Muñoz, F.; Hildebrandt, A.; Schacht, A.; Stürmer, B.; Bröcker, F.; Martín-Loeches, M.; Sommer, W. What makes the hedonic experience of a meal in a top restaurant special and retrievable in the long term? Meal-related, social and personality factors. *Appetite* **2018**, *125*, 454–465. [CrossRef]

30. Torri, L.; Tuccillo, F.; Bonelli, S.; Piraino, S.; Leone, A. The attitudes of Italian consumers towards jellyfish as novel food. *Food Qual. Prefer.* **2020**, *79*, 103782. [CrossRef]
31. Spence, C.; Piqueras-Fiszman, B. Using Surprise and Sensory Incongruity in a Meal. In *The Perfect Meal*; Spence, C., Piqueras-Fiszman, B., Eds.; John Wiley & Sons, Ltd.: Chichester, UK, 2014; pp. 215–242.
32. Vidigal, M.C.; Minim, V.P.; Simiqueli, A.A.; Souza, P.H.; Balbino, D.F.; Minim, L.A. Food technology neophobia and consumer attitudes toward foods produced by new and conventional technologies: A case study in Brazil. *LWT Food Sci. Technol.* **2015**, *60*, 832–840. [CrossRef]
33. Vanhonacker, F.; Kühne, B.; Gellynck, X.; Guerrero, L.; Hersleth, M.; Verbeke, W. Innovations in traditional foods: Impact on perceived traditional character and consumer acceptance. *Food Res. Int.* **2013**, *54*, 1828–1835. [CrossRef]
34. Chen, Q.; Anders, S.; An, H. Measuring consumer resistance to a new food technology: A choice experiment in meat packaging. *Food Qual. Prefer.* **2013**, *28*, 419–428. [CrossRef]
35. Den Uijl, L.C.; Jager, G.; De Graaf, C.; Waddell, J.; Kremer, S. It is not just a meal it is an emotional experience—A segmentation of older persons based on the emotions that they associate with mealtimes. *Appetite* **2014**, *83*, 287–296. [CrossRef] [PubMed]
36. Barrena, R.; Sánchez, M. Neophobia, personal consumer values and novel food acceptance. *Food Qual. Prefer.* **2013**, *27*, 72–84. [CrossRef]
37. Tuorila, H.; Hartmann, C. Consumer responses to novel and unfamiliar foods. *Curr. Opin. Food Sci.* **2020**, *33*, 1–8. [CrossRef]
38. Siegrist, M. Factors influencing public acceptance of innovative food technologies and products. *Trends Food Sci. Technol.* **2008**, *19*, 603–608. [CrossRef]
39. Roman, S.; Sánchez-Siles, L.M.; Siegrist, M. The importance of food naturalness for consumers: Results of a systematic review. *Trends Food Sci. Technol.* **2017**, *67*, 44–57. [CrossRef]
40. Tüzünkan, D.; Albayrak, A. Research about Molecular Cuisine Application as an Innovation Example in Instambul Restaurants. *Procedia Soc. Behav. Sci.* **2015**, *195*, 446–452. [CrossRef]
41. Głuchowski, A.; Czarniecka-Skubina, E.; Pielak, M.; Ołubiec-Opatowska, E. Sensory quality of molecular dishes and consumers' attitudes towards them. *Ital. J. Food Sci.* **2019**, *31*, 1–13. [CrossRef]
42. He, W.; Boesveldt, S.; de Graaf, C.; de Wijk, R.A. The relation between continuous and discrete emotional responses to food odors with facial expressions and non-verbal reports. *Food Qual. Prefer.* **2016**, *48*, 130–137. [CrossRef]
43. Yeomans, M.R.; Chambers, L.; Blumenthal, H.; Blake, A. The role of expectancy in sensory and hedonic evaluation: The case of smoked salmon ice-cream. *Food Qual. Prefer.* **2008**, *19*, 565–573. [CrossRef]
44. Mielby, L.H.; Frøst, M.B. Expectations and surprise in a molecular gastronomic meal. *Food Qual. Prefer.* **2010**, *21*, 213–224. [CrossRef]
45. Guiné, R.P.; Dias, A.; Peixoto, A.; Matos, M.; Gonzaga, M.; Silva, M. Application of molecular gastronomy principles to the development of a powdered olive oil and market study aiming at its commercialization. *Int. J. Gastron. Food Sci.* **2012**, *2*, 101–106. [CrossRef]
46. Guiné, R.P.; Ferreira, P.; Roque, A.R.; Pinto, H.; Tomás, A. Port Wine "Caviar": Product Development, Sensorial Analysis, and Marketing Evaluation. *J. Culin. Sci. Technol.* **2014**, *12*, 294–305. [CrossRef]
47. Traynor, M.P.; Burke, R.; O'Sullivan, M.G.; Hannon, J.A.; Barry-Ryan, C. Sensory and chemical interactions of food pairings (basmati rice, bacon and extra virgin olive oil) with banana. *Food Res. Int.* **2013**, *54*, 569–577. [CrossRef]
48. Traynor, M.; Moreo, A.; Cain, L.; Burke, R.; Barry-Ryan, C. Exploring Attitudes and Reactions to Unfamiliar Food Pairings: An Examination of the Underlying Motivations and the Impact of Culinary Education. *J. Culin. Sci. Technol.* **2020**, 1–23. [CrossRef]
49. Fraat, F.; Zainal, A. The Influence of Hedonic Characteristics on Chefs' Acceptance towards Molecular Asam Pedas. In *Regional Conference on Science, Technology and Social Sciences*; Springer: Singapore, 2016; pp. 1085–1094.
50. Kpossa, M.R.; Lick, E. Visual merchandising of pastries in foodscapes: The influence of plate colours on consumers' flavour expectations and perceptions. *J. Retail. Consum. Serv.* **2020**, *52*, 1–15. [CrossRef]
51. Piqueras-Fiszman, B.; Spence, C. Sensory and hedonic expectations based on food product-extrinsic cues: A review of the empirical evidence and theoretical accounts. *Food Qual. Prefer.* **2015**, *40*, 165–179. [CrossRef]
52. ISO. *ISO 13299:2016: Sensory Analysis—Methodology—General Guidance for Establishing a Sensory Profile*; ISO: Geneva, Switzerland, 2016.
53. ISO. *ISO 8586 2012: General Guidelines for the Selection, Training and Monitoring of Selected Assessors and Expert Sensory Assessors*; ISO: Geneva, Switzerland, 2012.
54. Pilner, P.; Hobden, K. Development of a scale to measure the trait of food neophobia in humans. *Appetite* **1992**, *19*, 105–120. [CrossRef]
55. Goldsmith, R.; Hofacker, C. Measuring consumer innovativeness. *J. Acad. Mark. Sci.* **1991**, *19*, 209–221. [CrossRef]
56. Huotilainen, A.; Pirttilä-Backman, A.M.; Tuorila, H. How innovativeness relates to social representation of new foods and to the willingness to try and use such foods. *Food Qual. Prefer.* **2006**, *17*, 353–361. [CrossRef]
57. Peryam, D.R.; Pilgrim, F.J. Hedonic scale method of measuring food preferences. *Food Technol.* **1957**, *11*, 9–14.
58. Ng, M.; Chaya, C.; Hort, J. Beyond liking: Comparing the measurement of emotional response using EsSense Profile and consumer defined check-all-that-apply methodologies. *Food Qual. Prefer.* **2013**, *28*, 193–205. [CrossRef]

59. Schouteten, J.J.; Gellynck, X.; De Bourdeaudhuij, I.; Sas, B.; Wender, L.P.B.; Perez-Cueto, F.J.A.; De Steur, H. Comparison of response formats and concurrent hedonic measures for optimal use of the EmoSensory®Wheel. *Food Res. Int.* **2017**, *93*, 33–42. [CrossRef] [PubMed]
60. Varela, P.; Ares, G. *Novel Techniques in Sensory Characterisation and Consumer Profiling*; CRC Press, Taylor & Francis Group: Boca Raton, FL, USA, 2014; p. 85.
61. Kostyra, E.; Rambuszek, M.; Waszkiewicz-Robak, B.; Laskowski, W.; Blicharski, T.; Poławska, E. Consumer facial expression in relation to smoked ham with the use of face reading technology. The methodological aspects and informative value of research results. *Meat Sci.* **2016**, *119*, 22–31. [CrossRef]
62. Kostyra, E.; Wasiak-Zys, G.; Rambuszek, M.; Waszkiewicz-Robak, B. Determining the sensory characteristics, associated emotions and degree of liking of the visual attributes of smoked ham. A multifaceted study. *LWT Food Sci. Technol.* **2016**, *65*, 246–253. [CrossRef]
63. Kemp, S.E.; Hollowood, T.; Hort, J. *Sensory Evaluation a Practical Handbook*; Wiley-Blackwell, A John Wiley & Sons, Ltd.: Oxford, UK, 2009; p. 120.
64. ISO. *ISO 8589 2007: Sensory Analysis—General Guidance for the Design of Test Rooms*; ISO: Geneva, Switzerland, 2007.
65. Jiang, Y.; King, J.M.; Prinyawiwatkul, W. A review of measurement and relationships between food, eating behavior and emotion. *Trends Food Sci. Technol.* **2014**, *36*, 15–28. [CrossRef]
66. Spinelli, S.; Jaeger, S.R. What do we know about the sensory drivers of emotions in foods and beverages? *Curr. Opin. Food Sci.* **2019**, *27*, 82–89. [CrossRef]
67. Santagiuliana, M.; Bhaskaran, V.; Scholten, E.; Piqueras-Fiszman, B.; Stieger, M. Don't judge new foods by their appearance! How visual and oral sensory cues affect sensory perception and liking of novel, heterogeneous foods. *Food Qual. Prefer.* **2019**, *77*, 64–77. [CrossRef]
68. Andersen, B.V.; Brockhoff, P.B.; Hyldig, G. The importance of liking of appearance, -odour, -taste and -texture in the evaluation of overall liking. A comparison with the evaluation of sensory satisfaction. *Food Qual. Prefer.* **2019**, *71*, 228–232. [CrossRef]
69. Favalli, S.; Skov, T.; Byrne, D.V. Sensory perception and understanding of food uniqueness: From the traditional to the novel. *Food Res. Int.* **2013**, *50*, 176–188. [CrossRef]
70. Santagiuliana, M.; van den Hoek, I.A.; Stieger, M.; Scholten, E.; Piqueras-Fiszman, B. As good as expected? How consumer expectations and addition of vegetable pieces to soups influence sensory perception and liking. *Food Funct.* **2019**, *10*, 665–680. [CrossRef] [PubMed]
71. Stolzenbach, S.; Bredie, W.L.; Byrne, D.V. Consumer concepts in new product development of local foods: Traditional versus novel honeys. *Food Res. Int.* **2013**, *52*, 144–152. [CrossRef]
72. Oliveira, S.; Fradinho, P.; Mata, P.; Moreira-Leite, B.; Raymundo, A. Exploring innovation in a traditional sweet pastry: Pastel de Nata. *Int. J. Gastron. Food Sci.* **2019**, *17*, 100160. [CrossRef]
73. Zellner, D.A.; Siemers, E.; Teran, V.; Conroy, R.; Lankford, M.; Agrafiotis, A.; Ambrose, L.; Locher, P. Neatness counts. How plating affects liking for the taste of food. *Appetite* **2011**, *57*, 642–648. [CrossRef] [PubMed]
74. Zellner, D.A.; Loss, C.R.; Zearfoss, J.; Remolina, S. It tastes as good as it looks! The effect of food presentation on liking for the flavor of food. *Appetite* **2014**, *77*, 31–35. [CrossRef] [PubMed]
75. King, S.C.; Meiselman, H.L.; Henriques, A. The effect of choice and psychographics on the acceptability of novel flavors. *Food Qual. Prefer.* **2008**, *19*, 692–696. [CrossRef]
76. Henriques, A.S.; King, S.C.; Meiselman, H.L. Consumer segmentation based on food neophobia and its application to product development. *Food Qual. Prefer.* **2009**, *20*, 83–91. [CrossRef]
77. Nasser El Dine, A.; Olabi, A. Effect of reference foods in repeated acceptability tests: Testing familiar and novel foods using 2 acceptability scales. *J. Food Sci.* **2009**, *74*, 97–106. [CrossRef]
78. Spence, C.; Harrar, V.; Piqueras-Fiszman, B. Assessing the impact of the tableware and other contextual variables on multisensory flavour perception. *Flavour* **2012**, *1*, 7. [CrossRef]
79. Cardello, A.V.; Pineau, B.; Paisley, A.G.; Roigard, C.M.; Chheang, S.L.; Guo, L.F.; Hedderley, D.I.; Jaeger, S.R. Cognitive and emotional differentiators for beer: An exploratory study focusing on "uniqueness". *Food Qual. Prefer.* **2016**, *54*, 23–38. [CrossRef]
80. Gunaratne, T.M.; Fuentes, S.; Gunaratne, N.M.; Torrico, D.D.; Gonzalez Viejo, C.; Dunshea, F.R. Physiological Responses to Basic Tastes for Sensory Evaluation of Chocolate Using Biometric Techniques. *Foods* **2019**, *8*, 243. [CrossRef] [PubMed]
81. Loss, C.R.; Zellner, D.; Migoya, F. Innovation influences liking for chocolates among neophilic consumers. *Int. J. Gastron. Food Sci.* **2017**, *10*, 7–10. [CrossRef]
82. Fibri, D.L.N.; Frøst, M.B. Consumer perception of original and modernised traditional foods of Indonesia. *Appetite* **2019**, *133*, 61–69. [CrossRef] [PubMed]
83. Danner, L.; Duerrschmid, K. Automatic facial expressions analysis in consumer science. In *Methods in Consumer Research*; Ares, G., Varela, P., Eds.; Woodhead Publishing: Cambridge, UK, 2018; Volume 2, pp. 231–252.
84. Danner, L.; Sidorkina, L.; Joechl, M.; Duerrschmid, K. Make a face! Implicit and explicit measurement of facial expressions elicited by orange juices using face reading technology. *Food Qual. Prefer.* **2014**, *32*, 167–172. [CrossRef]
85. Jaeger, S.R.; Vidal, L.; Kam, K.; Ares, G. Can emoji be used as a direct method to measure emotional associations to food names? Preliminary investigations with consumers in USA and China. *Food Qual. Prefer.* **2017**, *56*, 38–48. [CrossRef]

86. Crist, C.E.; Duncan, S.E.; Gallagher, D.L. Protocol for Data Collection and Analysis Applied to Automated Facial Expression Analysis Technology and Temporal Analysis for Sensory Evaluation. *J. Vis. Exp.* **2016**, *114*, 1–15. [CrossRef]
87. Burke, R.M.; Danaher, P.; Hurley, D. Creating bespoke note by note dishes and drinks inspired by traditional foods. *J. Ethn. Foods* **2020**, *7*, 1–7. [CrossRef]
88. Tan, H.S.G.; Tibboel, C.J.; Stieger, M. Why do unusual novel foods like insects lack sensory appeal? Investigating the underlying sensory perceptions. *Food Qual. Prefer.* **2017**, *60*, 48–58. [CrossRef]
89. Choe, J.Y.; Cho, M.S. Food neophobia and willingness to try non-traditional foods for Koreans. *Food Qual. Prefer.* **2011**, *22*, 671–677. [CrossRef]

Article

Generating New Snack Food Texture Ideas Using Sensory and Consumer Research Tools: A Case Study of the Japanese and South Korean Snack Food Markets

Rajesh Kumar [1], Edgar Chambers IV [1,*], Delores H. Chambers [1] and Jeehyun Lee [2]

1 Center for Sensory Analysis and Consumer Behavior, Kansas State University, 1310 Research Park Dr., Ice Hall, Manhattan, KS 66502, USA; krajesh@ksu.edu (R.K.); delores@ksu.edu (D.H.C.)
2 Department of Food Science and Nutrition, Pusan National University, Busan 46241, Korea; jeehyunlee@pusan.ac.kr
* Correspondence: eciv@ksu.edu

Citation: Kumar, R.; Chambers, E., IV; Chambers, D.H.; Lee, J. Generating New Snack Food Texture Ideas Using Sensory and Consumer Research Tools: A Case Study of the Japanese and South Korean Snack Food Markets. *Foods* **2021**, *10*, 474. https://doi.org/10.3390/foods10020474

Academic Editors: Claudia Ruiz-Capillas and Ana Herrero Herranz

Received: 4 December 2020
Accepted: 8 February 2021
Published: 22 February 2021

Publisher's Note: MDPI stays neutral with regard to jurisdictional claims in published maps and institutional affiliations.

Copyright: © 2021 by the authors. Licensee MDPI, Basel, Switzerland. This article is an open access article distributed under the terms and conditions of the Creative Commons Attribution (CC BY) license (https://creativecommons.org/licenses/by/4.0/).

Abstract: Food companies spend a large amount of money and time to explore markets and consumer trends for ideation. Finding new opportunities in food product development is a challenging assignment. The majority of new products launched in the market are either copies of existing concepts or line extensions. This study demonstrates how the global marketplace can be used for generating new texture concepts for snack foods. One hundred and twenty-three prepacked snack foods from South Korea (SK) and ninety-five from Japan (JP) were purchased for this study. Projective mapping (PM) was used to sort the snacks on a 2-dimensional map (texture and flavor). Sensory scientists grouped snacks on similarities and dissimilarities. PM results showed, 65% (JP) and 76% (SK) snacks were considered as hard textures, ranging from moderate to extremely hard. Sixty-five percent of JP snacks were savory, whereas 59% of SK snacks had a sweet flavor. The PM 2-dimensional map was used to find white spaces in the marketplace. Thirty-two diversified snacks from each country were screened and profiled using descriptive sensory analysis by trained panelists. Attributes such as sustained fracturability, sustained crispness, initial crispness, and fracturability were the main sensory texture characteristics of snacks. Results showed how descriptive analysis results can be used as initial sensory specifications to develop prototypes. Prototype refinement can be performed by doing multiple developmental iterations and consumer testing. The study showed how white spaces are potential opportunities where new products can be positioned to capture market space. Practical Application: The methodology produced in this study can be used by food product developers to explore new opportunities in the global marketplace.

Keywords: new product development; texture; snacks; ideation; white space; marketplace

1. Introduction

Food companies need to continue to innovate products to sustain market leadership. Current markets are overloaded with product offerings; thus, the challenge is to innovate new products and update existing products to gain new consumers [1]. The innovation of new products has a positive effect on the economic growth of companies [2]. Innovation helps to develop new market segments, expand current market segments and product portfolios, provide positive image building, and bring new consumers to food companies [3]. The rapid changes in technology, market trends, and consumer expectations (e.g., specific dietary, health, environmental sustainability, and packaging) is keeping the food industry under tremendous pressure to spend large amounts of money on new food product development (NPD) to either increase profits or survive [3–6].

Broadly, NPD consists of four stages, namely opportunity identification, development, optimization, and launch [1,7]. The success of NPD is directly related to several factors: (1) a unique product idea or opportunity; (2) large-scale predevelopment research; (3) superior knowledge of the market; and (4) a cross-functional team (management, scientist, marketing and launch) collaboration [2,7]. The combination of the first three factors truly determines the quality of the opportunity identification. At this stage, the idea and product developers unearth new areas of opportunities to fulfill the unmet needs of consumers [8,9]. Food companies use three primary sources for new product idea generation, i.e., the marketplace, within the company, and the environment outside the marketplace [1]. Global markets can be excellent places to explore new product ideas because those markets often provide products unknown to the developers [10].

Globalization has integrated regions, companies, markets, and societies from different countries and continents. The internationalization of markets has removed barriers for food availability and consumption and has allowed companies to explore foreign markets for product innovation and idea generation [10]. Food companies have successfully developed global food products by generating ideas and products and one country and moving those ideas or products to other countries; for example, beverages (e.g., Coca-Cola and Pepsi), tea (e.g., Lipton), coffee (e.g., Nescafe), cigarettes (e.g., Marlboro), or chewing gums (e.g., Wrigley). The inclusion of international markets in NPD for generating new product opportunities offers a great diversity of products, customers, and consumers. Food companies use data (consumer involvement, food trends, and environmental factors) most frequently in the opportunity identification and product design stage of NPD [11]. Thus, researchers and food companies need to find both novel and quality opportunities from the market [12]. These gaps (white spaces) could be potential unmet consumer needs that can be filled by developing products for these identified consumer needs [12].

The main task of NPD is to develop products that deliver desired benefits to their intended consumers. Developing consumer-centric products involves great risks and failures [2,3,13]. Fuller [1] identified two main early-stage risk components in NPD: (a) wrong investments in new products that would later fail in the market, and (b) overlooking a potentially successful new product, termed an opportunity loss. Dijksterhuis [14] explained five factors for a high number of new product failures: (1) the uncoordinated efforts of many different functions working on different aspects of consumer and product development; (2) lack of understanding of consumer behavior; (3) usage of outdated research models; (4) lack in seriousness towards behavioral sciences; and (5) high reliability on the notion that good-quality products automatically lead to high sales. Even after producing a large amount of literature on NPD, the failure rate is still very high. Between 2011 and 2013, 76% of the newly launched consumer goods did not survive one year on the market [15], 45% of products remained on the market for less than half a year [14], 75 to 95% of newly developed food and beverage products failed within one year of launch [16].

To increase the odds of NPD success, many researchers recognized the need to consider consumer behavior and choice-based ideas from external global markets [9,17–22]. Sensory science and consumer research provide techniques to identify white spaces in NPD, support research and development, and contribute to minimizing the decision uncertainty [23].

Researchers have identified the early stages in NPD as the most important activities for both product success and failure [1,24]. The early stages of NPD have sometimes been termed as the "fuzzy front-end" because they are looking to take vague ideas and provide some clarity in understanding actual needs. Unfortunately, they also have been called "fuzzy" for reasons such as ill-defined processes, ambiguities, confusion, and ad hoc decisions [18,24]. The early involvement of sensory and consumer research in NPD is recommended as an important success factor [2,23,25,26]. Thus, there is a need for a structured sensory science-based framework in the early stages of NPD for idea generation [3]. The use of techniques such as interviews, focus groups, behavioral observation, ethnography and other such qualitative measures plays an important part of the process for determining and documenting consumer needs [27–33]. In addition, quantitative measures of consumer understanding, attitudes, behaviors, and emotions as they relate to products provide additional information that may be critical to discerning potential product requirements [34–38].

A sensory method called projective mapping (PM) or "napping" is used as a tool to categorize products and discover white spaces among product groups. In PM, assessors position the products (samples) on a two-dimensional space according to the similarities and differences of product characteristics [5,39–42]. PM has been described as a natural, holistic, and spontaneous way for people to describe products. It has been successfully applied to various food products, e.g., orange juice [43], red sufu [44], wine [45], pork [46], peas, and sweetcorn [47]. The influence of extrinsic factors on a consumer's perception of foods such as smoked bacon [48], fermented dairy products [49], and chicken meat [50] as well as packaging [51] also has been studied with PM. Over the years, PM or "napping" has been shown to be efficient, timely, and cost-effective, to obtain a "big picture" overview of a category and is considered a rapid method for gauging some descriptive sensory attributes. The application of PM as a sensory tool for rapid product categorization and characterization for a large number of products is common [52].

The early stage of NPD includes brainstorming and ideation by looking at consumer and market trends. To develop new concepts, researchers and food companies obtain information from competitive food products in the market and then develop concepts for new products. Using descriptive sensory analysis gives an edge to the researcher in a better understanding of competitive products, and of the marketplace where the potential new product will be placed [53]. Descriptive sensory analysis is a classic sensory method used in NPD to profile products on all of its perceived sensory properties [54,55]. It involves the discrimination and description of both quantitative and qualitative sensory attributes by trained sensory panelists [53]. The descriptive analysis offers various applications such as help in understanding the relationship between sensory and instrumental measurements, the relationship between descriptive sensory and consumer preference measurements, product optimization and validation, product profiling, quality control (product comparison), sensory mapping and product matching, shelf life and packaging effect, etc. [53,56–60].

The descriptive profiling of foods helps to identify the main sensory attributes of food products which can be manipulated: (a) to create a profile of desirable sensory characteristics to help in the development and (b) to define early-stage specifications for a new product [53]. The key sensory attributes that are identified help to distinguish the importance of "tangible" product characteristics that form the basis of technical product specifications [18,25]. Sensory characteristics are measurable and can be manipulated, and therefore, characteristics obtained from a wide range of products can encourage the researcher to create a product with different and multiple sensory profiles [53,61]. Descriptive profiling methods have been used to profile many products including products such as bread [62], fresh and dried mushrooms [63], snacks and snack-like foods [64], potato varieties [65], mate tea [66], ground beef [67], and smoked food products [68]. Many sensory studies combined descriptive analysis results with consumer hedonics to determine why food products are liked by consumers [25,69,70]. The combination also helps to identify consumer segments and their specific sensory preferences for certain product characteristics, and also give insight into possible gaps in the marketplace [71,72].

Consumers describe a product's benefits by perceived intrinsic and extrinsic characteristics (e.g., the crispiness of potato chips [73], creaminess in dairy products [74,75], "health, good taste and convenience" [17]. Principal components analysis (PCA) plots generated on descriptive sensory profiling data provide an opportunity to access the positioning and comparison of products in the market space [53]. Using PCA plots, several white spaces (the open space between products) and product clusters can be identified with their identifying main sensory attributes [53]. Those sensory attributes are reported to be directly experienced by consumers to assess products' evaluation and significantly influence consumer product appraisal [76]. The "white spaces" suggest areas where new products could be developed to meet unmet needs [10,77–79]. However, the presence of white space does not necessarily mean that (a) products do not exist in that space but only that they were not part of the study, (b) just because a product is made to fill the space that the product will succeed, or (c) it is impossible to develop a product that fits the white space based on current technology.

A goal of this project was to highlight one strategic framework to find white spaces in the marketplace and then develop new snack texture concepts to fit the sensory concepts identified as white spaces. The specific objectives were to (a) find the new texture and flavor gaps in several large-scale markets; (b) identify key sensory texture characteristics of the Japan (JP) and South Korea (SK) snack foods; and (c) to demonstrate how unfamiliar marketplaces can be used in NPD for ideation. This study is a continuation and expansion of earlier work [10].

2. Materials and Methods

2.1. Materials

One hundred and twenty-three packaged snacks from Seoul and Busan, SK, and ninety-five packaged snacks from Kyoto, JP, were purchased in-country and shipped to the Center for Sensory Analysis and Consumer Behavior, Kansas State University (KSU), United States (US). Although a wide range of products other than those thought of as traditional snack foods are eaten as snacks [80], fresh fruits, candies and confectionary products often eaten as desserts were excluded from this study to focus on foods that were made and marketed primarily as snack foods. Trained sensory scientists and product developers from the US, China, India, and SK purchased snacks for this study, following a product procurement strategy recommended by Murley [10] to help ensure that the wide range of snack foods and types was represented. Package guidelines were followed for storage and handling.

2.2. Snacks Data Bank

Information related to each snack type such as product name, product description, manufacturer, package size, number of packages, ingredient list, and pictures (front and back) were collected to develop a snack data bank for each country (See the JapaneseSnacksDataBank.xlsx and SouthKoreanSnacksDataBank.xlsx at https://krex.k-state.edu/dspace/handle/2097/40897) (accessed on 2 January 2021). The collected data helped in product identification, product cataloging, and most importantly in knowledge generation about various snack foods such as packaging data, as well as the ingredient and nutritional data. Several authors concluded that knowledge generation on market products, and its proper integration with organizational learning are important aspects of NPD [3,13].

2.3. Projective Mapping

PM was used in its original concept as described by its authors with few modifications [42,81]. A subset of purchased snacks with the most diversified texture profile, new ingredients, and novel concepts were selected from each country for PM. Fifty-one snacks from JP and sixty-six from SK were included in the PM. The modalities used for PM were texture (hard to soft texture perception) and flavor (savory to sweet flavor perception). Snack foods were sorted for similarities and dissimilarities on the aforementioned modalities. The panel determined the key aspects for placement. The snacks were tasted blind with only a two-digit code and sorted into groups by six trained sensory scientists with prior experience in snack food evaluation.

Two 1 h training sessions were held to orient the panelists with products; training included tasting samples. PM was performed on a rectangular des covered k; the center of the desk was labelled for axis interaction and extreme ends were labeled exactly the same as represented in Figure 1. Panelists evaluated one sample at a time, discussed and reached consensus on positioning the samples. The number of samples evaluated in each session was restricted to ten samples. Additional sessions were held after a wait of at least 1 h. When all the samples for a country had been tested, discussed, and placed on the desktop, the panelists reviewed the placement and made any final modifications. At that point, the "x" and "y" coordinates of the desk were measured to provide the specific data for each sample. Water and unsalted crackers were used as palate cleansers. The products were grouped subjectively (based on the perceived texture and flavor evaluation).

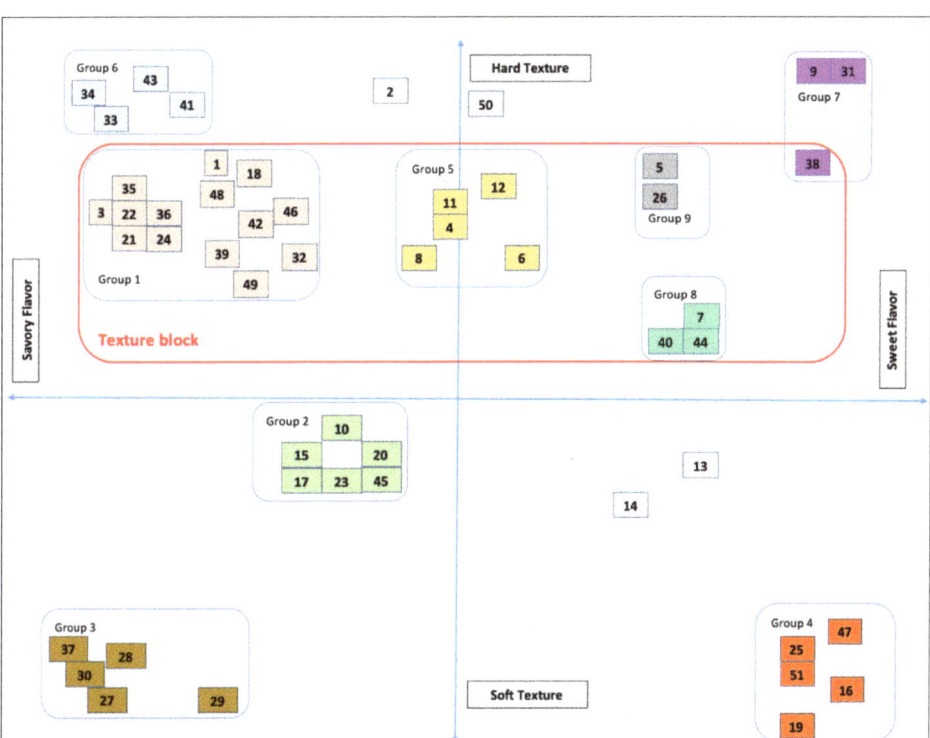

Figure 1. Projective mapping plot of the fifty-one JP snacks showing nine product groupings and outlying products (snacks are coded with 2-digit numbers and snacks with the same color are in the same group). The products' grouping was subjective.

2.4. Snacks Sensory Description

After PM the entire set of products and examining the results, 20 snacks from each country were selected to represent the entire map and were screened for descriptive sensory profiling. To increase the product pool size, 12 new snack products from each country were also added for descriptive profiling. The parameters used to screen snacks included the coverage of the map surface and the selection of diversified and novel textures, new ingredients, and novel concepts. The screened snack foods are listed in Table 1 (for JP) and Table 2 (for SK). In addition, three snacks (Stacy's pita original, Lay's classic potato chips and Tostitos original corn chips) widely available around the world also were included in the test to provide a "reference" set of products that could help anchor the maps. This also allows other researchers to help better understand the similarities and differences shown on the map, particularly because many would never have seen or tasted the products tested.

Table 1. List of the Japan (JP) snacks screened for descriptive profiling.

Serial Numbers	Snacks	Manufacturer [1]	PM Code * and Group Numbers **
1	3D corn bugle	-	23, group-2
2	Bourbon lubera rolls	Bourbon	13
3	Nagewa potato rings	Family Mart collection	4, group-5
4	Sesame wafer rolls	-	40, group-8
5	Seaweed coated crackers	-	34, group-6
6	Freeze-dried strawberries	Fukumi	16, group-4
7	Cheese-filled crackers	Family Mart collection	15, group-2
8	Plum meat snack	Seven Eleven	29, group-3
9	Baby star ramen	Oyatsu Company	33, group-6
10	Strawberry filled balls	Seven Eleven	25, group-4
11	Freeze-dried ice-cream cone	Glico	47, group-4
12	Cheese-filled rolls	Kirara	36, group-1
13	Squid chips	-	21, group-1
14	Unbranded rice crisps	-	41, group-6
15	Pasta shaped snack	Seven Eleven	1, group-1
16	Calbee Potato Sticks	Calbee	2
17	Pocky chocolate sticks	Glico	26, group-9
18	Pea crisps	Calbee	11, group-5
19	Sweet potato sticks	Family Mart collection	9, group-7
20	Unbranded seaweed crackers	-	35, group-1
21	Riska corn potage puffs	Riska	New
22	Bourbon rice crackers with cheese	Bourbon	New
23	Zaku curry filled snacks	-	New
24	Kameda nut clusters	Kameda	New
25	Renkon lotus root chips	Sokan group	New
26	Morianga bites	Morianga	New
27	Steamed plum seaweed	Family Mart Collection	New
28	Edamame crisps	Seven Eleven	New
29	Denroku crispy coated nuts	Denroku	New
30	Mayonnaise potato wedges	Seven Eleven	New
31	Soybean FL coated peanuts	Nuts.com	New
32	Peanut coated cotton candy balls	-	New
33	Lay's classic potato chips	Frito-Lay	Anchor
34	Tostitos original corn chips	Frito-Lay	Anchor
35	Stacy's pita original	Frito-Lay	Anchor

[1] Products without a manufacturer listed are snacks either sold "on the street" or in local "snack" shops in packages without a label. * PM code is a 2-digit number used in the projective mapping plot as an identifier for snack samples. ** Group numbers are provided to identify the snack sample association in projective mapping grouping. New = a product added to this study; Anchor = a common U.S. product used for comparison purposes.

Table 2. List of South Korea (SK) snacks screened for descriptive profiling.

Serial Numbers	Snacks	Manufacturer [1]	PM Code * and Group Numbers **
1	Orion turtle chips (Himalayan salt)	Orion	41, group-3
2	Orion Peanut Balls	Orion	44, group-4
3	Orion original potato chip	Orion	40
4	Taco chips	Lotte	28, group-1
5	Peanut crunchy bar	Koon brother	13, group-9
6	Pulmuone Crispy seaweed chips	Pulmuone	7, group-3
7	Dasang sweet potato sticks	Dasang	5, group-7
8	Soy sauce seaweed chips	Tempura Chips	60, group-2
9	Momali crown snack	Crown Co.	4, group-4
10	Haitai Rice Sticks	Haitai Calbee	52, group-1
11	Daiso orion potato chips	Orion	14
12	Florentin Coconut French dessert	Peacock	48, group-4
13	Heyroo Injeolmi snack	Heyroo	38, group-6
14	Heyroo Noodle snack	Heyroo	21, group-3
15	Heyroo sweet popcorn	Heyroo	43, group-4
16	Heyroo oranda clusters	Heyroo	46, group-5
17	Prawn snack	-	63, group-3
18	Laver Almond	Tom's farm	31
19	Mushroom snack	-	15, group-6
20	Kims crispy roasted laver chips	Dongwon Yangban	57, group-7
21	Seed filled cookie	Lotte	New
22	Seaweed rolls	Only price 2000	New
23	Squid rice balls	-	New
24	Roasted lotus seeds	Daily super nuts	New
25	Baby crab crunch	Farm & Dale	New
26	Soft somjulmi snack	Peacock	New
27	Seaweed crisps	Cheiljedang	New
28	Honey butter cashew-nut	Tom's Farm 1982	New
29	Yogurt cashew-nut	Murgerbon	New
30	Tofu snack	Hav'eat	New
31	NongHyup grain crisps	NongHyup	New
32	Chicken shaped snack	Lotte	New
33	Lay's classic potato chips	Frito-Lay	Anchor
34	Stacy's pita Original	Frito-Lay	Anchor
35	Tostitos original corn chips	Frito-Lay	Anchor

[1] Products without a manufacturer listed are snacks either sold "on the street" or in local "snack" shops in packages without a label. * PM code is a 2-digit number used in the projective mapping plot as an identifier for snack samples. ** Group numbers are provided to identify the snack sample association in projective mapping grouping. New = a product added to this study; Anchor = a common U.S. product used for comparison purposes.

2.5. Descriptive Profiling

Consensus methodology was used to develop sensory attributes, definitions, and references [55,82]. Panelists and the sensory analysts determined attributes for further rating by consensus. The final list of attributes was kept consistent for both JP and SK snacks. The snacks were profiled for flavor, amplitude, appearance, and texture attributes. However, because the flavors of many snack foods can be easily changed based on consumer preferences and many of the snacks tested come in many different flavors, only appearance and texture attribute data were considered in this analysis and are shown in this paper. The texture terms used in descriptive profiling were adopted from the snacks texture lexicon published by Kumar and Chambers [64].

Panelists used a scale ranging from 0 to 15.0 with 0.5 increments where 0 represents none and 15 extremely strong to profile snack samples. Each panelist independently allocated intensities to the attributes and then the intensities were discussed within the panel to reach a single consensus score for each attribute for each product. Three samples were evaluated in each session. Panelists cleaned their palates between samples with freshly cut cucumbers, mozzarella cheese (manufactured by Kroger, Cincinnati, OH, USA), hot water, and a washcloth for the cleaning of lips and hands. The descriptors list, definitions and reference standards are provided in Supplementary Table S1. Similar methodology has been used in other recent studies for the sensory profiling of various foods, e.g., [62–66,82,83].

2.6. Sample Preparation

The snacks used were all ready to eat and needed no preparation; they were served as they were. The samples were blind coded with three-digit codes, served in 8 oz (Styrofoam) and 3.25 oz (plastic) cups (based on the size and shape of the snacks) covered with a lid. One sample at a time was served to panelists in a randomized order.

2.7. Panelists

For the PM, six sensory analysts with experience in snack food evaluation served as the panel for the study. All of the analysts had training in PM techniques and worked as a group to produce a single joint map of snacks for each country. The panelists were trained to specifically focus on the texture and flavor stimuli. The assessors were tasked to screen the large pool of samples, position them on a 2-dimensional space by reaching a consensus on the general differences on texture perception. The objective of PM using a trained panelist was to layout an overall product space rather than generate data through scaling differences. After PM, the descriptive study was planned to identify the subtle difference and quantify descriptors among different panelists.

For the descriptive analysis, six highly trained descriptive sensory panelists were used for this study. Each panelist had more than 120 h of training in descriptive panel training and more than 1000 h of descriptive testing experience with various types of foods and beverages, including extensive testing on different snack type products. The panelist worked on evaluation techniques for appearance, texture, and flavor perception. The panelist received 9 h of additional orientation with both the JP and the SK snacks. The number of highly trained panelists who participated in this study was sufficient to differentiate the samples in the descriptive analysis [84–87] and similar panels have been used in other studies [66,88–92].

2.8. Data Analysis

Correlation-based principal component analysis (PCA) and agglomerative hierarchical clustering (AHC) were performed on the sensory descriptive data using data analysis software XLSTAT 2019.3.2.61545. To prevent data redundancy, attribute correlations were analyzed by the data analytical software R-studio version 4.0.0 (R Foundation for Statistical Computing, Vienna, Austria; https://www.R-project.org/) (accessed on 10 January 2020). Note that for consensus profiling, because there is no variance in scores, "significant" differences are not determined [55,82]. Instead, the size of intensity differences deemed "important" is determined in advance by researchers. In this case, differences were deemed important if they varied by ≥0.5 points, a typical level used in such studies.

3. Results

The sequential use of sensory tools produced information on the main sensory descriptors, the snacks market categorization based on sensory descriptors, existing snacks market space, and white spaces (potential opportunities). All that information was produced by the PM plots and subsequent PCA mapping along with the original data. The information can be used by a snack manufacturer to (a) have an overview of the snack markets (based on sensory parameters); (b) identify the major flavors, textures, and possible trends; (c) learn about a competitor's product positioning; (d) develop new concepts to bring to further sensory (including consumer) research; and (e) enhance their product snack portfolio. The results explain how this information can be generated using JP and SK snacks as examples.

3.1. Projective Mapping

The representative maps of the PM results are presented in Figure 1 (for JP) and Figure 2 (for SK). The snacks are coded with two-digit numbers for representation purposes.

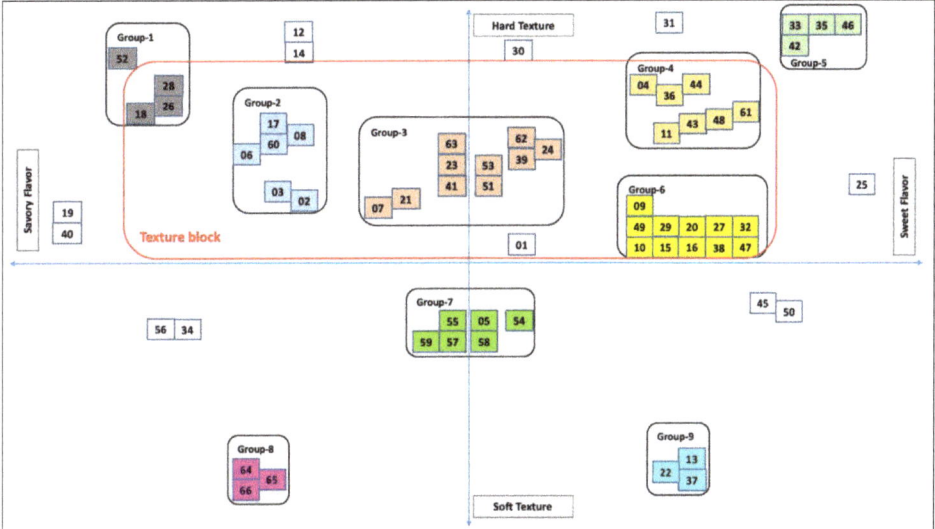

Figure 2. Projective mapping plot of the sixty-six SK snacks showing nine product groupings and outlying products (snacks are coded with 2-digit numbers and snacks with the same color are in the same group). The products' grouping was subjective.

3.1.1. Japanese Snacks

Fifty-one snacks with a variety of texture profiles were sorted into nine groups (Figure 1). The PM was primarily focused on the textural dimension from a hard to a soft texture. Because the snacks were seasoned with different types of flavors, sorting them based on flavor was much too difficult for a 2-dimensional space. The only flavor dimension that was considered was that from savory to sweet. All the products were analyzed visually, in the hand (tactile hand feel), and tested orally (for texture and flavor) by the sensory scientists who participated in the PM.

Out of 51 snacks, 33 snacks (64.71%) were considered as hard bite textures, ranging from moderately to extremely hard. The main texture descriptors were crispiness, crunchiness, sustained crispiness, sustained crunchiness, and hardness. The largest snacks group (group-1) had 14 products (for example, crackers, wafers, puffs, and rolls), representing 27.45% of total snacks. Similarly, group-6 had four snacks, grouped for extremely hard texture and strong savory flavor. Group-2 had six snacks (for example, corn trumpets, corn puffs, squid crackers, shrimp crackers, cheese-filled sticks, and unbranded grain crackers), representing a soft-bite texture with of the mild savory flavor category. A complete list of the JP snack food groups is presented in Table 3.

Table 3. Group identified in the projective mapping of the JP snacks. Group number, number of snacks in each group, texture, snack type, flavor, and snack names.

Groups	Number of Snacks	Texture and Flavor	Snacks Type and Flavor	Examples Snacks Names
Group-1	14	Moderate hard bite texture with mild to a strong savory flavor	Type—crackers, wafers, rolls, puffs Flavor—cheese, squid, savory	Ramen noodle snack, shrimp chips, seaweed crackers, squid snack, rice crackers, coated rice crackers, rice crackers, pasta shape fried snack, ginseng root chips, cheese-filled rolls
Group-2	6	Soft bite texture with a low savory flavor	Type—crackers, wafers, rolls, puffs Flavor—cheese, sweet, sesame	3D corn bugles, corn puffs, squid crackers, shrimp crackers, cheese-filled sticks, unbranded rice crackers
Group-3	5	Extremely soft-chewy with a strong savory flavor	Type—seafood and meat Flavor—seafood, fish	Dried squid, plum meat, dried fish, cheese with cod, spicy grilled kamaboko fish
Group-4	5	Extremely soft with a strong sweet flavor	Type—cake, freeze-dried, puffed balls Flavor—strawberry, chocolate, sweet	Baumkuchen cake, freeze-dried strawberries, strawberry filled puffed balls, freeze-dried strawberry ice-cream cone, chocolate sweet treats
Group-5	5	Moderate hard bite with a bland taste	Type—sticks, chips, crisps Flavor—bland, plain, salt	Fried rice crackers, potato rings, pea sticks, rice crackers, fried rice crackers with peanuts
Group-6	4	Extremely hard bite with a very strong savory flavor	Type—hard grain crackers Flavor—seaweed	Seaweed crackers, baby star ramen noodle snack, unbranded fried snack, unbranded crackers
Group-7	3	Extremely hard bite with a strong sweet flavor	Type—sticks, crackers Flavor -sweet	Unbranded baked crackers, soybean coated walnuts, sweet potato sticks
Group-8	3	Moderate hard bite with a mildly sweet flavor	Type—puffs, crackers, Flavor—sweet, chocolate	Chocolate-coated baked rice puffs, sesame wafer rolls, sugar granules coated crackers
Group-9	2	Moderate soft bite with a mildly sweet flavor	Type—sticks Flavor—sweet, chocolate, sesame	Rice crackers with sesame seeds, Pocky chocolate sticks

Group-1 represents the largest portion of JP snack foods from the selected snack pool. The results suggest that most JP snack foods are hard to bite texture snacks seasoned with various flavors such as savory, bland, and plain salt. Group-1 and -5 differed in terms of flavor intensities but were similar on textural dimensions. Collectively, snacks from groups-1, -5,-7,-8, and -9 formed a large hard texture block (highlighted with a red border) (Figure 1). The hard texture block accounted for 49% of the overall JP snacks market space. Hard texture snacks appeared to dominate the JP snacks market, which has a large number of existing products. The possible explanations could be (a) JP consumers prefer hard texture (crunchy and crispy) snacks; (b) our research team inadvertently collected more hard texture snacks and therefore limited the product pool; or (c) it is a true representative of the JP snacks market. Hence, for a new product developer, understanding the texture dimensions of JP snacks could be a potential framework or area of interest to explore either as copycat products (harder textures) or to create new textures (e.g., at the softer texture end of the spectrum). Of course, another niche area could be bringing new flavors into the existing texture spectrum where flavors may be lacking.

Thirty-three snacks (65%) were savory, including snacks seasoned only with plain salt. Other flavors (for example, seafood, seaweed, prawns, squid, crab, and fish) also were present in that grouping. Savory flavored snacks occupy the largest space in the JP snack market. Thus, for a product developer, a savory flavor could be an easy carry-over from one snack type to another, but also positions the product against a larger competitive set.

The broad range of textures and flavor, some of which were not found in tests conducted on snacks from other countries represent a new opportunity for manufacturers to transfer ideas from one country and culture to another. Taking ideas for new products from countries with a plethora of products often is an easy way to create new products for countries where existing products may be in more limited supply or exist in fewer sensory segments.

The gaps between the product grouping are the white spaces where no products were found to exist. Those empty spaces are potentially unexplored opportunities in the JP snack market and perhaps in other markets. The bottom half of the plot in Figure 1 represents the soft texture snacks space. More white space is available in soft texture snacks over hard texture snacks. This may be because (a) a smaller number of products are in the soft texture product pool (a potential opportunity), or (b) the JP consumer does not prefer soft texture snacks. If soft texture snacks are not as popular in various countries, they may not be a real opportunity. For JP, the further investigation of that snack segment is required in terms of consumer studies. For other countries, the opportunity for new snack development in the sweet category needs to be considered and further research with potential new products may be warranted. In addition, spaces that are not filled with many products also may be considered "white" spaces. For example, the space between group-1 and group-7 has only five products (i.e., group-5). Considering the number of products that exist in other areas of the map, more products could be developed to fill and position in this space.

The plot can be divided into four quadrants (Figure 1). The first quadrant (Q1) represents hard texture snacks with a sweet flavor, the second quadrant (Q2) is hard texture snacks with a savory flavor, the third quadrant (Q3) is soft texture snacks with savory flavor, and the fourth quadrant (Q4) is soft texture snacks with sweet flavor (Figure 1). Each quadrant produces different information. For example, Q4 and Q1 have the least number of snacks and more white spaces. A product developer can develop a wide range of new textures (hard to soft) with sweet flavors. The market space offered in these two quadrants is quite large. Similarly, other quadrants can be used to frame initial product concepts, either individually or in combination with other quadrants.

From a broader perspective, the plot can be divided into two halves. If a product developer is interested in new snack flavors, they can divide the plot on the vertical axis (Figure 1). For example, the left half of this plot, vertically divided, characterizes the savory flavor market space ranging from a hard to a soft texture. The right half of the plot represents the sweeter flavors market space with the same texture range from hard to soft. If the plot is divided into two halves on the horizontal axis, the top half contains all hard texture snacks with both sweet and savory flavors. The bottom half of the plot comprises all softer texture snacks spreading across savory and sweet flavor. There is a wide range of options that could be explored in soft texture with savory flavors. For example, there was no "soft texture, non-seafood" savory snack found in this study. Only 18 snacks were of soft texture, mainly groups-3 and -4. Group-3 consists of fish or seafood flavored soft chewy snack loaded with strong sour-savory flavors. In addition, group-4 snacks were soft textured sweet snacks but not chewy. Considerable white space is available across the savory-sweet flavor dimension with a soft texture profile that may help the developer in identifying additional products for the market.

One issue that must be considered is that many softer textured snacks were found when conducting the initial product search. However, many of those were in the form of freshly prepared "street snacks", such as fresh seafood or egg products that could not be sold in a shelf-stable manner given current technologies. Those products may be considered as inspiration for manufactured shelf-stable products but also represent a competitor that is not directly accounted for in this research.

3.1.2. South Korean Snacks

A total of sixty-six pre-packed snacks were sorted between the texture (hard to soft) and flavor dimensions (savory to sweet). Nine main groups were formed (Figure 2). Group-6 had eleven moderately hard texture snacks with a mild sweet flavor, group-3 had ten moderate hard texture snacks with a bland flavor, group-4 had seven sweet snacks with slightly harder texture than groups-6 and -3. Group-1 had four extremely hard texture snacks with an extremely strong savory flavor, and group-5 snacks had a similar texture but strong sweet flavor. Group-2 snack texture was similar to that of group-3. Group-7 snacks had bland flavors with a slightly softer texture compared to group-3. The other two groups representing soft texture snacks were groups-8 and -9. Both groups were similar in the texture dimensions, with group-8 snacks being savory and group-9 being sweet. A complete description of the groups, texture, flavor, and snack names are provided in Table 4.

Table 4. Group identified in the projective mapping of the SK snacks. Group number, number of snacks in each group, snack type, texture, flavor, and snack names.

Groups	Number of Snacks	Texture and Flavor	Snacks Type and Flavor	Examples Snacks Names
Group-1	4	Extremely hard bite with an extremely strong savory flavor	Type—chips, sticks Flavor—savory, corn, garlic, seaweed	Binggre smoky bacon chip with spicy beef flavor, Brito's snacks, Mexican taco chip, Mister free'd chia seed tortilla chips, Haitai spicy Rice cake sticks
Group-2	6	Moderate hard bite with a mild savory flavor	Type—potato chips, fish chips Flavor—seaweed, chicken, crab, savory	Nong shim cuttlefish roasted butter chips, crab-shaped baked snack, Nongshim chicken leg snack, Nongshim potato chips, Pulmuone seaweed chips
Group-3	10	Moderate hard bite with a bland or little sweet flavor	Type—chips, trail mix, crackers Flavor—soy, seaweed, bland, salt	Orion turtle chips, Peacock Seoul crispy rice chips, Pulmuone crispy seaweed snack, Soy Sauce Tempura Seaweed Snacks, Heyroo noodle snack, Prawn snack, Heyroo seaweed tofu snack, dried fish snack, Mum Mum rice rusks, ChungWoo Fermented Hardtack crackers, The Kims crispy laver chips
Group-4	7	Moderate hard bite with a mildly sweet flavor	Type—nuts, chips, crackers, Flavor—squid, coffee, sweet	Orion squid flavored peanut balls, crunchy and tasty deep anchovy fried, Momali shinchon (crown) snack, Peacock Florentin coconut French dessert, Heyroo sweet popcorn kernel covered with sweet butter scent, coffee coated peanut, Nobrand coconut sticks
Group-5	4	Extremely hard bite with an extremely sweet flavor	Type—puffs, chips Flavor—sweet, peanut	Nobrand seashell-shaped snack, Haitai matdongsan peanut crunch, Heyroo oranda snacks, Amigo chips
Group-6	11	Slight hard bite with a mildly sweet flavor	Type—chips, sticks, crisps, crackers, rolls Flavor—sweet, rice, seaweed, fish	Fried butter potato chips, Orion potato sticks, Orion Gosomi Sweet Cookie Cracker, heyroo injeolmi traditional Rice Cake Snack Crispy Coated by Bean powder, Heyroo egg snacks, Shinhwa seasoned dried fish meat, Haitai calbee sweet potato chips, Crown rice crackers, Market O nature mushroom snack, Big roll grilled seaweed roll: classic flavor, Pulmuone snack chip

Table 4. Cont.

Groups	Number of Snacks	Texture and Flavor	Snacks Type and Flavor	Examples Snacks Names
Group-7	6	Slight soft bite with a bland or little sweet flavor	Type—sticks, chips, crisps Flavor—seaweed, sweet, sesame	Roasted sweet potato chew snack with pineapple flavor, coconut seaweed baby snack, seaweed snack with white sesame, The Kims crispy roasted laver chips, Team Korea crispy laver snack, K-fish seaweed chips
Group-8	3	Very soft chewy texture with a mild savory flavor	Type—Jerky, dried meat Flavor—seafood and meat	Roast horse mackerel, baked cheese dried squid, hot pork jerky
Group-9	3	Very soft texture with a mild to very sweet flavor	Type—grain bars, crisps Flavor—sweet, banana	Mybizcuit peanut crunchy bar, Premium Grain bars, Kiddylicious banana crispy

Group-3 snacks were bland or seasoned with plain salt. Group-7 snacks were seaweed flavored with a slightly soft texture. Overall, 12 snacks, mainly from groups-1 and -2 were seaweed flavored. Group-8 snacks were savory chewy meat/seafood snacks. Group-9 snacks were savory with a soft texture. Thirty-nine (59%) snacks were sweet-flavored or lingered with a sweet taste. Among sweet-flavored snacks, thirty-one (47%) had a slight to moderate hard texture and only eight snacks were soft textured. The PM results obtained from the pooled products showed that the SK market had more sweet snacks over savory.

PM results showed that fifty (75.8%) snacks were in the hard-textured space, varying from slightly hard to moderately hard. Only nine snacks were of extremely hard texture. PM results indicate that the SK snack market space is mainly constituted of slightly to moderately hard texture snacks. The texture dimensions of the SK snacks market were similar to the JP snacks market but with slightly less hard textures. The white space in soft texture products either with savory or sweet flavor is due to the small number of snacks available in that segment. Overall, slight to moderate hard texture with low-intensity sweet flavor can be said to be the best description of the SK snack market. The texture dimension of SK snacks mainly varied from moderately hard to slightly hard with most being sweet flavored. On the other hand, the texture dimension of JP snacks varied from moderately hard to extremely hard and seasoned with savory flavors.

The PM results helped to identify the existing snack food positioning in the market space. This enabled researchers to do a product segmentation and explore white spaces for new opportunities. The developers can look at PM plots as a whole, or as individual quadrants, or half plots to find new product opportunities.

3.2. Descriptive Profiling

3.2.1. Japanese Snacks

Thirty-three texture descriptors were used to profile thirty-five snacks. The PCA plot obtained from the descriptive data is presented in Figure 3. The product variability explained by the first two principal components (PCs) was 44.07% of the total variability. The main differentiating texture attributes were PC1 (initial crispness, fracturability, roughness of mass, sustained fracturability, sustained crispness, cohesiveness, dissolvability, puffiness and firmness) and PC2 (dissolvability, surface shine, porous, cohesiveness, surface roughness, roughness of surface, and puffiness). One set of snacks featured high-intensity scores of PC1 attributes, the other set of products highlighted strong intensities of dissolvability, powdery, porous, and chalky mouthfeel. Another large set of snacks close to the center of the PCA plot represented low intensities of attributes such as adhesive, cohesive of mass, waxy mouthfeel, gritty, mealy, uniformity of bite, and uniformity of surface.

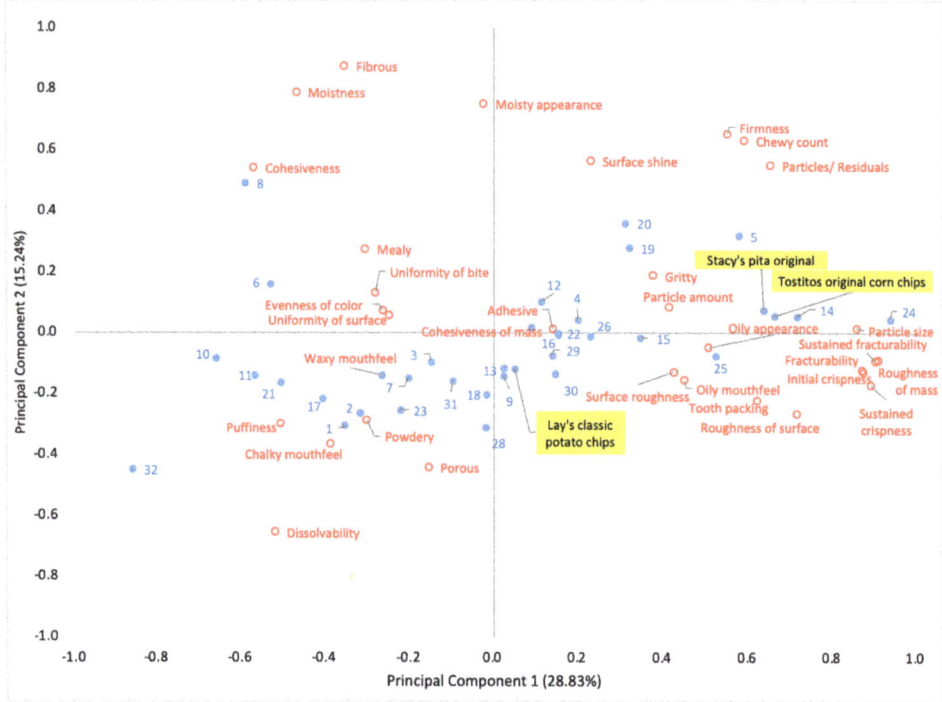

Figure 3. Principal Componensts Analysis (PCA) plot representing the descriptive texture profiling results of JP snacks. The numbers (including dots) highlighted in blue color represent the snack type as listed in Table 1, and the text (including dots) in red color denotes texture attributes. Three US snacks—Stacy's pita original, Lay's classic potato chips, and Tostitos original corn chips (highlighted in yellow color)—were used to compare texture dimensions with JP snacks.

The PCA plot provided a space where new products of certain textures could be developed. For example, there is a scarcity of snacks that are fibrous, cohesive, mealy, moist, having waxy mouthfeel, etc. Similarly, a large white space can be seen around descriptors such as firmness, chew count, gritty, etc. The developer can utilize descriptive data to incubate new texture profiles to fulfill empty texture spaces by introducing new prototypes. The analytical descriptive profiling data can be used as a reference guide to shape new prototypes for further development [24,56]. Of course, white spaces such as the one mentioned in the firm, chewy, gritty area may be undeveloped because that product "concept" may not be appetizing for consumers. However, some products, such as meat jerky, may fit with some aspects of that concept. We also imagine that some high protein products made from plants might fall into that category and whether they are successful or not may depend on accentuating characteristics that might be desirable (firm, chewy) in certain contexts, while reducing characteristics that usually are less desirable (e.g., gritty). Overall, the descriptive sensory profiling can help to design the prototype, determine prototype requirements, and define the key sensory specifications [18].

3.2.2. South Korean Snacks

The PCA plot representing texture descriptive results is shown in Figure 4, with three main snack clusters being noted. The largest groups of snacks had moderate intensities mainly described by the cohesiveness of mass, uniformity of the surface, mealy, chalky mouthfeel, moistness, and adhesive. The second group of snacks was profiled by cohesiveness, doughy, evenness of color, puffiness, and dissolvability. The third group of snacks with strong intensities of texture attributes was marked by PC1. The snacks with strong intensities are represented on the edges of the PCA plot, whereas the snacks with low intensities of textures attributes are located near the center of the PCA plot (Figure 4). The first (PC1) and second principal (PC2) components explained 40.42% of the total variability. The texture attributes contributing to PC1 were dissolvability, cohesiveness, roughness of mass, initial crispness, fracturability, sustained crispness, and roughness of surface. The texture attributes for PC2 were roughness of surface, dissolvability, firmness, fracturability, and initial crispness.

Figure 4. PCA plot representing the descriptive texture profiling results of SK snacks. The numbers (including dots) highlighted with blue color represent the snack type as listed in Table 2, and the text (including dots) in red color denotes texture attributes. Three US snacks—Stacy's pita original, Lay's classic potato chips, and Tostitos original corn chips (highlighted in yellow color)—were used to compare texture dimensions with SK snacks.

Large white spaces between and within snack groups are present. For example, the white space around Stacy's pita original chips shows the unavailability of a similar product in the SK snack market. Similarly, white space around the Peacock Florentin Coconut French dessert, prawn snack, and Heyroo noodle snack shows where new texture concepts could be developed to fill these spaces. The developers can use the tested products as references to quantify texture descriptors.

4. Discussion

This research work adopted a market assessment and product category appraisal approach for new product ideation [90]. This research work applied sensory tools to deliver a pool of new texture concepts. The developer can narrow down the list of new concepts after evaluating consumer response and technical feasibility. The discussion below explains how a step-by-step process can be used to funnel new ideas.

Step 1: Pre-development Homework (Preliminary Market Assessment, Which Markets and Why?)

Detailed preliminary homework was conducted to explore the JP snack market [10], and similar work was done for the SK market, except that an in-country sensory professional was used to help the process move more quickly. The critical sections covered in the pre-development homework includes an assessment of the JP snack market potential, desired snacks market portfolio, the size, feasibility, and area of interest. The other pertinent segments were market selection, location, information acquisition, innovation trends, funds, skilled teams (manpower), product procurement strategy, product shipment, timelines, climate, travel, lodging, boarding, storage, and shipment, etc. Pre-development work is considered important in NPD [3,10,13,18]. During the early stages of NPD, researchers aim to search for novel ideas (for example, texture, ingredient, shape, size, packaging, convenience, and flavor) [1,93,94]. Many researchers reasoned that earlier stage work such as market exploration is most beneficial for the NPD process [1,11,95].

Step 2: Market-Driven Product Assessment

A deep understating of the nature of the market, competitive index, and consumer trends are essential for new product ideation and success [96]. Failure to understand market orientation, assessment, and leaving consumers out of the development process could lead to disasters for innovators. The notion of deep market research to discover white space is supported in several studies [12,18,78]. Researchers undertook a detailed market assessment of the markets which included the participation of local consumers from both countries. A multi-stage market assessment process includes different teams exploring different zones of the market, product procurement strategy, consumer interviews, daily sensory evaluation by sensory scientists, information collection, and shipping enough quantities from the market for further investigation [10].

Once snacks were procured, sensory tools such as a 2-dimensional PM were applied to sort the products into groups. The snacks were segmented for texture and flavor modalities. Sixty-five percent of JP snacks had hard textures (ranging from extremely hard to moderately hard). Results indicate that a big block of snacks across the flavor dimension accounted for 49% of the snacks market space. PM results are a close representation of the JP snacks market space.

The PM tool helped to portray each country's existing snack market texture and flavor outlooks. PM enables the researcher to perform a product segmentation and explore the white spaces in the market. New ideas can fill the white spaces by testing with consumers through models, mock-ups, product concepts, and actual prototypes [18,19,97]. Once the new product concepts are extracted, they should be tested to explore insights on consumer relevance [24]. The initial inputs from the consumers on the needs, likings, and preferences can help to screen and envisage these concepts. A thorough market assessment is a key step in NPD [1,96]. Developers also can use any other sensory dimensions to sort products based on their interests. For example, scientists who work on product renovation or novel ingredients can also use PM as a tool to identify an ingredient's market space.

In the SK snacks, PM results showed that 75% of snacks are hard textured, varying from slightly hard to moderately hard. Fifty-nine percent of SK snacks are sweet flavored or had a sweet aftertaste. Among the sweet-flavored snacks, 47% were hard textured and only eight snacks were soft textured. PM results obtained from pooled products show that SK consumers eat more sweet-flavored snacks than savory.

The overwhelming presence of hard bite texture snacks in the JP and SK market also reflects the product characteristics that currently drive consumer interest. This also advances the need to explore detailed texture attributes that form a product profile. Once foundational characteristics such as the hard bite texture are framed, then the developer seek to measure these texture attributes via descriptive analysis. By identifying what texture attributes form product characteristics (for example crispness, fracturability, firmness in case of snacks) the developers can cement inputs for the subsequent technical prototype developmental stage [2,8,9,26,98].

Large white spaces were discovered on the soft texture axis for both countries. In addition, considerable white space is available across the savory–sweet flavor dimension with a soft texture profile in the JP market (Figure 1). This may be related to the lack of creative product development and marketing in that space, lack of consumer interest in those textures or flavors, or a lack of technology to satisfactorily produce such products. We believe that based on the products encountered, the white space likely results from a lack of product development in that space and the lack of technology to produce products suitable for that space. A number of freshly prepared "street snacks" with a softer, sweeter profile were available in JP, but those products would be difficult or impossible to reproduce today because of distribution and shelf-life issues. For example, a seemingly popular snack of prepared seafood and egg that was both sweet and soft in texture could not be mass-produced and sold because of shelf-life issues based on both sensory changes over time and food safety issues. The expansion of this area through both technology innovation and creative product design and development could bring new textures and flavors into the existing white spaces. Because 65% of the JP snacks evaluated were savory, the potential opportunity to create sweet or sweet and savory snacks is great.

Step 3: Opportunity Definition (Distinct, Early Features, Requirements, and Product Specifications)

Another essential part of NPD is defining the project scope, target market, as well as product features, attributes, and specifications [18,99]. The PCA plots generated from the sensory profiling of snacks can be used as guidelines to frame the sensory profile of new concepts and the direction of potential new product definitions and specifications. Descriptive profiling provided essential elements of the existing snacks such as appearance (color), shape, flavor, and texture attributes (physical components). These key attributes and components can be manipulated in iterative or "structured ways" to come up with new product configurations [54,94]. For example, the attributes of the PC1 and PC2 contributed the most to explaining the total variability from a list of key texture and appearance attributes. The strengths of these texture attributes are measurable and manipulatable to predict and develop new product candidates. Since texture has been identified as an important function of snack foods from which derive consumer desired benefits [64,100], and serves as the base of many snack food development projects, knowing those existing attributes is key information, both to provide reference points for "me-too" products and for companies interested in finding new opportunities. The descriptive analysis helped to quantify product attributes and translate them into measurable product characteristics [55].

The white spaces between snack groups identified by their texture attributes represent the gaps where new prototypes can be placed. The existing snacks' (near to white space) key sensory specifications could be used as a starting point for prototype development. Developers can tweak the key sensory texture intensities by using consumer feedback. Sensory profiles of prototype products can be plotted on the same PCA plot to verify texture positioning. For example, there is a scarcity of snacks that are fibrous, cohesive, mealy, moist, and have a waxy mouthfeel for the JP market (Figure 3). Similarly, wide product space is available for snacks with other key sensory attributes such as firmness, chew count, and gritty.

For the SK snacks, large white spaces were found between and within each snack group (Figure 2). For example, the white space around Stacy's pita original chips shows the unavailability of a similar product in the SK snack market. The developers can use the tested products as reference products to quantify texture specifications. Throughout the NPD process, the prototypes should be compared with the target product for the key attributes and other desirable sensory characteristics identified in descriptive profiling. The inclusion of either target or main competitive products makes it easier for developers to evaluate whether the newly developed prototypes adhere to the desired product concept [101].

Descriptive analysis is valuable for the replacement of essential components. A product developer can either replace essential components (for example, ingredient, flavors, or base material) of the product with something novel or close to the immediate background of the product that can accomplish the same necessary function. For example, the replacement of oil with plant sterols in mayonnaise. The plant sterols not only fulfill the functional requirement of providing structure and flavor carrying ability but also added health benefits by reducing serum cholesterol [24,102]. Once the desired product is fully developed, multiple consumer studies must be carried out to evaluate hedonics towards the newly developed product(s) and comparison must be made with current or competitive products. The foremost benefit of performing the descriptive analysis throughout the NPD is a detailed understanding of products. In addition, descriptive analysis is usually more cost-effective than consumer studies.

A product developer can also make several copies of an existing snack component and alter them in creative ways. For example, the development of purple corn tortilla chips on the line of regular yellow corn tortilla chips. Another creative way would be increasing the plant protein component of existing products for delivering more protein within existing product texture space. A smart developer can include several ideas (for example, environmentally sustainable ingredients, novel ingredients, plant proteins, less processing, and natural) to create niche product spaces but maintain similar texture profiles.

Step 4: Opportunity for Fine-Tuning (Iterative, Prototype Development, Test, Feedback, and Revise Iterations)

In rapidly changing consumer needs, it is not always possible to identify consumer needs and obtain correct product definitions. Developers should use iterative steps to build prototypes to fulfill identified white spaces. Sometimes, consumer requirements change in the time that passes between the beginning and end of development. Thus, the original product definition no longer satisfies consumer requirements.

Often consumers are not clear or fail to articulate what they need in the product until they see the product [103,104]. Thus, it is difficult to obtain an accurate product definition in the early stages of product development if the developer solely depends on explicit consumer inputs for idea generation. Because of limited exposure, consumer inputs are believed to restrict new ideas [26]. Instead, the product definition should be driven by presenting successive versions of the prototypes to consumers for feedback and verification. Iterative development is a dynamic process to capture accurate product definition by presenting a series of deliberative iterative prototypes to consumers. Therefore, the iterative development of prototypes is fluid, captures changing information, and floats the final products close to consumer requirements [18,26].

Information such as what consumers like or dislike and the value consumers see in prototypes should be gathered. The developer can revise, reset or plan the next (future) iteration about the benefits required, propositions, and product design based on gathered feedback.

Step 5: Opportunity Feasibility (Marketing, R&D, Engineering, Production)

Only those prototypes that address the needs of the consumer will be most likely to succeed and should be offered to product and other technology specialists to develop into a tangible product. Similarly, marketing must be involved to determine how products and market needs can be paired and promoted to produce successful launch and sales data.

4.1. Implications

In the current scenario of globalization and high competition, the methodology and results produced in this study could serve as a market-based source for innovation. The market-based approach is a form of open innovation that uses markets as a source for external knowledge [3]. The inclusion of this methodology in the innovation process can be beneficial for the industry. More specifically, it is beneficial for revenues from incremental innovation, reduced time to market, to achieve marginal improvements to existing products, and tremendously impact new product performance [3,105,106]. However, to capitalize market knowledge for innovation, food companies and developers are required to increase absorption capacity [107,108]. The adoption of this methodology could give developers an edge in the hyper competitive environment, capture merging and fast changing consumer needs, and other requirements that can be included in innovation process.

4.2. Limitations

The NPD methodology used in this study to generate new texture concepts could be used for other food product categories. However, researchers are advised to do rigorous homework before applying the suggested methodology to other markets or market categories. This study suggests the utility of sensory methods in market assessment and ideation. However, the adoption of these methods does not guarantee the success of new products, especially if the market has not been evaluated thoroughly. The prototypes developed by using this methodology only confront consumers with products developed within the existing framework of the market tested. This can be noted in two ways: (1) the products selected drive the PM and the descriptive profiling. If the products are not representative of the market or category tested, the results will have limited application or could even be misleading; (2) current products do not necessarily lead to "outside-the-box" innovation. It may be difficult to understand unfulfilled needs by examining the prototypes based on the existing marketplace. This is where the creativity of the product development team including the food scientists, sensory scientists, engineers, and marketing specialists need to come together to "imagine" and then create new product concepts and prototypes for testing.

Results projecting products from one market onto another also are not always successful depending on the similarity in preferences and consumer segments between countries. For example, one study showed that the same segments of consumers existed in multiple countries for a product (pomegranate juice) [109]. However, the proportion of consumers in those segments was completely different in the US and Spain, suggesting that a juice developed for the Spanish market may not be successful in the US. On the other hand, this could be the result of products not being readily available in certain countries or the difference in consumption rates among countries. Testing with consumers who regularly eat certain products in a category is quite different from testing with consumers who are new to the product type. Thus, prototypes developed using the JP and SK snack market framework could be a potential opportunity for the US market or maybe too far out of the current repertory of snack products to be successful. Testing with consumers for acceptance within the current framework and within an "altered framework" is required when testing completely new products.

4.3. Future Research

It is important to conduct comprehensive studies that analyze the impact of a market source innovation model on the performance of new products, incremental income and the cost associated with them. Although these are recommended methods, they need to be examined in a critical light based on case-studies and other use research. In addition, future studies must be conducted to explain what kinds of human capabilities the industry must develop and use to capitalize on external market knowledge.

5. Conclusions

The world is changing rapidly, i.e., more global, less predictable, and more abstruse. The product developers' task is full of multi-faceted challenges. A plethora of literature has been published to deal with these challenges. For example, "open innovation, agile development, design thinking for ideation, stage-gate development, and lean product development". The developers require more creative techniques than "just ask consumers what they want" to increase the chances of success in competitive markets.

This paper showed one method of how new product concepts can be developed using sensory science tools such as product categorization, PM, and descriptive profiling. This research approach for novel and distinctive market opportunities displays an innovative, practical side of NPD research as a compliment. This study also identified the foremost sensory attributes of the JP and SK snack foods that drive consumer benefits. The proposed methodology can be used by food manufacturers to develop new product ideas from unfamiliar markets.

The results of this study can help developers learn to find white spaces in the marketplace and fill these spaces by designing prototypes. The developers can use tested products (close to white spaces) for initial specifications and then build several concepts for consumer assessment. This study is unique in its approach because it allows developers to use sensory methods to put several new ideas on the table for refinement and consumer feedback. The significance of product innovation is critical to business prosperity and consumer satisfaction, however, the keys to success remain indefinable.

Supplementary Materials: The following are available online at https://www.mdpi.com/2304-8158/10/2/474/s1, Table S1: Definitions and references.

Author Contributions: Conceptualization, R.K., E.C.IV, D.H.C.; methodology, R.K., E.C.IV, J.L.; software, R.K., E.C.IV; validation, R.K.; formal analysis, R.K, J.L.; investigation, R.K., E.C.IV, D.H.C., J.L.; resources, E.C.IV; data curation, R.K., E.C.IV; writing—original draft preparation, R.K.; writing—review and editing, R.K., E.C.IV, D.H.C., J.L.; visualization, R.K., J.L.; supervision, E.C.IV, D.H.C.; project administration, E.C.IV; funding acquisition, E.C.IV. All authors have read and agreed to the published version of the manuscript.

Funding: This research is based on work supported by the National Institute of Food and Agriculture, U.S. Department of Agriculture, Hatch under accession number 1016242.

Institutional Review Board Statement: The study was approved by the Committee on Research with Human Subjects at Kansas State University with protocol #7297.3.

Informed Consent Statement: Informed consent was obtained from all participants in the descriptive sensory study and a notice of informed consent was provided to all consumer survey participants who were anonymous to the researchers.

Acknowledgments: We are indebted to our colleagues at Frito-Lay USA for their inspiration and comments on the portions of this project and their help in product procurement. We are also grateful to the sensory scientists and panelists from Kansas State University who participated in the panels for this project.

Conflicts of Interest: The authors declare no conflict of interest.

References

1. Fuller, G.W. *New Food Product Development*; CRC Press: Boca Raton, FL, USA, 2016; ISBN 9780429062711.
2. Guiné, R.P.F.; Ramalhosa, E.C.D.; Valente, L.P. New Foods, New Consumers: Innovation in Food Product Development. *Curr. Nutr. Food Sci.* **2016**, *12*, 175–189. [CrossRef]
3. Santoro, G.; Vrontis, D.; Pastore, A. External knowledge sourcing and new product development. *Br. Food J.* **2017**, *119*, 2373–2387. [CrossRef]
4. Bresciani, S. Open, networked and dynamic innovation in the food and beverage industry. *Br. Food J.* **2017**, *119*, 2290–2293. [CrossRef]
5. Della Corte, V.; Del Gaudio, G.; Sepe, F. Innovation and tradition-based firms: A multiple case study in the agro-food sector. *Br. Food J.* **2018**, *120*, 1295–1314. [CrossRef]
6. Merieux NutriSciences; Lascom. How to Facilitate Your Product Development in a Global Regulatory Environment. A white paper. Food and Beverage. 2018. Available online: http://www.lascom.com/wp-content/uploads/2018/05/White-Paper_Merieux-NutriSciences-Lascom_How-to-facilitate-product-development-in-a-global-regulatory-environment.pdf (accessed on 10 January 2020).
7. Stewart-Knox, B.; Parr, H.; Bunting, B.; Mitchell, P. A model for reduced fat food product development success. *Food Qual. Prefer.* **2003**, *14*, 583–593. [CrossRef]
8. Banović, M.; Krystallis, A.; Guerrero, L.; Reinders, M.J. Consumers as co-creators of new product ideas: An application of projective and creative research techniques. *Food Res. Int.* **2016**, *87*, 211–223. [CrossRef] [PubMed]
9. De Pelsmaeker, S.; Gellynck, X.; Delbaere, C.; Declercq, N.; Dewettinck, K. Consumer-driven product development and improvement combined with sensory analysis: A case-study for European filled chocolates. *Food Qual. Prefer.* **2015**, *41*, 20–29. [CrossRef]
10. Murley, T.; Kumar, R.; Chambers, E.; Chambers, D.; Ciccone, M.; Yang, G. A process for evaluating a product category in an unfamiliar country: Issues and solutions in a case study of snacks in Japan. *J. Sens. Stud.* **2020**, *35*, e12574. [CrossRef]
11. Horvat, A.; Granato, G.; Fogliano, V.; Luning, P.A. Understanding consumer data use in new product development and the product life cycle in European food firms—An empirical study. *Food Qual. Prefer.* **2019**, *76*, 20–32. [CrossRef]

12. Johnson, M.W. *Reinvent Your Business Model: How to Seize the White Space for Transformative Growth*; Harvard Business Press: Brighton, MA, USA, 2018; ISBN 978-1-6336-9646-4.
13. Jagtap, S.; Duong, L.N.K. Improving the new product development using big data: A case study of a food company. *Br. Food J.* **2019**, *121*, 2835–2848. [CrossRef]
14. Dijksterhuis, G. New product failure: Five potential sources discussed. *Trends Food Sci. Technol.* **2016**, *50*, 243–248. [CrossRef]
15. Nielson. Breakthrough innovation report. 2014. Available online: https://www.nielsen.com/us/en/insights/report/2014/breakthrough-innovation-report/ (accessed on 10 January 2020).
16. Kemp, S.; Hort, J. Trends in sensory science. *Food Sci. Technol.* **2015**, *29*, 36–39.
17. Asioli, D.; Varela, P.; Hersleth, M.; Almli, V.L.; Olsen, N.V.; Næs, T. A discussion of recent methodologies for combining sensory and extrinsic product properties in consumer studies. *Food Qual. Prefer.* **2017**, *56*, 266–273. [CrossRef]
18. Cooper, R.G. The drivers of success in new-product development. *Ind. Mark. Manag.* **2019**, *76*, 36–47. [CrossRef]
19. Costa, A.I.A.; Jongen, W.M.F. New insights into consumer-led food product development. *Trends Food Sci. Technol.* **2006**, *17*, 457–465. [CrossRef]
20. Grujić, S.; Odžaković, B.; Ciganović, M. Sensory analysis as a tool in the new food product development. In Proceedings of the II International Congress Food Technology Quality and Safety, Novi Sad, Serbia, 28–30 October 2014; pp. 325–330.
21. Ryynänen, T.; Hakatie, A. "We must have the wrong consumers"—A case study on new food product development failure. *Br. Food J.* **2014**, *116*, 707–722. [CrossRef]
22. Simeone, M.; Marotta, G. Towards an integration of sensory research and marketing in new food products development: A theoretical and methodological review. *Afr. J. Bus. Manag.* **2010**, *4*, 4207–4216.
23. Talavera, M.; Chambers, E. Using sensory sciences help products succeed. *Br. Food J.* **2017**, *119*, 2130–2144. [CrossRef]
24. MacFie, H.J.H. Index. In *Consumer-Led Food Product Development*; Elsevier: Amsterdam, The Netherlands, 2007; pp. 593–613, ISBN 9781845690724.
25. Crofton, E.C.; Scannell, A.G.M. Snack foods from brewing waste: Consumer-led approach to developing sustainable snack options. *Br. Food J.* **2020**. [CrossRef]
26. Cuny, C.; Petit, C.; Allain, G. Capturing implicit texture-flavour associations to predict consumers' new product preferences. *J. Retail. Consum. Serv.* **2020**. [CrossRef]
27. Godin, L.; Laakso, S.; Sahakian, M. Doing laundry in consumption corridors: Wellbeing and everyday life. *Sustain. Sci. Pract. Policy* **2020**, *16*, 99–113. [CrossRef]
28. Mahama-Musah, F.; Vanhaverbeke, L.; Gillet, A. The impact of personal, market- and product-relevant factors on patronage behaviour in the automobile tyre replacement market. *J. Retail. Consum. Serv.* **2020**, *57*, 102206. [CrossRef]
29. Mora, M.; Romeo-Arroyo, E.; Torán-Pereg, P.; Chaya, C.; Vázquez-Araújo, L. Sustainable and health claims vs sensory properties: Consumers' opinions and choices using a vegetable dip as example product. *Food Res. Int.* **2020**, *137*, 109521. [CrossRef]
30. Zocchi, D.M.; Piochi, M.; Cabrino, G.; Fontefrancesco, M.F.; Torri, L. Linking producers' and consumers' perceptions in the valorisation of non-timber forest products: An analysis of Ogiek forest honey. *Food Res. Int.* **2020**, *137*, 109417. [CrossRef]
31. Donelan, A.K.; Chambers, D.H.; Chambers, E.; Godwin, S.L.; Cates, S.C. Consumer Poultry Handling Behavior in the Grocery Store and In-Home Storage. *J. Food Prot.* **2016**, *79*, 582–588. [CrossRef]
32. Pawera, L.; Khomsan, A.; Zuhud, E.A.M.; Hunter, D.; Ickowitz, A.; Polesny, Z. Wild food plants and trends in their use: From knowledge and perceptions to drivers of change in West Sumatra, Indonesia. *Foods* **2020**, *9*, 1240. [CrossRef]
33. Talavera, M.; Sasse, A.M. Gathering consumer terminology using focus groups—An example with beauty care. *J. Sens. Stud.* **2019**, *34*, 12533. [CrossRef]
34. Hoppu, U.; Puputti, S.; Mattila, S.; Puurtinen, M.; Sandell, M. Food consumption and emotions at a salad lunch buffet in a multisensory environment. *Foods* **2020**, *9*, 1349. [CrossRef]
35. Bryant, C.; van Nek, L.; Rolland, N.C.M. European markets for cultured meat: A comparison of germany and france. *Foods* **2020**, *9*, 1152. [CrossRef]
36. Chambers, E.; Tran, T.; Chambers, E. Natural: A $75 billion word with no definition—Why not? *J. Sens. Stud.* **2019**, *34*, e12501. [CrossRef]
37. Murley, T.; Chambers, E. The influence of colorants, flavorants and product identity on perceptions of naturalness. *Foods* **2019**, *8*, 317. [CrossRef] [PubMed]
38. Doungtip, P.; Sriwattana, S.; Kim, K.T. Understanding Thai consumer attitudes and expectations of ginseng food products. *J. Sens. Stud.* **2020**, *35*, 12553. [CrossRef]
39. Aguiar, L.A.; Melo, L.; de Lacerda de Oliveira, L. Validation of rapid descriptive sensory methods against conventional descriptive analyses: A systematic review. *Crit. Rev. Food Sci. Nutr.* **2019**, *59*, 2535–2552. [CrossRef] [PubMed]
40. Cartier, R.; Rytz, A.; Lecomte, A.; Poblete, F.; Krystlik, J.; Belin, E.; Martin, N. Sorting procedure as an alternative to quantitative descriptive analysis to obtain a product sensory map. *Food Qual. Prefer.* **2006**, *17*, 562–571. [CrossRef]
41. Pagès, J.; Cadoret, M.; Lê, S. The sorted napping: A new holistic approach in sensory evaluation. *J. Sens. Stud.* **2010**, *25*, 637–658. [CrossRef]
42. Risvik, E.; McEwan, J.A.; Rødbotten, M. Evaluation of sensory profiling and projective mapping data. *Food Qual. Prefer.* **1997**, *8*, 63–71. [CrossRef]

43. Zhang, T.; Lusk, K.; Mirosa, M.; Oey, I. Understanding young immigrant Chinese consumers' freshness perceptions of orange juices: A study based on concept evaluation. *Food Qual. Prefer.* **2016**, *48*, 156–165. [CrossRef]
44. He, W.; Chung, H.Y. Comparison between quantitative descriptive analysis and flash profile in profiling the sensory properties of commercial red sufu (Chinese fermented soybean curd). *J. Sci. Food Agric.* **2019**, *99*, 3024–3033. [CrossRef] [PubMed]
45. Brand, J.; Kidd, M.; van Antwerpen, L.; Valentin, D.; Næs, T.; Nieuwoudt, H.H. Sorting in combination with quality scoring: A tool for industry professionals to identify drivers of wine quality rapidly. *S. Afr. J. Enol. Vitic.* **2018**, *39*, 163–175. [CrossRef]
46. González-Mohíno, A.; Antequera, T.; Pérez-Palacios, T.; Ventanas, S. Napping combined with ultra-flash profile (UFP) methodology for sensory assessment of cod and pork subjected to different cooking methods and conditions. *Eur. Food Res. Technol.* **2019**, *245*, 2221–2231. [CrossRef]
47. Cliceri, D.; Dinnella, C.; Depezay, L.; Morizet, D.; Giboreau, A.; Appleton, K.M.; Hartwell, H.; Monteleone, E. Exploring salient dimensions in a free sorting task: A cross-country study within the elderly population. *Food Qual. Prefer.* **2017**, *60*, 19–30. [CrossRef]
48. Saldaña, E.; Martins, M.M.; Behrens, J.H.; Valentin, D.; Selani, M.M.; Contreras-Castillo, C.J. Looking at non-sensory factors underlying consumers' perception of smoked bacon. *Meat Sci.* **2020**, *163*, 108072. [CrossRef] [PubMed]
49. Soares, E.K.B.; Esmerino, E.A.; Ferreira, M.V.S.; da Silva, M.A.A.P.; Freitas, M.Q.; Cruz, A.G. What are the cultural effects on consumers' perceptions? A case study covering coalho cheese in the Brazilian northeast and southeast area using word association. *Food Res. Int.* **2017**, *102*, 553–558. [CrossRef] [PubMed]
50. Katiyo, W.; Coorey, R.; Buys, E.M.; Kock, H.L. Consumers' perceptions of intrinsic and extrinsic attributes as indicators of safety and quality of chicken meat: Actionable information for public health authorities and the chicken industry. *J. Food Sci.* **2020**, *85*, 1845–1855. [CrossRef] [PubMed]
51. Thomas, S.; Chambault, M. *Integrating the Packaging and Product Experience in Food and Beverages*; Elsevier: Amsterdam, The Netherlands, 2016; ISBN 9780081003565.
52. Mayhew, E.; Schmidt, S.; Lee, S.Y. Napping—Ultra flash profile as a tool for category identification and subsequent model system formulation of caramel corn products. *J. Food Sci.* **2016**, *81*, S1782–S1790. [CrossRef] [PubMed]
53. Valentin, D.; Cholet, S.; Nestrud, M.; Abdi, H. Projective mapping and sorting tasks, Chapter 15. In *Descriptive Analysis in Sensory Evaluation*; Kemp, S.E., Hort, J., Hollowood, T., Eds.; John Wiley & Sons, Ltd.: Chichester, UK, 2018; pp. 535–559, ISBN 978-0-470-67139-9.
54. Lawless, H.T.; Heymann, H. *Sensory Evaluation of Food*; Food Science Text Series; Springer: New York, NY, USA, 2013; ISBN 978-1-4419-6487-8.
55. Chambers, E. Consensus Methods for Descriptive Analysis. In *Descriptive Analysis in Sensory Evaluation*; Kemp, S., Hort, J., Hollowood, T., Eds.; John Wiley & Sons, Ltd.: Chichester, UK, 2018; pp. 213–236, ISBN 978-0-470-67139-9.
56. Chambers, E. Analysis of sensory properties in foods: A special issue. *Foods* **2019**, *8*, 291. [CrossRef]
57. Luchsinger, S.E.; Kropf, D.H.; García Zepeda, C.M.; Chambers IV, E.; Hollingsworth, M.E.; Hunt, M.C.; Marsden, J.L.; Kastner, C.L.; Kuecker, W.G. Sensory analysis and consumer acceptance of irradiated boneless pork chops. *J. Food Sci.* **1996**, *61*. [CrossRef]
58. Muñoz, A.M.; Chambers, E.I. Relating measurements of sensory properties to consumer acceptance of meat products. *Food Technol.* **1993**, *47*, 128–131.
59. Suwonsichon, S. The importance of sensory lexicons for research and development of food products. *Foods* **2019**, *8*, 27. [CrossRef] [PubMed]
60. Yang, J.; Lee, J. Application of sensory descriptive analysis and consumer studies to investigate traditional and authentic foods: A review. *Foods* **2019**, *8*, 54. [CrossRef]
61. Van Kleef, E.; van Trijp, H.C.M.; Luning, P. Consumer research in the early stages of new product development: A critical review of methods and techniques. *Food Qual. Prefer.* **2005**, *16*, 181–201. [CrossRef]
62. Tran, T.; James, M.N.; Chambers, D.; Koppel, K.; Chambers, E. Lexicon development for the sensory description of rye bread. *J. Sens. Stud.* **2019**, *34*, e12474. [CrossRef]
63. Chun, S.; Chambers, E.; Han, I. Development of a Sensory Flavor Lexicon for Mushrooms and Subsequent Characterization of Fresh and Dried Mushrooms. *Foods* **2020**, *9*, 980. [CrossRef] [PubMed]
64. Kumar, R.; Chambers, E. Lexicon for multiparameter texture assessment of snack and snack-like foods in English, Spanish, Chinese, and Hindi. *J. Sens. Stud.* **2019**, *34*, e12500. [CrossRef]
65. Sharma, C.; Chambers, E.; Jayanty, S.S.; Sathuvalli Rajakalyan, V.; Holm, D.G.; Talavera, M. Development of a lexicon to describe the sensory characteristics of a wide variety of potato cultivars. *J. Sens. Stud.* **2020**, *35*, e12577. [CrossRef]
66. de Godoy, R.C.B.; Chambers, E.; Yang, G. Development of a preliminary sensory lexicon for mate tea. *J. Sens. Stud.* **2020**, *35*, e12570. [CrossRef]
67. Laird, H.; Miller, R.K.; Kerth, C.R.; Chambers, E. The Flavor and Texture Attributes of Ground Beef. *Flavor Texture Attrib. Gr. Beef* **2017**, *1*, 7. [CrossRef]
68. Jaffe, T.R.; Wang, H.; Chambers, E. Determination of a lexicon for the sensory flavor attributes of smoked food products. *J. Sens. Stud.* **2017**, *32*, e12262. [CrossRef]
69. Culbert, J.A.; Ristic, R.; Ovington, L.A.; Saliba, A.J.; Wilkinson, K.L. Influence of production method on the sensory profile and consumer acceptance of Australian sparkling white wine styles. *Aust. J. Grape Wine Res.* **2017**, *23*, 170–178. [CrossRef]

70. Lee, J.; Chambers, E.; Chambers, D.H.; Chun, S.S.; Oupadissakoon, C.; Johnson, D.E. Consumer acceptance for green tea by consumers in the United States, Korea and Thailand. *J. Sens. Stud.* **2010**, *25*, 109–132. [CrossRef]
71. Bowen, A.J.; Blake, A.; Tureček, J.; Amyotte, B. External preference mapping: A guide for a consumer-driven approach to apple breeding. *J. Sens. Stud.* **2019**, *34*, e12472. [CrossRef]
72. Sharma, C.; Jayanty, S.S.; Chambers, E.; Talavera, M. Segmentation of potato consumers based on sensory and attitudinal aspects. *Foods* **2020**, *9*, 161. [CrossRef]
73. Salvador, A.; Varela, P.; Sanz, T.; Fiszman, S.M. Understanding potato chips crispy texture by simultaneous fracture and acoustic measurements, and sensory analysis. *LWT Food Sci. Technol.* **2009**, *42*, 763–767. [CrossRef]
74. Antmann, G.; Ares, G.; Salvador, A.; Varela, P.; Fiszman, S.M. Exploring and explaining creaminess perception: Consumers' underlying concepts. *J. Sens. Stud.* **2011**, *26*, 40–47. [CrossRef]
75. Frøst, M.B.; Janhøj, T. Understanding creaminess. *Int. Dairy J.* **2007**, *17*, 1298–1311. [CrossRef]
76. De Pelsmaeker, S.; Dewettinck, K.; Gellynck, X. The possibility of using tasting as a presentation method for sensory stimuli in conjoint analysis. *Trends Food Sci. Technol.* **2013**, *29*, 108–115. [CrossRef]
77. Adriana, L.S.; Mauricio, C.; Delores, C.; Loreida, T.; Kadri, K.; Edgar, C.; Yu, H. Benefits, Challenges, and Opportunities of Conducting a Collaborative Research Course in an International University Partnership: A Study Case Between Kansas State University and Tallinn University of Technology. *J. Food Sci. Educ.* **2019**, *18*, 78–86. [CrossRef]
78. Corley, J. Finding the white space in natural personal care: How Olivina identified a niche (natural male grooming) and fast-tracked from formulation to shelf in nine months. *Glob. Cosmet. Ind.* **2017**, *185*, 34–39.
79. Thompson, A. Kinder Joy, M&M'S Caramel top IRI's list of 2018 New Product Pacesetters. *Candy Ind.* **2019**, *184*, 10–11.
80. Phan, U.T.X.; Chambers, E. Application of An Eating Motivation Survey to Study Eating Occasions. *J. Sens. Stud.* **2016**, *31*, 114–123. [CrossRef]
81. Pagès, J. Collection and analysis of perceived product inter-distances using multiple factor analysis: Application to the study of 10 white wines from the Loire Valley. *Food Qual. Prefer.* **2005**, *16*, 642–649. [CrossRef]
82. Chambers, E.; Chambers, D.H. Chapter 1 | Consensus Profile Methods Derived from the Flavor Profile Method. In *Descriptive Analysis Testing for Sensory Evaluation*, 2nd ed.; Bleibaum, R., Ed.; ASTM International: West Conshohocken, PA, USA, 2020; pp. 1–18.
83. Griffin, L.E.; Dean, L.L.; Drake, M.A. The development of a lexicon for cashew nuts. *J. Sens. Stud.* **2017**, *32*, e12244. [CrossRef]
84. Chambers, E.; Bowers, J.A.; Dayton, A.D. Statistical designs and panel training/experience for sensory analysis. *J. Food Sci.* **1981**, *46*, 1902–1906. [CrossRef]
85. Chambers, E.; Smith, E.A. Effects of testing experience on performance of trained sensory panelists. *J. Sens. Stud.* **1993**, *8*, 155–166. [CrossRef]
86. Chambers, D.H.; Allison, A.M.A.; Chambers IV, E. Training effects on performance of descriptive panelists. *J. Sens. Stud.* **2004**, *19*, 486–499. [CrossRef]
87. Otremba, M.M.; Dikeman, M.E.; Milliken, G.A.; Stroda, S.L.; Unruh, J.A.; Chambers IV, E. Interrelationships among evaluations of beef longissimus and semitendinosus muscle tenderness by Warner-Bratzler shear force, a descriptive-texture profile sensory panel, and a descriptive attribute sensory panel. *J. Anim. Sci.* **1999**, *77*, 865–873. [CrossRef] [PubMed]
88. Belisle, C.; Adhikari, K.; Chavez, D.; Phan, U.T.X. Development of a lexicon for flavor and texture of fresh peach cultivars. *J. Sens. Stud.* **2017**. [CrossRef]
89. Chambers, E.; Lee, J.; Chun, S.; Miller, A.E. Development of a Lexicon for Commercially Available Cabbage (Baechu) Kimchi. *J. Sens. Stud.* **2012**, *27*, 511–518. [CrossRef]
90. Muñoz, A.M.; Chambers IV, E.; Hummer, S. A multifaceted category research study: How to understand a product category and its consumer responses. *J. Sens. Stud.* **1996**, *11*. [CrossRef]
91. Suwonsichon, S.; Chambers Iv, E.; Kongpensook, V.; Oupadissakoon, C. Sensory lexicon for mango as affected by cultivars and stages of ripeness. *J. Sens. Stud.* **2012**, *27*, 148–160. [CrossRef]
92. Thompson, K.R.; Chambers, D.H.; Chambers IV, E. Sensory characteristics of ice cream produced in the U.S.A. and ITALY. *J. Sens. Stud.* **2009**, *24*, 396–414. [CrossRef]
93. Grunert, K.G. (Ed.) *Consumer Trends and New Product Opportunities in the Food Sector*; Wageningen Academic Publishers: Wageningen, The Netherlands, 2017; ISBN 978-90-8686-307-5.
94. Simms, C.; Trott, P. Packaging Dependent Products: How do Firms in the Packaged Food Sector Manage the Development of new Packaging Opportunities? In Proceedings of the 12th European Conference on Innovation and Entrepreneurship, Academic Conferences International Limited, Paris, France, 21–22 September 2017; pp. 611–619.
95. Wind, J.; Mahajan, V. Issues and Opportunities in New Product Development: An Introduction to the Special Issue. *J. Mark. Res.* **1997**, *34*, 1–12. [CrossRef]
96. Cooper, R.G. Best practices and success drivers in new product development. In *Handbook of Research on New Product Development*; Edward Elgar Publishing: Cheltenham, UK, 2018; ISBN 978-1-78471-814-5.
97. Costa, G.M.; Paula, M.M.; Costa, G.N.; Esmerino, E.A.; Silva, R.; Freitas, M.Q.; Barão, C.E.; Cruz, A.G.; Pimentel, T.C. Preferred attribute elicitation methodology compared to conventional descriptive analysis: A study using probiotic yogurt sweetened with xylitol and added with prebiotic components. *J. Sens. Stud.* **2020**. [CrossRef]

98. Moussaoui, K.A.; Varela, P. Exploring consumer product profiling techniques and their linkage to a quantitative descriptive analysis. *Food Qual. Prefer.* **2010**, *21*, 1088–1099. [CrossRef]
99. Cooper, R.G. *Winning at New Products: Creating Value through Innovation*; Basic Books: New York, NY, USA, 2017; ISBN 0465093329.
100. Kumar, R.; Chambers, E. Understanding the terminology for snack foods and their texture by consumers in four languages: A qualitative study. *Foods* **2019**, *8*, 484. [CrossRef]
101. O'sullivan, M. Index. In *A Handbook for Sensory and Consumer-Driven New Product Development*; Elsevier: Amsterdam, The Netherlands, 2017; pp. 325–337.
102. Goldenberg, J.; Mazursky, D. *Creativity in Product Innovation*; Cambridge University Press: Cambridge, UK, 2002; ISBN 9780521800891.
103. Reid, S.E.; de Brentani, U. The Fuzzy Front End of New Product Development for Discontinuous Innovations: A Theoretical Model. *J. Prod. Innov. Manag.* **2004**, *21*, 170–184. [CrossRef]
104. Savela-Huovinen, U.; Muukkonen, H.; Toom, A. Sensory expert assessor's learning practices at workplace: Competencies and contexts in sensory evaluation. *J. Sens. Stud.* **2018**, *33*. [CrossRef]
105. Knudsen, M.P. The relative importance of interfirm relationships and knowledge transfer for new product development success. *J. Prod. Innov. Manag.* **2007**, *24*, 117–138. [CrossRef]
106. Capitanio, F.; Coppola, A.; Pascucci, S. Product and process innovation in the Italian food industry. *Agribusiness* **2010**, *26*, 503–518. [CrossRef]
107. Chen, Y.; Vanhaverbeke, W.; Du, J. The interaction between internal R&D and different types of external knowledge sourcing: An empirical study of Chinese innovative firms. *R D Manag.* **2016**, *46*, 1006–1023. [CrossRef]
108. Zobel, A.K. Benefiting from Open Innovation: A Multidimensional Model of Absorptive Capacity. *J. Prod. Innov. Manag.* **2017**. [CrossRef]
109. Koppel, K.; Chambers, E.; Vázquez-Araújo, L.; Timberg, L.; Carbonell-Barrachina, T.A.; Suwonsichon, S. Cross-country comparison of pomegranate juice acceptance in Estonia, Spain, Thailand, and United States. *Food Qual. Prefer.* **2014**, *31*. [CrossRef]

Article

Trends of Using Sensory Evaluation in New Product Development in the Food Industry in Countries That Belong to the EIT Regional Innovation Scheme

Katarzyna Świąder [1,*] and Magdalena Marczewska [2]

1 Department of Functional and Organic Food, Institute of Human Nutrition Sciences, Warsaw University of Life Sciences (SGGW–WULS), 159C Nowoursynowska Street, 02-776 Warsaw, Poland
2 Department of Organization and Management Theory, Faculty of Management, University of Warsaw, 1/3 Szturmowa Street, 02-678 Warsaw, Poland; mmarczewska@wz.uw.edu.pl
* Correspondence: katarzyna_swiader@sggw.edu.pl; Tel.: +48-225-937-047

Citation: Świąder, K.; Marczewska, M. Trends of Using Sensory Evaluation in New Product Development in the Food Industry in Countries That Belong to the EIT Regional Innovation Scheme. *Foods* **2021**, *10*, 446. https://doi.org/10.3390/foods10020446

Academic Editors: Claudia Ruiz-Capillas and Ana Herrero Herranz

Received: 23 December 2020
Accepted: 16 February 2021
Published: 18 February 2021

Publisher's Note: MDPI stays neutral with regard to jurisdictional claims in published maps and institutional affiliations.

Copyright: © 2021 by the authors. Licensee MDPI, Basel, Switzerland. This article is an open access article distributed under the terms and conditions of the Creative Commons Attribution (CC BY) license (https:// creativecommons.org/licenses/by/ 4.0/).

Abstract: Sensory evaluation plays an important role in New Product Development (NPD) in food industry. In the present study, the current trends of using sensory evaluation in NPD in the food industry in countries that belong to EIT Regional Innovation Scheme (RIS) were identified. The research was conducted in the first quarter of 2020. Computer assisted self-interviewing (CASI) technique for survey data collection was used. The sample included 122 respondents representing RIS countries that are the EU Member States and European Horizon 2020 Associated Countries that are classified as modest and moderate innovators according to European Innovation Scoreboard. The analysis presented in the paper allowed to describe the methods of sensory evaluation that can be used to support NPD in the food industry, identify the trends of using sensory evaluation in NPD in the food industry companies in RIS countries. The research results showed that almost 70% of companies apply sensory evaluation methods in NPD. The larger the company, the more often the methods of sensory evaluation are used in NPDs. Almost 60% of companies employing 51–100, 101–1000 and more than 5000 people, respectively declare the use of expert (analytical) test. However, regardless of size, most companies prefer consumer (affective) test to expert tests. Based on the results, it seems that the potential of usage sensory evaluation methods is not yet fully exploited in the food industry.

Keywords: sensory evaluation; sensory analytical test; affective test; food industry; Regional Innovation Scheme (RIS); new product development (NPD)

1. Introduction and Background

One of the basic conditions for the development of the company and its long-term success is its innovation [1–3]. Innovative companies are able to respond to the challenges of the world around much faster and more effectively than non-innovative ones [1,4]. Companies, both small and medium enterprises (SMEs) and large corporations, have begun to consider innovation as an integral part of their strategy in order to create a lasting competitive advantage and to adapt the products or services offered to the needs of consumers, which has led to a greater need for teams mainly involved in the development of new products [1].

New product development (NPD) is "the process of designing a new product, producing it and bringing it to market" [5]. The definitions of the new product proposed by Fuller [6] are as follows "A product not previously manufactured by a company and introduced by that company into its marketplace or into a new marketplace" or "The presentation or rebranding by a company of an established product in a new form, a new package or under a new label into a market not previously explored by that company". NPD is based on customized procedures, models and is supported by appropriate tools;

thus, it brings significant benefits in terms of production costs, product quality and supply chain availability, which is crucial for success and business development [5]. Moreover, it is one of the ways to increase the company's profitability [6]. However, we must bear in mind that the success of a new product depends on many factors and in some cases the development of a new product may involve a high risk [5,7,8].

There are many different types of new food products on the market, such as line extensions [6,8] (e.g., new flavors for an ice-cream), repositioned existing product [6] (e.g., oil as one of the main constituents of mayonnaise or vegetable paste), new form or size of existing product [6] (e.g., instant oatmeal in ready to eat cup), reformulation of existing product [6] (e.g., sugar-reduced or sugar-free cakes), repackaging of existing product [6] (e.g., infant food containers changed from glass to squeezed), innovative products [6] (e.g., the plant-based analogue of tuna meat), creative products [6] (e.g., 3D printing food). However, one must remember that consumers purchase products regardless of their category, but rather because they meet their needs [9]. The growing awareness of consumers and their needs for natural, safe and healthy food [10] as well as consumers preferences for personalized, customized food products have led to diversity in the food market [5]. These changes led to new food trends such as functional food [10,11] with their healthful properties and nutritional value [10,12], novel food or use of nanotechnology in food sector [11].

It is not difficult to develop a recipe for a new product, but it is difficult to develop a new product that meets the expectations of the assumed number of consumers and is profitable to sell [8]. Every manufacturer's dream is to have their new product on the shelf as soon as possible and to sell it with great success, but the pathway from idea to shelf can be long and winding and leads through different stages of the process of designing new products. New product process proposed by Cooper [13] assumes the following stages: (1) new product idea; (2) costumer defined, internally defined, idea refined; (3) project formally opened; (4) development; (5) prototyping; (6) field trials; (7) launch. O'Sullivan [8] presents NPD process from the inception to the shelf divided into the following steps: (1) ideation; (2) project pre-planning; (3) validation of proof of concept; (4) process optimization and up-scaling; (5) commercialization; (6) pre and post-approval and shelf-life testing. The commonly used framework listing the main stages in the NPD process is based on four stages: (1) opportunity identification; (2) product design and development; (3) testing; (4) introduction and launch [14]. Azanedo [5] points out, however, that currently in the development of new food products (NFPD), the NPD models used need to be modified in order to better and more effectively support the food industry. Moreover, the study suggests that the most attention should be paid to areas such as meeting consumer demands, consumer preferences and sensory characteristics of products, taking into account the seasonality of ingredients and the traceability and safety of final products, and the large-scale production of food with respect for the environment and the supply and distribution of local ingredients.

In the process of designing food products, it is important to recognize the needs of the customer, direct the designed product towards them, and then communicate and explain the value of this designed product to the consumer [9]. Consumers buy image, comfort, nutrition, using their senses, sensory sensitivity, they buy sensory properties. That is why sensory methods are an important, integral tool that should be used in NPD process. When designing products, the most important quality feature of a product is its direct relationship to satisfaction, perception and ultimate acceptance by the consumer of the sensory qualities of the product [8,15].

Sensory evaluation and new product development are strongly linked. Sensory analysis methods can be used at many stages of the design process to assess the quality of the product and the expectations of consumers and their reactions to the product. Following the framework indicating the importance of sensory evaluation in NPD [5,8,15–21] this empirical research aims to identify the trends of using sensory evaluation in New Product Development in the food industry companies in countries that belong to EIT Regional

Innovation Scheme (RIS) by addressing research questions listed below. There are only a few studies tackling application of sensory evaluation methods in food industry companies that focus on general trends and compare the use of sensory evaluation methods between companies of different size and those that value the most diverse stages of the NPD process. Thus, this paper seeks to fill this research gap by addressing the following research questions:

RQ1 What methods of sensory evaluation can be used to support NPD in the food industry?

RQ2 What are the trends of using sensory evaluation in NPD in the food industry companies in countries that belong to EIT Regional Innovation Scheme (RIS)?

RQ3 What sensory evaluation methods are used by companies that tend to value the most a specific stage of the NPD process?

RQ4 What are the differences in applying sensory evaluation methods among companies of different sizes from RIS countries?

The first research question will be answered based on desk research results and literature review, whereas the answers to the questions 2-4 will be based on the results of empirical research.

The article proceeds as follows. First the research background along with the research aims are presented in the introductory part. The next section of the paper is devoted to sensory evaluation methods, their use and importance in NPD process with the focus on food industry. The third part of the paper describes methods used for the analysis, along with data selection and extraction process. It is followed by the presentation of results of a pilot study aimed to identify the trends of using sensory evaluation in New Product Development in the food industry in countries that belong to EIT Regional Innovation Scheme (RIS) and its discussion. The paper ends with concluding remarks and future research directions.

2. Sensory Evaluation in New Product Development

Sensory tests have been used since people began to assess everything that can be used by them and to distinguish between good and bad, from water and food, starting with and ending with weapons and other objects. The increase in trade, on the other hand, has inspired the formal application of sensory testing significantly [17].

The history of "sensory" analysis also dates back to the wars, when efforts were made to provide the American forces with acceptable food [17]. The early 1900s gave rise to a professional taster and consultant in emerging industries food, beverages, and cosmetics. The term "organoleptic examination" was then used to describe allegedly objective sensory characteristics. However, these tests were still often subjective rather than objective [17].

International interest in food and agriculture in the mid-1960s and on into the 1970s, the energy crisis, food production and raw material costs, competition and market internationalization have created opportunities for sensory evaluation [22]. The course of events has made sensory evaluation a recognized scientific specialty [17,22]. Sensory evaluation is defined as "a scientific discipline used to evoke, measure, analyze and interpret reactions to those characteristics of foods and materials as they are perceived by the senses of sight, smell, taste, touch and hearing" [18,22,23]. Sensory evaluation, like other scientific methods in which we take measurements, is based on taking measurements in a precise and accurate way, considering the sensitivity and aiming at avoiding false-positive results [24]. In order to be considered a reliable method, sensory analysis must be based on the skills of a sensory analyst to optimize definition of the problem (what should be measured) and test design (produce the desired accuracy of results), instrumentation (selected and trained panelists) and interpretation of results [17,18].

When assessing the characteristics of a food product, we first assess its appearance, then its odor, texture/consistency and flavor/taste [17]. The reaction to a sensory stimulus, on the other hand, can be divided into three different dimensions: qualitative perception, quantitative perception, and hedonic reaction [25]. In order to obtain that information, we must use analytical or affective methods during the sensory evaluation [16,17,22,25].

The purpose of analytical tests is to assess in detail the sensory quality of a product, while affective tests are used to measure the acceptability or preference of a product by consumers [8,25]. The basic goal while choosing sensory evaluation methods is to match the right test with the right question that we want to answer. Among the analytical tests (Table 1) that are mainly evaluated by the panel experts, we can use the discrimination test to determine if there are sensory differences or similarities between products, without describing their nature. We can use the Triangle test, Duo-trio test, Two out of five test. As far as the nature of the differences between products is known, we can use a grading test such as paired comparison test, to position different products according to their sensory characteristics. A ranking test can be used to assess noticeable differences between several products depending on the intensity of the difference, and a scoring test may be used to assess the specific intensity of the sensory characteristics of products. In the analytical test, a descriptive test (called sensory profiling) is very often used to describe and evaluate both the intensity and quality of perceived product characteristics, i.e., Quantitative Descriptive Analysis®, Texture Profile® [8,18,22,25,26].

Table 1. Types of most popular analytical and affective tests used in sensory evaluation.

	Type of Test	**Question**
ANALYTICAL TEST	Discrimination test: Triangle Duo-trio Two out of five	Which sample is different? Which sample is different from the reference sample? Which 3 samples are the same type?
	Grading test: Paired-comparison test Ranking test Scoring test	Which sample is most (sweet, bitter, etc.)? List the samples in increasing order of intensity for a selected attribute (sweet, bitter, etc.)? How (sweet, bitter, etc.) the sample is?
	Descriptive test: Quantitative Descriptive Analysis® Flavor Profile® Texture Profile® Spectrum™ Descriptive Analysis Free-Choice Profiling	Are the products different and how do they differ?
AFFECTIVE TEST	Paired comparison test	Which do you prefer? Which do you like most?
	Ranking test	Rank this product by preference?
	Hedonic scoring test	Asses the degree of pleasure/liking given by products on the scale?

Source: Own elaboration based on [8,18,22,25–27].

Apart of above-mentioned methods, there are new one called rapid sensory evaluation methods, that are more flexible, simple and easy to perform and can be used with semi trained assessors or naive assessors such as: flash profiling, ultraflash profiling, ranking test, napping, free sorting, optimized descriptive profiling, ideal profile method, check-all-that-apply, temporal dominance of sensation [8].

Sensory acceptance of the product by the consumer, its hedonic reaction, can be assessed using an affective test (Table 1). This may be a paired comparison test, in which the consumer chooses the sample he or she prefers or likes most from two or more, or a ranking test, in which the consumer rank the product according to his or her preferences, whereas in order to determine the scale of preference among the products or the degree of pleasure/liking the product gives, a hedonic scoring test with scales can be used [22,25,27]. An example of a qualitative affective sensory test is the focus group, a rapid method to test the product and packaging concepts and ideas [8].

Sensory evaluation of a product, including both the analytical sensory evaluation carried out by a panel of experts and the affective test carried out on consumers, allows to obtain more information about the product being analyzed, its quality and to verify factors influencing its acceptability by consumers, which facilitates work on improving the quality of the product or its reformulation [10].

It is quite common practice in food companies to use inappropriate sensory analysis methods for specific research purposes [8].

Both affective tests and analytical (expert) sensory tests can be use on each step of new product development (Table 2). During ideation, the initial project planning and validation of proof-of-concept affective test, such as focus groups, can be used, but also methods such as free elicitation, information acceleration (IA), Kelly repertory grid, laddering, lead user technique and Zaltman metaphor elicitation technique (ZMET) are recommended. The stage where both the affective test and the sensory test can be applied is process optimization and up-scaling, where the sensory acceptance test (affective test) can be applied, as well as the analytical test: a descriptive test such as the Quantitative Descriptive Analysis®and a rapid test such as the Ranking Descriptive Analysis. A very important aspect during the commercialization of a product is to carry out sensory Acceptance Tests. Carrying out Consumer Tests is also very important during pre- and post-approval tests and product durability tests [8].

Table 2. Use of sensory evaluation methods on each step of new product development.

Stages of New Product Development	Applied Sensory Evaluation Methods	
	Affective Test	Analytical Test
Ideation	Focus Groups, Free Elicitation, IA *, Kelly Repertory Grid, Laddering, Lead User Technique, ZMET *	-
Project Pre-Planning		
Validation of Proof of Concept		
Process optimization and up-scaling	Sensory Acceptance Testing	Descriptive test: QDA®*
		Rapid test: RDA *
Commercialization	Sensory Acceptance Testing	-
Pre- and Post-Approval and shelf-life testing	Consumer Testing ($n > 100$)	-

IA *, information acceleration; ZMET, Zaltman metaphor elicitation technique; QDA®, Quantitative Descriptive Analysis; RDA, Ranking Descriptive Analysis. Source: Own elaboration based on [8].

Besides new product development, sensory evaluation can also be used in other product development activities, such as product prototype evaluation [8,28]; product concept fit [8,28]; pilot plant scale-up; cost reduction study by substituting or modifying ingredients [28]; process change [29,30]; ingredients changes for example caused by reduction of salt [20,21,31], sugar [21,32,33], or fat [21,31] or purchase specifications change [19] as well as product improvement and optimization of product formula [34,35].

Moreover, sensory evaluation methods are used to supports marketing and marketing research activities [19,28], beginning with new product development and assessment of market potential, continuing through tracking product performance, and contributing to special assignments such as developing tests. They can be used in sensory marketing as data to support or challenge advertising claims [36]. Sensory marketing defined as "marketing that engages the senses of consumers and influences their perception, judgement and behavior" examines how acoustic, tactile and olfactory sensory stimuli influence decision making processes and the formation of consumer attitudes and can thus be used for advertising design and effectiveness [36].

Sensory evaluation is also used to compare the quality of competing products [37]. Consumer test can be used to define the most important characteristics of food affecting purchasing decisions, identify preferences and to know consumers when purchasing food products [12,37].

Sensory evaluation methods can be used in several food industry departments; however, they are mainly used for quality control and product research and development (R&D) in big companies [17,19] so their potential is not yet fully exploited in the food industry.

Sensory evaluation methods can be used for shelf-life assessment of food products [38–40] and new technologic that can extend product durability and quality such as pulsed electric fields [29,30]. Changes in the sensory characteristics of food products affect the determination of their shelf-life, and the freshness of a product's safety and quality are characteristics to which consumers are now paying increasing attention [40].

Expert tests such as sensory descriptive analysis and consumer test can be well used to investigate exotic, authentic, ethnic, or artisanal foods [23] such as green tea [41,42], soy sauce [43,44], kimchi [45,46], tofu [47,48], dates [49,50] as well as innovative one such as tea-infused yoghurts [10] or plant-based yogurts made from almond, cashew, coconut, hemp or soy [51].

There are many possible ways to apply sensory evaluation in NPD in the food industry. Companies can use analytical and affective (hedonic) sensory tests and choose from a variety of sensory methods that can be used at different stages of the NPD process and allow the different characteristics of food products to be studied and consumer reactions to these products and their expectations.

3. Materials and Methods

The study was conducted as a part of the project "Summer school on the New Product Development for the food industry" (2020 edition) financed by the EIT Food under Horizon2020, which aimed to address contemporary challenges related to NPD in the food industry. The research was carried out as a pilot study (part of the project) to identify the trends of using sensory evaluation in New Product Development in the food industry in countries that belong to EIT Regional Innovation Scheme (RIS); thus, the research sample was composed of respondents representing these countries. The EIT's Innovation Communities are strengthening the innovation ecosystem in parts of Central, Eastern and Southern Europe and have been set up to increase the number and business maturity of start-ups coming from these regions [52].

The EIT RIS countries are the EU Member States and European Horizon 2020 Associated Countries who are classified as modest and moderate innovators according to European Innovation Scoreboard [53]. These are Bulgaria, Croatia, Cyprus, Czech Republic, Estonia, Greece, Hungary, Italy, Latvia, Lithuania, Malta, Poland, Portugal, Romania, Slovakia, Slovenia and Spain, as well as Albania, Armenia, Bosnia and Herzegovina, Faroe Islands, Georgia, Moldova, Montenegro, Republic of North Macedonia, Serbia, Turkey and Ukraine. Modest and moderate innovators show an innovation performance below the EU average. Since NPD is one of the dimensions of countries' innovation performance, it seems important to analyze it, especially in the context of countries which are not best performers, in order to identify the ways to boost its development, as well as explore and design new potential pathways to support it.

The data used in this research were collected through computer assisted self-interviewing (CASI). This technique allows to collect data from respondents who complete the survey questionnaire via computer without any external assistance. The use of such research method is based on the assumption that respondents can read and understand the questions well enough to give precise answers [54]. The questionnaires were conducted in the first quarter of 2020, before the CODIV-19 pandemic.

The sample included 122 respondents representing RIS countries. All participants have agreed to participate in the study and received a link to the questionnaire for them only. If respondents provided their questionnaires with missing/unreliable data, they were excluded from the sample analyzed ($n = 8$).

The respondents were employed in companies from the food industry and performed various jobs related to NPD. Most of them were food technologists, project/program managers or C-level executives. The characteristics of the respondents are presented in

(Table 3). Respondents differ in gender, age and length of professional experience in the food sector.

Table 3. The characteristics of the respondents (n = 122).

Characteristic		n *	%
Gender	Male	73	59.8%
	Female	49	40.2%
Age group	18–24	20	16.4%
	25–34	77	63.1%
	35–44	22	18.0%
	45–54	2	1.6%
	55 or more	1	0.8%
Professional experience in food sector	less than 1 year	36	29.5%
	between 1 and 2 years	27	22.1%
	between 2 and 5 years	31	25.4%
	between 5 and 10 years	16	13.1%
	more than 10 years	12	9.8%
Area of expertise	Food science/chemistry/technology	23	18.9%
	Food safety/quality	16	13.1%
	Product development in the food sector	15	12.3%
	Food production/manufacturing/processing	13	10.7%
	Entrepreneurship/business startup/development/acceleration in agri-food or life sciences	11	9.0%
	Marketing/consumer behavior/market research, preferably in the food sector	10	8.2%
	Nutrition/food related health	8	6.6%
	Other	7	5.7%
	Agriculture/agricultural technologies	6	4.9%
	Consumer testing/sensory science	4	3.3%
	Food-health nexus	3	2.5%
	Food waste/side stream valuation	2	1.6%
	New business models	2	1.6%
	Bioeconomy/resource stewardship/sustainability	1	0.8%
	** STEM/STEAM/science education	1	0.8%

* number of respondents. ** STEM, Science, Technology, Engineering and Math; STEAM, Science, Technology, Engineering, Arts and Math.

The respondents were employees of food industry companies of various sizes, where SMEs constituted almost three quarters of the sample. Further, almost 40% of the sample were micro enterprises with fewer than 10 employees and around 20% represented small enterprises. Table 4 presents details on sample characteristics.

Table 4. Sample characteristics ($n = 122$).

Characteristic		n *	%
Number of employees in organization	1–10	48	39.3%
	11–50	23	18.9%
	51–100	12	9.8%
	101–1000	21	17.2%
	1001–5000	6	4.9%
	Above 5000	12	9.8%
Country **	ES	19	15.6%
	IT	17	13.9%
	GR	14	11.5%
	PL	14	11.5%
	HU	8	6.6%
	TR	7	5.7%
	PT	6	4.9%
	BG	4	3.3%
	HR	4	3.3%
	CZ	3	2.5%
	EE	3	2.5%
	LT	3	2.5%
	ME	3	2.5%
	RO	3	2.5%
	RS	3	2.5%
	AL	2	1.6%
	SI	2	1.6%
	LV	2	1.6%
	SK	2	1.6%
	UA	2	1.6%
	GE	1	0.8%

** AL, Albania; BG, Bulgaria; CZ, Czech Republic; EE, Estonia; ES, Spain; GE, Georgia; GR, Greece; HR, Croatia; HU, Hungary; IT, Italy; LT, Lithuania; LV, Latvia; ME, Montenegro; PL, Poland; PT, Portugal; RO, Romania; RS, Serbia; SI, Slovenia; SK, Slovakia; TR, Turkey; UA, Ukraine. * Number of respondents.

The questionnaire was composed of questions of different types, i.e., single-choice questions and open-ended questions. Most of the questions had already defined answers to choose from; however, in many cases, there was space left for participants to write additional comments. The questionnaire used in the study was related to various aspects of NPD and sensory analysis, including average length of NPD project, most important stages of NPD, application of sensory evaluation in NPD, sensory methods used in NPD, use of consumer tests to verify consumer preferences or acceptance for the developed product and use of expert tests performed by a panel of trained experts to determine the quality of products. The obtained data were analyzed considering the sample of companies from RIS countries in general, and also taking into account the specificity of companies of different sizes and comparing the importance of a specific stage of NPD and its impact on the use of sensory evaluation. Sample survey questions and answers are presented in (Table 5).

Table 5. Sample survey questions and answers.

General Questions	
Question	Answer
Country	(open ended question)
Gender	Male Female
Age group	18–24 25–34 35–44 45–54 55 or more
Please indicate your area of expertise	Agriculture/agricultural technologies; Bioeconomy/resource stewardship/sustainability; Consumer testing/sensory science; Entrepreneurship/business start-up/development/acceleration in agri-food or life sciences; Education/andragogy, in particular in entrepreneurship or food systems; Food-health nexus; Food production/manufacturing/processing; Food science/chemistry/technology; Food safety/quality; Food systems/food value chains; Food waste/side stream valuation; Marketing/consumer behavior/market research, preferably in the food sector; New business models; Nutrition/food related health; Product development in the food sector; Science communication/public engagement of science/citizen science; * STEM/STEAM/science education; Trust/transparency; Other
How long is your professional experience in the food sector?	less than 1 year; between 1 and 2 years; between 2 and 5 years; between 5 and 10 years; more than 10 years
How many employees are there in your entire organization?	1–10 11–50 51–100 101–1000 1001-5000 Above 5000
Which role best describes your current position in the company?	C-Level Executive (*CEO, CTO, etc.) Development Leadership: VP/Director Development Direct Manager: Team leader/Group Leader Development Team Member: Architect/Developer/*QA Project/Program Manager System Engineer Product Manager/Product Owner DevOps Engineer External Consultant/Trainer Food technologist Sensory and Consumer Manager Other (please specify below)
Industry specific questions	
Question	Answer
According to the best of your knowledge, what is the average length of NPD project in your organization? Please take into account the time from ideation to commercialization.	Less than 1 year 1–2 years 2–4 years More than 4 years

Table 5. Cont.

General Questions	
Which stages of food design are the most important in the company you work for? Please select all that apply.	a. Creating a new product, idea b. Developing a product recipe, selection and safety of raw materials, its health-promoting properties c. Packaging, its appearance, functionality, impact on product durability and the environment d. Labeling in accordance with legal requirements e. Ensuring the safety of produced food f. Sensory quality of the product and its acceptability by consumers g. Product distribution h. Product marketing and advertising i. Obtaining funds/grants j. Other (please specify)
Does the company you work for apply a sensory evaluation test in the product development?	Yes/No
What kind of sensory methods are used by the company you work for in the process of food product design? Please select all that apply.	a. Discrimination test b. Descriptive analysis c. Consumer test d. Other (please specify) e. I don't know what specific methods were used
Does the company you work for perform consumer tests to verify consumer preferences/acceptances for the developed product?	Yes/No
What is the average number of consumer test participants that the company you work for performs?	a. $\leq$30 consumers b. 31–50 consumers c. 51–99 consumers d. $\geq$100 consumers
Does the company you work for use expert tests performed by a panel of trained experts to determine the quality of products?	Yes/No
If company you work for does not use expert tests performed by a panel of trained experts to determine the quality of products, please indicate why not. Please select all that apply.	a. The company does not employ such experts b. Such research is too expensive c. Such research is not necessary when designing products d. The sensory quality of products is assessed by company owners, management board or employees and friends e. The company has no knowledge or experience in this field f. Other (please specify)

* STEM, Science, Technology, Engineering and Math; STEAM, Science, Technology, Engineering, Arts and Math; CEO, chief executive officer, CTO, chief technical officer, QA, Quality Assurance.

4. Results and Discussion

On the basis of the studies carried out, it is worth pointing out that the NPD projects in the analyzed food sector companies are relatively short (Table 6), i.e., 34.4% of the companies claim that they manage to develop and introduce an entire project within a year, and 41.8% need less than 2 years to complete it. The companies argue that due to market dynamics and growth they have to act fast in order to be competitive.

One of the main factors influencing product development is the speed with which the product is placed on the market. If this process took too long, the research previously carried out may no longer correspond to reality, e.g., demographic consumer segments initially identified as potential buyers and optimistic users may have changed their minds about wanting to buy the product [8]. Moreover, because the development of a new product may involve a high risk [5,7,8] it is worth running the NPD process relatively fast.

According to Dijksterhuis [55], 50 to 75% of newly developed products placed on the market are disposed of, which is far from the assumed financial targets. More than

90% (some say it is even 98%) of all NPDs in the food and drink industry fail, while the remaining 10% (or rather 2%) have been extremely successful, and the final prize is huge [8]. The main research problem behind the high failure rate of new products is the lack of understanding of consumers' motivation and choice. Therefore, the research on consumer behavior should be used more effectively to address this problem [55].

Table 6. Average length of New Product Development (NPD) project.

Average Length of NPD Project	n *	%
Less than 1 year	42	34.4%
1–2 years	51	41.8%
2–4 years	22	18.0%
More than 4 years	7	5.7%

* Number of respondents.

Sensory evaluation seems to be an important element of NPD in RIS countries. 67.2% of analyzed companies claim to apply sensory evaluation methods while working on new products.

Sensory food science is a discipline that is increasingly used and needed in order to better understand the factors influencing consumer preferences. Sensory evaluation is an essential tool for use by the food industry now and in the future, when, due to social and industrial needs that are consumer-oriented, their use will increase in the future [18].

There are differences in applying sensory evaluation among companies from RIS countries, which value the most different stages of NPD. It is not surprising that the use of consumer assessment tests, expert tests and other sensory evaluation methods is most common among companies, which see "sensory quality of the product and its acceptability by consumers" as the most important stage of NPD process (Table 7). However, "developing a product recipe, selection and safety of raw materials, its health-promoting properties" is seen as crucial in the NPD process by more than one fourth of the researched companies and within this group sensory evaluation, along with consumer assessment and expert tests are also seen as valuable. Table 7 presents the use of various sensory evaluation methods by companies that tend to value the most a specific stage of the NPD process. E.g., 74.4% of companies that see as the most important stage of NPD "developing a product recipe, selection and safety of raw materials, its health-promoting properties" apply sensory evaluation in NPD, 66.7% use consumer tests, and 46.2% use expert tests. This means that there are some companies that use both types of tests, and other focus only on one specific test.

In theory, sensory analysis is applied at many stages of new product development. Affective tests, e.g., focus group, can be used at the product concept stage, while from the assessment of a prototype, affective methods such as examining product preference or acceptability can be used. Analytical research, e.g., discrimination tests, can be used to optimize the product, e.g., nutritional optimization of the product related to the reduction or exchange of sucrose in the product. However, to determine the quality of the product prototype and the differences between the variants, it is worth using Quantitative Descriptive Analysis (QDA®). In order to obtain more complete information about the product, descriptive analysis can be carried out in parallel with sensory affective analysis and consumer testing and instrumental measurements. Before a product is placed on the market, Multivariate Data Analysis (MVA) can be used to correlate sensory descriptive analysis with factors affecting the consumer [8].

There are three most popular sensory methods used in NPD in the food sector in RIS countries (Figure 1), and discrimination test is the most popular one (37.8%). However, companies usually use more than one method in order to verify, compare and combine the results of sensory methods used.

Table 7. Most important stages of NPD and use of sensory evaluation.

	% of Companies Indicating Selected Stage as the Most Important NPD Stage	Application of Sensory Evaluation in NPD	Use of Consumer Tests	Use of Expert Tests
Developing a product recipe, selection and safety of raw materials, its health-promoting properties	32.0%	74.4%	66.7%	46.2%
Creating a new product, idea	24.6%	66.7%	70.0%	40.0%
Ensuring the safety of produced food	13.9%	70.6%	52.9%	41.2%
Sensory quality of the product and its acceptability by consumers	9.0%	81.8%	72.7%	63.6%
Packaging, its appearance, functionality, impact on product durability and the environment	5.7%	71.4%	57.1%	42.9%
Product marketing and advertising	5.7%	57.1%	57.1%	42.9%
Obtaining funds/grants	2.5%	33.3%	66.7%	0.0%
Product distribution	2.5%	0.0%	33.3%	0.0%
Labeling in accordance with legal requirements	0.8%	100.0%	100.0%	100.0%
Other	3.3%	25.0%	25.0%	25.0%

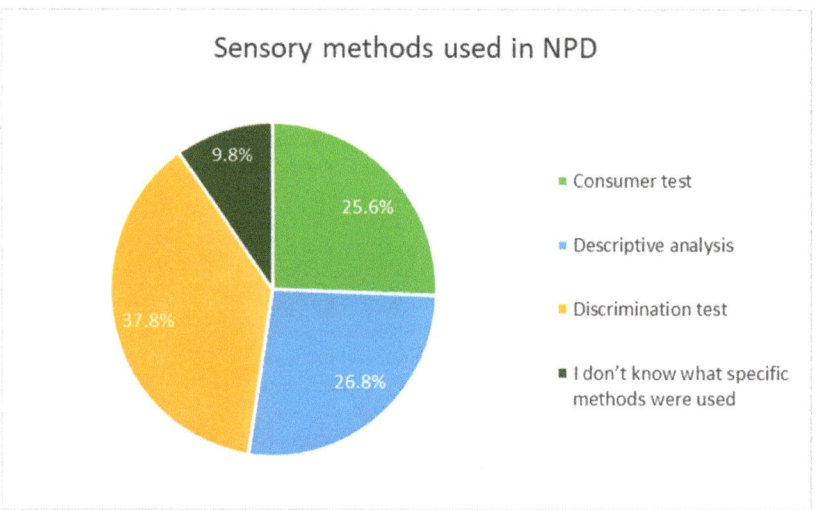

Figure 1. Sensory methods used in NPD.

The discriminatory test is a powerful sensory evaluation method in terms of its sensitivity, which provides reliable and important results that, because of its effectiveness, have saved companies a considerable amount of time, money and effort [22].

Sensory methods can be used at many stages of the NPD, so it is important to use them at many stages, but you can see that discriminatory methods are most often used, and the use of methods depends on the purpose of the research.

Companies from the sample tend to value consumers opinions. Of the companies, 63.1% use consumer tests (affective tests) to verify consumer acceptance or preferences for the developed product. Interestingly, the average number of consumer test participants differs significantly between the companies form the sample. On the one hand, almost 33% of companies collect information from a small number of consumers, i.e., ≤30. On the other hand, over 33% conduct consumer assessment on relatively big samples of more than 100 consumers (Figure 2). As confirmed by other researchers, consumer acceptance testing can be used during the product development and optimization based on 25–75 individuals, while a large number of consumers (more than 100) is used in consumer tests before product lunch [8].

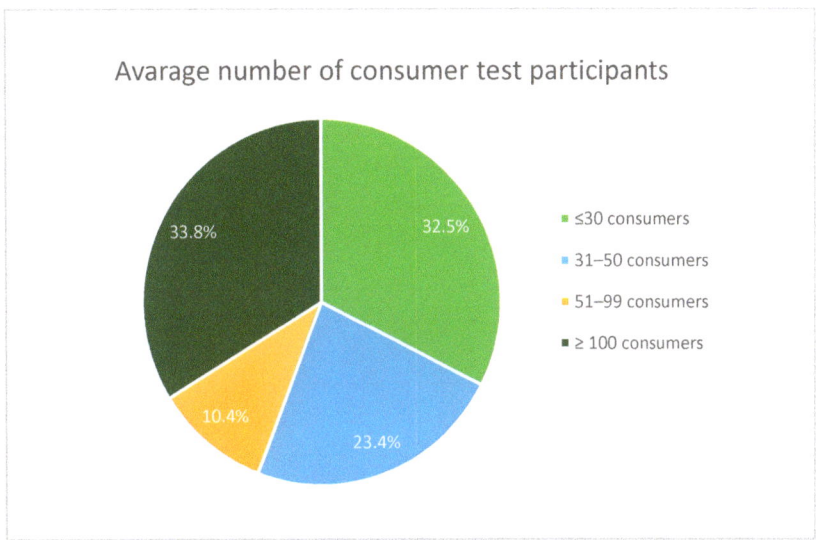

Figure 2. Average number of consumer test participants.

Interest in consumer (hedonic) research in both basic psychophysics and applied and consumer food research has increased significantly in recent years. Research on the identification of differences in the hedonic response to chemical stimuli has become the basis for a better understanding of the role of sensory, perceptual, cognitive and genetic factors influencing consumer food preferences and choice. Now that the consumer market has become more crowded and competitive, applied product research is not only investigating which products are more popular with consumers than others, but it has become more important than ever to discover the basic segmentation of consumers [56].

Compared to other types of sensory evaluation methods, expert tests performed by a panel of trained experts to determine the quality of products seem to be rather unpopular among companies representing food industry companies from RIS countries. Only 42.6% of these companies perform expert tests. There are several primary reasons mentioned by the food industry companies from RIS countries why such tests are not popular. First, these companies do not employ such experts and are not willing to outsource such service. Second, these companies have no knowledge or experience in this field, thus preferring to use other, better known methods of sensory evaluation. Third, the sensory quality of products is assessed by company owners, management board or employees and friends. Among secondary reasons for not using such tests, companies list high expenses and lack of need. However, looking at the answers given by the companies from the sample, along with additional justifications, it seems that companies representing the food industry in RIS countries do not have sufficient knowledge in the field to use expert tests.

The use of sensory evaluation methods differs among companies of various sizes from RIS countries. In general, the bigger the company, the more popular and used sensory evaluation methods in NPD (Table 8). Most of the companies, regardless of size, prefer consumer assessment tests over expert tests. However, almost 60% of companies employing 51–100, 101–1000 and above 5000, respectively, declare to use expert tests. These results seem to confirm the research results presented by other authors stating that the bigger the company, the wider knowledge and application of sensory evaluation methods.

Table 8. Use of sensory evaluation in companies from RIS countries by company size.

Company size (number of employees in organization)	Application of Sensory Evaluation in NPD		Use of Consumer Assessment Tests to Verify Consumer Preferences for the Developed Product		Use of Expert Tests Performed by a Panel of Trained Experts to Determine the Quality of Products	
		%		%		%
1–10	Yes	52.08%	Yes	54.17%	Yes	33.3%
	No	47.92%	No	45.83%	No	66.7%
11–50	Yes	47.83%	Yes	39.13%	Yes	30.4%
	No	52.17%	No	60.87%	No	69.6%
51–100	Yes	83.33%	Yes	66.67%	Yes	58.3%
	No	16.67%	No	33.33%	No	33.3%
	-	-	-	-	N/A	8.3%
101–1000	Yes	95.24%	Yes	80.95%	Yes	57.1%
	No	4.76%	No	19.05%	No	42.9%
1001–5000	Yes	100.00%	Yes	100.00%	Yes	50.0%
	No	0.00%	No	0.00%	No	50.0%
Above 5000	Yes	83.33%	Yes	91.67%	Yes	58.3%
	No	16.67%	No	8.33%	No	41.7%

Consumer assessment tests run by companies of different sizes vary in average number of consumer test participants. Almost 60% of companies employing 1–10 employees run these tests on the sample of ≤30 consumers. Companies employing 11–50 and 51–100 employees prefer samples of 31–50 or more than 100 consumers; more than 40% of companies employing 101–1000 people conduct these tests with more than 100 consumers, whereas 33.3% of those employing 1001–5000 run tests on the samples of ≤30 and more than 100 consumers, respectively. Not surprisingly, almost 82% of the biggest companies tend to research more than 100 consumers at a time.

When it comes to specific sensory methods used in NPD by companies of different sizes, consumer acceptance tests are the most popular among those employing 1–10 and 1001–5000 employees, whereas discrimination tests are more widely used by those employing 11–50, 101–1000 and above 5000 people.

Sensory evaluation is beginning to be applied in many food companies, and research results of other authors also confirm that its adoption depends, among others, on the size of the company [19]. For example, in the case of large companies such as Puleva Biotech S.A., Spain, sensory testing is carried out daily in several departments, i.e., quality control, research and development and marketing. In the case of small companies, the situation is completely different. Small companies do not have the structure, personnel and/or qualifications to carry out sensory research, although they are aware of its existence and effectiveness. Medium-sized companies try to include sensory evaluation as one of the modern tools to improve their efficiency and thus their income [19].

All in all, sensory food research can contribute to understanding consumer response to emerging trends in food production, processing and consumption. In order to make better use of sensory research, it is necessary to allow access to appropriate university training programmes, funding for fundamental research and multidisciplinary cooperation [18].

5. Conclusions

The analysis presented above, based on desk research and CASI, allowed to answer the research questions (RQ) outlined in the introductory part of the paper.

Thereby RQ1 allowed to describe the methods of sensory evaluation that can be used to support NPD in the food industry. Companies can benefit from the many achievements of the scientific discipline of sensory evaluation, and they can use analytical and affective sensory tests. They have many different sensory methods at their disposal, which can be used at different stages of the NPD process, from the idea and conception of the product to its launch and subsequent approval, which allows them to assess both the quality of the product and the factors influencing consumers and their purchasing decisions.

RQ2 addressed the trends of using sensory evaluation in NPD in the food industry companies in RIS countries. The research results showed that almost 70% of companies apply sensory evaluation methods in NPD; however, among them there are the following three most popular ones: discrimination test, descriptive analysis and consumer test. Moreover, the companies generally value consumers opinions and more than 63% of them uses consumer assessment tests to verify the match between the products they offer and expectations of consumers. Nevertheless, looking at the answers given by the companies from the sample, along with their additional justifications, it seems that companies representing food industry in RIS countries do not have sufficient knowledge in the field to use expert tests, which may lead to the use of unsuitable methods of sensory analysis to achieve the research objectives set companies.

RQ3 was aimed at identifying sensory evaluation methods that are used by companies, which mostly tend to value a specific stage of the NPD process. First, it allowed to identify that more than one fourth of the companies from the sample see "developing a product recipe, selection and safety of raw materials, its health-promoting properties" as the most important stage of the NPD process, almost one fifth values the most "creating a new product, idea", whereas around 10%, respectively, treat "ensuring the safety of produced food" and "sensory quality of the product and its acceptability by consumers" as highly important. Second, the analysis allowed to link specific sensory methods used with companies valuing different staged of NPD. Almost 82% of companies that value the most "sensory quality of the product and its acceptability by consumers" apply sensory evaluation methods in NPD; 73% use consumer assessment tests, and 64% use expert tests. The knowledge of sensory evaluation methods and their application is really high in this group of companies, compared to the whole sample. Among other groups of companies identified based on the stage of NPD process they value, 70% of companies focused on "creating a new product, idea" use consumer assessment tests and 67% of those focused on "developing a product recipe, selection and safety of raw materials, its health-promoting properties". The knowledge and application of different sensory evaluation methods is diverse among companies representing different groups; however, interestingly, companies focused on "obtaining funds/grants" and "product distribution" seem to disregard expert tests.

RQ4 allowed to characterize the differences in applying sensory evaluation methods among companies of different sizes from RIS countries. Sensory evaluation is increasingly being used in many food companies, and the use of these methods depends to a large extent on the size of the company. Among the companies analyzed in the RIS countries, it can be seen that the larger the company, the more often the methods of sensory evaluation are used in NPDs. Almost 60% of companies employing 51–100, 101–1000 and more than 5000 people, respectively, declare the use of expert (analytical) test. However, regardless of size, most companies prefer consumer (affective) test to expert tests. Consumer tests are most popular among companies with 1–10 and 1001–5000 employees, while discrimination tests included in analytical test are more frequently used by companies with 11–50, 101–1000 and over 5000 employees.

All in all, it seems that the potential of usage sensory evaluation methods is not yet fully exploited in the food industry. Since this pilot study has been carried out on a sample including many countries from a specific group (RIS countries), a further study is planned to analyze a specific group of companies from selected RIS countries in order to take a closer look at specific sensory analysis methods used in the development of selected food products.

Drawing on research results presented above, future research directions in investigating the importance of sensory evaluation in NPD in the food industry may also include the following topics: focus on analytical test or affective test, one sensory method, one method and one country only; focus on SMEs from specific branch of food industry in RIS countries and/or comparison with other countries; focus on size and experience of food industry in RIS countries; focus on specific stage of NPD and the use of sensory evaluation.

Author Contributions: Study conception and design, K.Ś., M.M.; methodology, K.Ś., M.M.; performed research, K.Ś., M.M.; analyzed the data, K.Ś., M.M.; interpreted the data, K.Ś., M.M.; writing, K.Ś., M.M. All authors have read and agreed to the published version of the manuscript.

Funding: Funding from the Rector of WULS-SGGW as part of the system of financial support for scientists and teams.

Institutional Review Board Statement: Not applicable.

Informed Consent Statement: Not applicable.

Data Availability Statement: Data is contained within the article.

Acknowledgments: The research presented in the paper has been conducted as a part of the project ID 20228 that has received funding from EIT Food, the European Knowledge and Innovation Community (KIC) on Food, as part of Horizon 2020.

Conflicts of Interest: The authors declare no conflict of interest. The funders had no role in the design of the study; in the collection, analyses or interpretation of data; in the writing of the manuscript or in the decision to publish the results.

References

1. Vignali, G.; Bigliardi, B.; Bottani, E.; Montanari, R. Successful new product development in the food packaging industry: Evidence from a case study. *Int. J. Eng. Sci. Technol.* **2011**, *2*. [CrossRef]
2. Balkin, D.B.; Markman, G.D.; Gomez-Mejia, L.R. Is Ceo Pay in High-Technology Firms Related to Innovation? *Acad. Manag. J.* **2000**, *43*, 1118–1129. [CrossRef]
3. Lyon, D.; Ferrier, W. Enhancing performance with product-market innovation: The influence of the top management team. *J. Manag. Issues* **2002**, *14*, 452–469. Available online: https://www.jstor.org/stable/40604404 (accessed on 19 December 2020).
4. Jimenez, J.; Valle, R.S.; Espallardo, M.H. Fostering innovation. The role of market orientation and organizational learning, Europ. *J. Innov. Manag.* **2008**, *11*, 389–412. [CrossRef]
5. Azanedo, L.; Garcia-Garcia, G.; Stone, J.; Rahimifard, S. An Overview of Current Challenges in New Food Product Development. *Sustainability* **2020**, *12*, 3364. [CrossRef]
6. Fuller, G.W. *New Food Product Development: From Concept to Marketplace*, 3rd ed.; CRC Press, Broken Sound Parkway NW: Boca Raton, FL, USA, 2011.
7. Gao, J.; Bernard, A. An overview of knowledge sharing in new product development. *Int. J. Adv. Manuf. Technol.* **2017**, *94*, 1545–1550. [CrossRef]
8. O'Sullivan, M.G. Innovative tech-nologies for the food and beverage industry. In *A handbook for Sensory and Consumer-Driven New Product Development*; Elsevier: Cambridge, MA, USA, 2017.
9. Okoye, I.N.H. How Do You Explain A New Product Category? Product Knowledge Explains It! Europ. *J. Bus. Manag.* **2015**, *7*, 18. Available online: https://www.iiste.org/Journals/index.php/EJBM/article/view/23195 (accessed on 19 December 2020).
10. Świąder, K.; Florowska, A.; Konisiewicz, Z.; Chen, Y.-P. Functional Tea-Infused Set Yoghurt Development by Evaluation of Sensory Quality and Textural Properties. *Foods* **2020**, *9*, 1848. [CrossRef]
11. Santeramo, F.; Carlucci, D.; De Devitiis, B.; Seccia, A.; Stasi, A.; Viscecchia, R.; Nardone, G. Emerging trends in European food, diets and food industry. *Food Res. Int.* **2018**, *104*, 39–47. [CrossRef]
12. Kraus, A. Development of functional food with the participation of the consumer. Motivators for consumption of functional products. *Int. J. Consum. Stud.* **2015**, *39*, 2–11. [CrossRef]
13. Cooper, R.G. Perspective: The Stage-Gate®Idea-to-Launch Process—Update, What's New, and NexGen Systems. *J. Prod. Innov. Manag.* **2008**, *25*, 213–232. [CrossRef]
14. Bagchi, D.; Nair, S. *Developing New Functional Food and Nutraceutical Products*, 1st ed.; Elsevier: London, UK, 2017.
15. O'Sullivan, M.G.; Kerry, J.P.; Byrne, D.V. Use of sensory science as a practical commercial tool in the develop-ment of consumer-led processed meat products. In *Processed Meats*; Kerry, J.P., Kerry, J.F., Eds.; Woodhead Publishing Ltd.: Southston, UK, 2011.
16. Cruz, A.G.; Cadena, R.S.; Walter, E.H.; Mortazavian, A.M.; Granato, D.; Faria, J.A.; Bolini, H.M. Sensory Analysis: Relevance for Prebiotic, Probiotic, and Synbiotic Product Development. *Compr. Rev. Food Sci. Food Saf.* **2010**, *9*, 358–373. [CrossRef] [PubMed]
17. Meilgaard, M.; Civille, G.V.; Carr, B.T. *Sensory Evaluation*, 4th ed.; CRC Press: Boca Raton, FL, USA, 2007.
18. Tuorila, H.; Monteleone, E. Sensory food science in the changing society: Opportunities, needs, and challenges. *Trends Food Sci. Technol.* **2009**, *20*, 54–62. [CrossRef]

19. Carbonell-Barrachina, Á.A. Application of sensory evaluation of food to quality control in the Spanish food industry. *Pol. J. Food Nutr. Sci.* **2007**, *57*, 71–76.
20. Raithatha, C. The role of sensory perception and sensory evaluation in the development of reduced sodium foods. *Agro Food Ind. Hi Tech* **2014**, *25*, 48–52.
21. Romagny, S.; Ginon, E.; Salles, C. Impact of reducing fat, salt and sugar in commercial foods on consumer acceptability and willingness to pay in real tasting conditions: A home experiment. *Food Qual. Prefer.* **2017**, *56*, 164–172. [CrossRef]
22. Stone, H.; Sidel, J. *Sensory Evaluation Practices*, 3rd ed.; Academic Press: Berkeley, CA, USA, 2004.
23. Yang, J.; Lee, J. Application of Sensory Descriptive Analysis and Consumer Studies to Investigate Traditional and Authentic Foods: A Review. *Foods* **2019**, *8*, 54. [CrossRef] [PubMed]
24. Choi, S.E. Chapter 3: Sensory Evaluation. In *Food Science: An Ecological Approach*, 2nd ed.; Jones and Bartlett Publisher: Sudbury, MA, USA, 2013.
25. Caugant, M. *Sensory Evaluation, Guide of Good Practice*; Actia: Paris, France, 2001.
26. Lawless, H.T.; Heymann, H. *Sensory Evaluation of Food, Food Science Text Series*; Springer: New York, NY, USA, 2010.
27. Resurrection, A.V.A. *Consumer Sensory Testing for Product Development*; An Aspen Publication: Gaithersburg, MD, USA, 1998.
28. Moskowitz, H.R.; Chandler, J. Notes on consumer oriented sensory evaluation. *J. Food Qual.* **1979**, *2*, 269–276. [CrossRef]
29. Mosqueda-Melgar, J.; Raybaudi-Massilia, R.M.; Martín-Belloso, O. Microbiological shelf life and sensory evaluation of fruit juices treated by high-intensity pulsed electric fields and antimicrobials. *Food Bioprod. Process.* **2012**, *90*, 205–214. [CrossRef]
30. Walkling-Ribeiro, M.; Noci, F.; Cronin, D.A.; Lyng, J.G.; Morgan, D.J. Shelf life and sensory attributes of a fruit smoothie-type beverage processed with moderate heat and pulsed electric fields. *LWT* **2010**, *43*, 1067–1073. [CrossRef]
31. Arnarson, A.; Olafsdottir, A.; Ramel, A.; Martinsdottir, E.; Reykdal, O.; Thorsdottir, I.; Thorkelsson, G. Sensory analysis and consumer surveys of fat- and salt-reduced meat products and their use in an energy-reduced diet in overweight individuals. *Int. J. Food Sci. Nutr.* **2011**, *62*, 872–880. [CrossRef] [PubMed]
32. Oliveira, D.; Antúnez, L.; Giménez, A.; Castura, J.C.; Deliza, R.; Ares, G. Sugar reduction in probiotic chocolate-flavored milk: Impact on dynamic sensory profile and liking. *Food Res. Int.* **2015**, *75*, 148–156. [CrossRef] [PubMed]
33. Mahato, D.K.; Keast, R.; Liem, D.G.; Russell, C.G.; Cicerale, S.; Gamlath, S. Sugar Reduction in Dairy Food: An Overview with Flavoured Milk as an Example. *Foods* **2020**, *9*, 1400. [CrossRef] [PubMed]
34. Hough, G.; Sánchez, R.; Barbieri, T.; Martínez, E. Sensory optimization of a powdered chocolate milk formula. *Food Qual. Prefer.* **1997**, *8*, 213–221. [CrossRef]
35. Badwaik, L.S.; Prasad, K.; Seth, D. Optimization of ingredient levels for the development of peanut based fiber rich pasta. *J. Food Sci. Technol.* **2014**, *51*, 2713–2719. [CrossRef] [PubMed]
36. Krishna, A.; Cian, L.; Sokolova, T. The power of sensory marketing in advertising. *Curr. Opin. Psychol.* **2016**, *10*, 142–147. [CrossRef]
37. Beriain, M.J.; Sanchez, M.; Carr, T.R. A comparison of consumer sensory acceptance, purchase intention, and willingness to pay for high quality United States and Spanish beef under different information scenarios. *J. Anim. Sci.* **2009**, *87*, 3392–3402. [CrossRef]
38. Kilcast, D. 4 - Sensory evaluation methods for shelf-life assessment. In *The Stability and Shelf-Life of Food*; Woodhead Publishing Series in Food Science; Technology and Nutrition: Sawston, UK; Cambridge, UK, 2000.
39. Hough, G.; Garitta, L. Methodology for sensory shelf-life estimation: A review. *J. Sens. Stud.* **2012**, *27*, 137–147. [CrossRef]
40. Giménez, A.; Ares, F.; Ares, G. Sensory shelf-life estimation: A review of current methodological approaches. *Food Res. Int.* **2012**, *49*, 311–325. [CrossRef]
41. Lee, O.H.; Lee, H.S.; Sung, Y.E.; Lee, S.M.; Kim, K.O. Sensory characteristics and consumer acceptability of various green teas. *Food Sci. Biotechnol.* **2008**, *17*, 349–356.
42. Lee, J.; Chambers, D.H. Descriptive Analysis and U.S. Consumer Acceptability of 6 Green Tea Samples from China, Japan, and Korea. *J. Food Sci.* **2010**, *75*, S141–S147. [CrossRef] [PubMed]
43. Imamura, M. Descriptive terminology for the sensory evaluation of soy sauce. *J. Sens. Stud.* **2016**, *31*, 393–407. [CrossRef]
44. Heo, J.; Lee, J. US consumers' acceptability of soy sauce and bulgogi. *Food Sci. Biotechnol.* **2017**, *26*, 1271–1279. [CrossRef]
45. Cho, J.-H.; Lee, S.-J.; Choi, J.-J.; Chung, C.-H. Chemical and sensory profiles of dongchimi (Korean watery radish kimchi) liquids based on descriptive and chemical analyses. *Food Sci. Biotechnol.* **2015**, *24*, 497–506. [CrossRef]
46. Jang, S.; Kim, M.; Lim, J.; Hong, J. Cross-Cultural Comparison of Consumer Acceptability of Kimchi with Different Degree of Fermentation. *J. Sens. Stud.* **2016**, *31*, 124–134. [CrossRef]
47. Kamizake, N.K.K.; Silva, L.C.P.; Prudencio, S.H. Impact of soybean aging conditions on tofu sensory characteristics and acceptance. *J. Sci. Food Agric.* **2018**, *98*, 1132–1139. [CrossRef]
48. Kim, Y.-N.; Muttakin, S.; Jung, Y.-M.; Heo, T.-Y.; Lee, D.-U. Tailoring Physical and Sensory Properties of Tofu by the Addition of Jet-Milled, Superfine, Defatted Soybean Flour. *Foods* **2019**, *8*, 617. [CrossRef]
49. Al-Farsi, M.; Alasalvar, C.; Morris, A.; Baron, A.M.; Shahidi, F. Compositional and Sensory Characteristics of Three Native Sun-Dried Date (Phoenix dactyliferaL.) Varieties Grown in Oman. *J. Agric. Food Chem.* **2005**, *53*, 7586–7591. [CrossRef] [PubMed]
50. Ismail, B.; Haffar, I.; Baalbaki, R.; Henry, J. Development of a total quality scoring system based on consumer preference weightings and sensory profiles: Application to fruit dates (Tamr). *Food Qual. Prefer.* **2001**, *12*, 499–506. [CrossRef]
51. Grasso, N.; Alonso-Miravalles, L.; O'Mahony, J.A. Composition, Physicochemical and Sensorial Properties of Commercial Plant-Based Yogurts. *Foods* **2020**, *9*, 252. [CrossRef]

52. EIT RIS Report. 2019. Available online: https://eit.europa.eu/sites/default/files/eit_ris_report_2019.pdf (accessed on 19 December 2020).
53. EIT Scorebord. 2020. Available online: https://ec.europa.eu/growth/industry/policy/innovation/scoreboards_en (accessed on 19 December 2020).
54. Lavrakas, P. *Encyclopedia of Survey Research Methods*; SAGE Publications Pvt Ltd.: Thousand Oaks, CA, USA, 2008; Vols. 1–0.
55. Dijksterhuis, G. New product failure: Five potential sources discussed. *Trends Food Sci. Technol.* **2016**, *50*, 243–248. [CrossRef]
56. Lim, J. Hedonic scaling: A review of methods and theory. *Food Qual. Prefer.* **2011**, *22*, 733–747. [CrossRef]

Article

Part Meat and Part Plant: Are Hybrid Meat Products Fad or Future?

Simona Grasso [1,*] and Sylvia Jaworska [2]

1. Institute of Food, Nutrition and Health (IFNH), School of Agriculture Policy and Development, University of Reading, Reading RG6 6EU, UK
2. Department of English Language and Applied Linguistics, University of Reading, Reading RG6 6AH, UK; s.jaworska@reading.ac.uk
* Correspondence: simona.grasso@ucdconnect.ie

Received: 10 November 2020; Accepted: 17 December 2020; Published: 17 December 2020

Abstract: There is a growing interest in flexitarian diets, which has resulted in the commercialisation of new hybrid meat products, containing both meat and plant-based ingredients. Consumer attitudes towards hybrid meat products have not been explored, and it is not clear which factors could affect the success of such products. This study is the first to overview of the UK hybrid meat product market and to explore consumer's attitudes towards hybrid meat products in 201 online reviews, using tools and techniques of corpus linguistics (language analysis). In the positive reviews, consumers emphasised the taste dimension of the hybrid meat products, seeing them as healthier options with good texture and easy to prepare. The negative reviews related to the poor sensory quality and not to the concept of hybridity itself. Using a multidisciplinary approach, our findings revealed valuable insights into consumer attitudes and highlighted factors to consider to market new hybrid meat products effectively.

Keywords: meat product; consumers; flexitarianism; corpus linguistics; online product reviews

1. Introduction

High levels of meat consumption are associated with perceived health, social and environmental concerns resulting in calls to reduce the quantity of meat we consume [1]. To achieve a partial substitution of animal proteins in the diet with more sustainable plant proteins, long-term dietary transitions rather than short phases need to be established [2].

Studies have found that to create an effective dietary change, new practices should not diverge too much from consumers' previous behaviour [3]. Food choice has been recognised as a complex process that goes beyond sensory properties and involves many factors that can be grouped into the characteristics of the consumer, the product and the specific context in which the choice is made [4]. Factors related to consumer behaviour, which might limit consumer transition to alternative protein sources, are convenience and minimal cooking skills [5].

It is difficult to fully shift from a meat-centric diet to strict vegetarianism or veganism because of positive beliefs and attachments to meat and meat-centric societal constructs, however switching to a flexitarian or semi-vegetarian diet (mainly plant-based, with limited meat consumption) is less strict and can still have a positive impact [6].

A survey by the Humane Research Council [7] on 11,399 Americans found that 5 out of 6 people who become vegans or vegetarians eventually went back to eating meat. The authors suggest that it would be more important to persuade the majority of the population to reduce meat consumption rather than convincing a small percentage to give up meat completely [8].

Flexitarianism is increasing in popularity amongst consumers, with a market research study in the UK [9] reporting that while around 90% of consumers eat red meat or poultry, more than a third (34%) of eaters and buyers of meat and poultry have regular days when they avoid meat.

Within this context, the concept of hybrid meat products, that is meat products in which a proportion of meat has been partially replaced by more sustainable protein sources, might be suitable to bridge the gap between meat and meat-free products, while providing convenience, and allowing consumers to continue using foods as they would conventionally do [10].

Hybrid meat products could open new business opportunities for the food industry [11], and indeed very recently, hybrid meat products have started appearing in the UK market [12]. The meat industry might be answering the growing flexitarian consumer needs, but the launch of hybrid meat products might also be representing a moment for change and an attempt from meat manufacturers to gain additional market share over new popular plant-based alternative protein sources [8]. As Hicks et al. [13] point out, "it would be efficient and wise for the meat industry to build a strategy around the flexitarian demographic, to ensure their needs are met and to keep them consuming meat, rather than risk losing them to veganism".

We will now discuss the difference between hybrid meat products and meat extenders, provide a literature overview on consumer attitudes towards hybrid meat products and introduce the corpus linguistics (language analysis) techniques that will help achieve the aims of the current study.

Many processed meat products available in the market are already somehow "hybrid" as they often do not contain 100% meat [8]. For example in the UK, according to the Meat Products Regulation [14], only 42% of pork is needed to label sausages as pork sausages and the pork meat used can contain 30% fat and 25% connective tissue. A variety of functional ingredients have been traditionally added to processed meats, including fillers (plant substances with high carbohydrate content), extenders (non-meat compounds with considerable protein content), and binders (substances with high-protein content able to bind both water and fat) [15]. Indeed, plant-based ingredients from soy and wheat have been used by the meat industry to achieve cost savings [16], as well as for their functional properties: fat emulsification, gelling capability, and water binding [17].

The difference between hybrid meat products and meat products with plant-based functional ingredients (extenders, fillers and binders) is in the purpose of the mix of meat and plant proteins [8]. Usually plant-based functional ingredients are used traditionally for economic and technological reasons, in hybrid meat products this concept is pushed further to include positive connotations on the meat "extension", including healthiness, lower environmental impact and generally the idea of decreasing meat consumption [8].

Several research articles have shown that although challenging, it is technologically feasible to manufacture hybrid meat products such as burgers, meatballs and sausages with acceptable sensory quality [10,11,18,19]. However, consumer attitudes towards hybrid meat products have been investigated in a limited number of studies [8]. A study by de Boer et al. [20] compared hybrid meat products vs. alternative protein snacks such as insects, lentils and seaweed. The most popular snack was the hybrid one (chosen by 54% of 1083 participants). The authors concluded that it would be valuable to combine animal and plant-based protein and that hybrid meat products could be acceptable to lowly involved consumers who will not actively search for more environmentally friendly proteins. Similarly, previous work by the same authors found that hybrid meat products could be acceptable to many consumers, especially those who are weakly involved, because they may seem more familiar to them [5,8].

While these studies offer initial invaluable findings that could be used to develop more popular hybrid meat products, more research is still needed to understand sensory aspects and specifically consumer attitudes towards those products. Having a better understanding of which factors might have an impact on consumer acceptability would allow the effective formulation and marketing of existing and future hybrid meat products. A more holistic and multidisciplinary approach could offer richer and more nuanced insights into the stance and views of consumers, including a range of perceived advantages and disadvantages. Consumers' online reviews provide a unique opportunity to do so and allow the researcher to tap into consumers' authentic responses and opinions on dimensions that are relevant to them, but might not have been included and tested in previous research. Because online reviews are essentially texts, they require text analytics derived from linguistics. We explored consumer's attitudes towards hybrid-meat products in online reviews by utilising tools and techniques of corpus linguistics that allowed quantitative identification of the most frequent words and key terms across larger textual data sets. Frequency counts and key terms are useful in that they can highlight the distinctive (salient) words and two-word combinations in a given data set (a corpus of texts), which in turn point to dominant stances and attitudes shared by producers of the texts. The tools and techniques of corpus linguistics were applied to study a corpus of 201 online reviews in order to identify the dominant stances and opinions expressed by consumers who bought and consumed hybrid-meat products.

The aim of this study was therefore to (1) review the presence of hybrid meat products in the UK market, and (2) extract UK online consumer reviews on hybrid meat products and gather preliminary consumer insights utilising tools and techniques of corpus linguistics.

2. Materials and Methods

2.1. Hybrid Meat Products in the UK Market and Review Collection

A web search was conducted to understand the presence of hybrid meat products in the UK market. If a specific retailer was found to have launched a range, then the retailer website was further investigated to get more details on the characteristics and availability of each product. Product details were compiled into a list. Some of these products contained customer product reviews, and all available reviews were collected. This made up a total of 201 reviews from the websites of Waitrose, Ocado and Sainsbury's for three hybrid meat products. The products were Waitrose pork, chickpea and spinach sausages (79 reviews), Waitrose harissa chicken cauliflower rice & chickpea meatballs (106 reviews) and Sainsbury's Love Meat & Veg! Mediterranean beef meatballs (16 reviews). All reviews are publicly available on the retailer's website and were downloaded into an excel spreadsheet, noting down the product name, review title, review comment and score from 1 to 5. The reviews were divided into two groups: reviews with a score equal to or above 3.5 were included in the corpus of positive reviews, whereas reviews with a score below 3.5 and below formed the corpus of negative reviews. Table 1 shows the size of the two corpora, with the majority of reviews being positive (80%).

Table 1. Size of the corpora of online product reviews on three UK hybrid meat products.

Corpus	No. and % of Reviews	No. of Words
Positive reviews	161 (80%)	4223
Negative reviews	40 (20%)	1148

Because the reviews were short and often included only a few words (not complete sentences), the sizes of the corpora are relatively small. Nonetheless, they were still large enough to perform frequency counts and key term analysis.

2.2. Statistical Approach for Frequency Counts and Key Term Analysis

Both positive and negative corpora were uploaded onto the linguistic software program Sketch Engine, which performed frequency counts and an extraction of key terms. Frequency counts of language items in reviews can be a useful indicator of preferences in that frequent items can signal preferred lexical choices which in turn can point to attitudes and stances. Yet, the most frequent items in English are grammatical words (e.g., articles, prepositions) that as such are used to form grammatical constructions and do not hold a lexical meaning. Because we were interested in attitudes towards the hybrid-meat products, and attitudes are likely to be revealed in the ways in which the consumers describe and evaluate the products, the analysis was focused on adjectives. Adjectives are parts of speech that function primarily as descriptors denoting a whole range of features and dimensions such as size, colour, quantity, texture, taste, judgment and affect [21]. Since all these dimensions can be relevant to hybrid-meat products, adjectives were selected as good indicators of consumers' attitudes towards specific features of the products.

Once the corpora were uploaded onto Sketch Engine, a parser was applied which tagged each word in a given corpus with its parts of speech (verbs, nouns, adjectives, etc.). In this way, adjectives were identified, and their frequencies retrieved. Because adjectives can describe a range of dimensions, subsequently all adjectives retrieved from the two corpora were grouped into their semantic domains. This allowed us to determine which dimensions of the products (e.g., texture, taste) were particularly emphasised and how they were evaluated by those who liked and disliked them. The grouping of adjectives into semantic domains was conducted first independently by the two researchers; disagreements and ambiguous meanings (e.g., the adjective 'hot' can be used to describe temperature or the level of spiciness) were resolved by checking the meanings of the adjectives in context, that is, how they were used in the reviews.

In order to gain insights into other salient themes and issues mentioned by the consumers, we also retrieved distinctive multiword items from both corpora, also known as key terms. Key terms are simply distinctive combinations of two or three words which appear more frequently in the studied corpus as compared to a reference corpus and, additionally, match the typical format of terminology in the language, that is, they are lemmatised. Key terms are good indicators of the content and distinctive topics of the studied corpus. For the purpose of this analysis, we used the EnglishTenTen corpus (available on Sketch Engine) as a reference corpus because it is a large compilation of general English collected from online sources. The key terms were retrieved using keyness scores calculated as follows:

$$\frac{fpm_{focus} + n}{fpm_{ref} + n}$$

Fpm_{focus} stands for normalised frequency (per million) of the term in the focus corpus (in our case in positive or negative reviews), while fpm_{ref} is the normalised frequency (per million) of the term in the reference corpus. N is the simple maths parameter added to account for the problem that we cannot divide by zero. Retrieved key terms were then grouped into semantic domains using the same procedure as above. The next section summarises the main findings that emerged from the search of hybrid meat products in the UK market and the results of the corpus linguistic analysis.

3. Results and Discussion

3.1. Hybrid Meat Products in the UK Market

Table 2 shows a list of hybrid meat products launched in the UK with information on the brand, launch date, range, prices, availability and percentage of plant-based ingredients used in the formulation (where available). Hybrid meat products were launched in four UK supermarkets (Marks and Spencer, Waitrose, Tesco, Sainsbury's and Aldi) between 2017 and 2020. More details on each retailer launch are discussed here.

Table 2. The brand, launch date, range, prices, availability and % of plant-based ingredients used in hybrid meat products launched in the UK.

Brand	Launch Date	Range Launched and Prices	Product Still Available?	% of Plant-Based Ingredients Used (If Not on Product Name Already)
Retailer brand—Marks and Spencer	6 Jan 2020	Hidden Veggies range:		Hidden Veggies range: one of your five a day per portion (vegetable content unknown)
		Beef and carrot mince (£3.50)	Yes	
		Hidden Veggies chicken and vegetable meatballs	Yes	
		Hidden Veggies beef and mushroom burgers	Yes	
Retailer brand—Tesco	April 2019	Meat & Veg range:		
		Tesco Meat & Veg 4 Beef, Carrot & Onion Burgers (454 g/£2.50)	Yes	35% average vegetable blend: vegetable blend (38%) of carrot and onion
		Tesco Meat & Veg 8 Beef, Red Pepper & Carrot Koftas (600 g/£3.00)	Yes	vegetable blend (30%) of carrot, onion, red pepper
		Tesco Meat & Veg Lamb, Carrot & Onion Mince (500 g/£4.00)	Yes	vegetable blend (38%) of carrot and onion
		Tesco Meat & Veg 12 Beef, Carrot & Onion Meatballs (336 g/£4.00)	Yes	vegetable blend (31%) of carrot, onion, butternut squash
		Tesco Meat & Veg Lean Beef, Carrot and Onion Mince (250 g/£2.19, 500 g/£3.39, 750 g/£4.50)	Yes	vegetable blend (31%) of carrot, onion, butternut squash
Retailer brand—Waitrose	10 Jan 2018	6 Pork, Butternut Squash, Quinoa & Kale Sausages £3.29/400 g	No	25% vegetables
		12 mini Pork, Butterbean, Lentil & Garlic Toulouse Style Sausages £3.29/400 g	Yes	30% vegetables
		6 Spanish Style Pork, Chickpea, Spinach & Tomato Sausages £3.29/400 g	Yes	35% vegetables
		12 Harissa Chicken, Cauliflower Rice & Chickpea Meatballs £3.29/360 g	Yes	25% vegetables
		20 Asian-Style Beef Meatballs with Beans £3.29/300 g	No	35% vegetables
		2 Caribbean Inspired/Style Spiced Pork, Sweet Potato & Black Turtle Bean Burgers £3.29/270 g	No	Unknown
		Waitrose Cumberland Chipolata with Mixed Pulses £3.29/375 g	No	20% vegetables
		Waitrose Beef Meatballs with Mixed Pulses £2.49/300 g	No	50% vegetables
		Waitrose Beef Mince with Mixed Pulses £3.99/454 g	No	50% vegetables
Retailer brand—Sainsbury's	early 2018	Sainsbury's Love Meat and Veg range all £2.50/350 g pack:		50% meat and 50% vegetables
		Mediterranean Beef Meatballs	Yes	
		Chicken Sausages with Feta, Spinach and Peas	No	
		Pork Sausages with Kale and Butternut Squash	No	
		Pork Sausages with Kidney Beans, Sweet Potato and Smoked Paprika	No	
		Pork Sausages with Roasted Red Pepper, Sundried Tomato and Quinoa	No	
Retailer brand—Aldi	early 2017	Flexitarian "Full of Beans" chilled mince (£1.99/400 g)	No	50% (haricot beans)
		BBQ Flexitarian burger	No	

Table 2. Cont.

Brand	Launch Date	Range Launched and Prices	Product Still Available?	% of Plant-Based Ingredients Used (If Not on Product Name Already)
Private label—Finnerbrogue Artisan	2016	#Funky Flexitarian range: Spicy lamb'alafal chipolatas	No	47% vegetables and legumes
		Smokey pork n'bombay beet bangers	No	
		Lightly curried cauli'nation chicken chipolatas	No	
		Beef, tomato n'basil bangers	No	
Private label—Debbie & Andrew's brand (ABP group)	Jan 2017	Flexilicious range: Chilli Con Carne Beef Sausages	No	40% vegetables and legumes
		Super Sausages 6 Chorizo Style Pork & Bean 400 g	No	
Private label—MOR Sausages	2017	Moroccan Spiced Pork & Red Pepper Sausages	No	% of plant-based ingredients not specified: Pork (55%)
		Mediterranean Chicken with Sundried Tomato & Basil Chipolatas	No	Chicken (60%)
		Pork, Super Green Veg & Lentil Sausages	No	Pork (55%)
		Pork, Beetroot & Bramley Apple Sausages	No	Pork (52%)
Private label—Kerry	2017	The Crafty Carnivore range: Smoky Chipotle Pork Sausages with Sweet Potatoes and Red Pepper	No	43% vegetables and legumes
		Harissa Spiced Pork Sausages with Butternut Squash and Red Pepper	No	40% vegetables and legumes
Restaurant chain—BrewDog	October 2019	Burger	No—special in October only	50% plant-based "Beyond Meat"
Restaurant chain—Byron	2018	Classic Flex Burger	No	30% mushrooms

According to our search, in 2017 Aldi was the first UK retailer to launch an own-label hybrid meat product. The retailer launched the flexitarian "Full of Beans" chilled mince in 2017 [22] and a burger with haricot beans. Because of its flexitarian name, the burger attracted negative press and social media attention [23], and neither of the two products are currently available on the retailer's website. It is interesting to note that the plain red packaging used by Aldi did not convey any special message regarding the non-meat ingredients, it just highlighted the characteristics of the meat component "Scotch BBQ flexitarian burgers-reared to higher welfare standards from farms we know and trust".

Waitrose and Sainsbury's both launched their hybrid meat product ranges in 2018 (Table 2). Waitrose, a UK retailer targeting upper-middle class consumers, launched a range of 9 hybrid meat products with 20–50% vegetables in 2018 and at least 3 of them are still available for purchase on the retailer website today [24]. The sausages were developed "for shoppers looking to reduce their meat intake" and carry the green Waitrose 'Good Health' label, designed to make it easier for shoppers to make healthier choices. Sainsbury's in 2018 launched a range called "Love Meat and Veg". The range aimed to help consumers to reduce their meat-eating habits and explore the switch to a higher vegetable intake targeting flexitarians. They contain 50% meat and 50% vegetables and include the range shown in Table 2. However, on the Sainsbury's website, only Mediterranean Beef Meatballs seem to be currently available. Both of these retailers made the ranges look different from the meat versions and plant-based version, using colourful packaging with vegetables and attractive names and these are all factors that have been shown to affect food choice [4].

Tesco in 2019 introduced the "Meat & Veg" range [25] which comprises products made from beef or lamb. The retailer claims that this range "helps make scratch cooking easier, removing the need to buy vegetables separately to make the base of popular dishes such as bolognese, lasagne or meatballs" and "the range champions vegetables as flavour enhancers to provide sweetness to home-cooked dishes". The focus of these products seems to be on delivering convenience and flavour to consumers, rather than highlighting the lower meat content or the health characteristics. All the products in the range are currently available to purchase on the Tesco website.

The last retailer to launch hybrid meat product is Marks and Spencer [26]. The range includes 3 products, and they have been formulated to deliver "1 of your 5 a day" per portion. Here the focus is on vegetables as healthy ingredients but also as flavourful compounds ("an easy way to your five a day" and "more veggies, more flavour").

A total of 4 private labels also launched hybrid meat products into the UK market between 2016 and 2017. These were ABP Food Group under the Debbie & Andrew's brand, Walkers Sausage Co. under the brand MOR sausages, Kerry under the brand The Crafty Carnivore and Finnerbrogue Artisan under the brand #Funky Flexitarian. Debbie & Andrew's launched the range "Flexilicious" in 2017 [27]. The flexilicious sausages consisted of 40% beef, 40% vegetables and legumes, 10% herbs and seasonings along with 10% gluten-free crumbs and water. They launched on Amazon Fresh, Asda, Morrisons, Ocado, Sainsbury's and Tesco, but they were not available for purchase when our search was conducted. MOR sausages launched in 2017 [28] and contained 52–60% meat and the inclusion of a variety of vegetables. Although the website suggests that MOR sausages are stocked in Tesco and Morrisons, these products were not found on the retailers' websites. Kerry's The Crafty Carnivore range of 2 sausages and Finnerbrogue Artisan's #Funky Flexitarian range of 4 sausages launched in 2016 were not successful either.

Finally, in the foodservice sector, 2 examples of hybrid meat products are available. Byron burger, a chain with 53 stores in the UK, launched the "Classic Flex" in 2018, made of 70% British beef and 30% mushrooms [29], however, this item is not on the current menu. In October 2019, Brew Dog launched a patty with 50% UK beef and 50% plant-based "Beyond Meat" [30], however, this was not on the online menu, and it was only offered as a "special" for the month of October.

In total, 38 hybrid meat products were launched in the UK in 2016–20, and 12 of these products seem to be still currently available for purchase [8]. These numbers should not surprise, as it is well known that most new foods fail in the market [31]. The most popular hybrid meat products launched onto the market were sausages, with 20 products launched, followed by meatballs with 7 launches, burgers with 6 launches and mince with 5 launches. The base meats used vary: beef and pork were the most popular (16 and 15 products respectively), followed by chicken (5 products) and lamb (2 products). The amount of meat to non-meat ingredient ratio changes widely, from 25% of vegetables, up to 50% of vegetables. The type of non-meat ingredients used also vary, and they are usually a blend of different spices, fruits and vegetables, however, mushrooms have also been used as an ingredient on their own, as mushrooms have been shown to be effective in maximising umami taste in meat formulations [32].

Interest in processed meat products with plant-based ingredients is increasing, with Mintel [12] reporting in a recent survey on 1678 consumers that on average 27% of processed meat buyers would be interested in buying processed meat products with added vegetables. This number increases to 35% for consumers who eat meatballs and to 31% for consumers who eat burgers.

However, looking at the launch dates and current availability in the market, this search shows that so far, attempts to bring to the market hybrid meat products have had mixed results. Some of the earlier launches did not seem to have managed to maintain a place in the market, and perhaps they might have been received with confusion or were not understood by consumers. The newer retailer launches by Tesco, Sainsbury's, Waitrose and Marks and Spencer, carry a more targeted message to consumers [8]. Overall, it seems that the most recent launches do not mention flexitarianism and stress more the flavour, healthiness and convenience of these meat products, including messages such as "5-a-day", the convenience of having vegetables already in minced meat and the use of vegetables as flavour enhancers. For these new launches to be successful, it is important to codevelop and codesign new foods with consumers, and the new product development literature stresses the importance of incorporating consumer insights into the new product development process for foods [31].

3.2. Consumer Attitudes and Evaluations in Hybrid Meat Reviews

Table 3 shows all the adjectives identified and retrieved from the corpus of positive reviews and their semantic classification. As can be seen, consumers emphasised mostly the taste dimension of hybrid meat products. Taste is quite a subjective domain, but many of those who purchased the products evaluated them as tasty, delicious and spicy. These results are in accordance with those by Reed et al. [33], who found that in commercial food product reviews, taste-associated words were mentioned more than words associated with other factors such as price, food texture, customer service, nutrition or smell. After taste, there was a heavy use of general positive descriptors, which is not surprising given the positive scores. Interestingly, several consumers saw the products as healthier options (19 mentions in total), low in fat (mentioned in 3 reviews) and as a way to increase the intake of vegetables (mentioned in 7 reviews). It can therefore be deduced that the hybrid meat products are appreciated by consumers who are health conscious. This is an interesting finding related to hybrid meats, as traditional meat products in the literature have scored quite low in terms of perceived healthiness by consumers [34]. Consumers also seem to positively value the texture of the products and their quick and easy preparation at home. Processed meats can be a convenient type of food product [31] and Grunert [35] suggested that there might be a synergy between the desire for healthiness and the demand for convenience in functional food products.

Table 3. Semantic domains of adjectives in positive online reviews.

Semantic Domain/Dimensions	Adjectives	Total Freq.
Taste	tasty (76), delicious (34), spicy (13), yummy (4), flavoursome (3), succulent (2), well-seasoned (2), strong (2), tangy (1), scrumptious (1), subtle (1), versatile (1)	140
General positive descriptors	great (39), good (21), nice (21), different (12), lovely (7), excellent (4), new (4), unusual (4), popular (4), perfect (2), pleasant (2), superb (2), traditional (2), wonderful (2), special (2)	128
Health	healthy (19), fresh (7), extra [veg] (5), low [in] (3), balanced (2)	36
Texture	moist (7), dry (5), meaty (4), fatty (2), mushy (2), soft (2), hard (2), light (2), tough (2)	28
Manner of processing	quick (7), easy (7), homemade (2), simple (2), standalone (1)	19
People	[e.g., my 3 year] old (4), whole (4)	8
Colour	brown (3), red [meat] (3), white (1)	7
Temperature of processing	cold (2), hot (2), warm (1)	5
General negative descriptors	sceptical (2), suspicious (1), unpleasant (1)	4
Portion size	small (2)	2

A total of 4 positive reviews included negative adjectives such as 'sceptical' and 'suspicious', but these were to emphasise consumers' initial suspicions which disappeared once the products were tasted. The following extracts are indicative of this stance:

(1) "Bought off the back of good reviews, was a bit sceptical that a meatball with cauliflower, etc. would actually be nice but I thought these were superb".
(2) "When I put these in the pan I was sceptical, because the cauliflower smell was very strong. As it happens these are tasty and quite light for a meatball".

These four reviews can be related to the concepts of willingness to try, change-seeking and consumer innovativeness as a way to relate to new foods [36], which in this case led to a positive outcome.

The top key terms revealed similar themes. The largest category was two-word combinations highlighting the healthiness of the products, with 19 mentions in total (see Table 4). Consumers seem to view hybrid meats as a healthy choice, as a way to reduce meat consumption and increase the intake of vegetables. The following extract from the corpus are indicative of this stance:

(3) "These sausages tasted good and are a great way to get extra veg into the kids".
(4) "We love that we can enjoy healthy choice sausages".
(5) "Healthy sausage, pleased with these sausages as lower in fat but still tasty!".

Table 4. Key terms in positive reviews.

Semantic Domain	Key Terms	Total Freq.
Health	right amount (3), healthy choice (2), meat consumption (2), veg content (2), red meat (2), healthy sausage (2), fat content (1), low meat content (1), healthy tasty dinner (1), healthy eating (1), nice balance (1), low carb (1)	19
Novelty/Surprise	first time (8), pleasant surprise (2), great alternative (1), standalone alternative (1), tasty alternative (1), delicious change (1), tasty change (1), tasty new experiment (1)	16
General positive terms	great way (3), regular addition (2), great find (2), great idea (2), excellent sausage (2), great combination (1)	14
Taste/texture	good flavour (3), great texture (3)	9
People	whole family (6)	6

The positive reviews indicate that hybrid meats might be seen as an opportunity to lower meat intake and increase vegetable intake while maintaining acceptable taste (as seen in Table 3). It has been reported in the literature that consumers are unwilling to compromise on taste when it comes to healthier food options [37], therefore, hybrid meat products should be designed to deliver both on taste and health.

The second most distinctive domain was that of novelty and positive surprise. Many consumers who took to the online forum to review the products emphasised that this was something new in their shopping basket almost treated as a form of experiment, which met the expectations:

(6) "It was with a pleasant surprise with all the veg how juicy they where (sic) and very tasty great combination of flavours very good and something a little different from standard",
(7) "A tasty change found this to be a delicious change from ordinary meatballs".
(8) "This is the first time we purchased this sausage will certainly be purchasing it again".
(9) "Interesting new combination tasty new experiment, hopefully will continue to be stocked".

These reviews highlight the interest of these consumers in purchasing something new and show how the differentiation strategy in these hybrid meat products were well received by consumers.

Consumers also emphasised that the sausages were enjoyed by the whole family, which made the hybrid meat products a convenient alternative.

In terms of the negative reviews, it is not surprising to find the category of negative descriptors as the most prominent ones (see Table 5). The dissatisfaction with the products seemed to be mostly due to a lack of taste and texture as evidenced with the use of adjectives such as 'flavourless', 'tasteless', 'bland', 'mushy' and 'dry'. It is therefore not necessarily the concept of hybridity itself but rather the specificity of the product's sensory quality that did not seem to match expectations. Consumers purchased the hybrid-meat products, which suggests that there was a willingness to try these new foods because they were perceived as healthy.

Table 5. Semantic domains of adjectives in negative online reviews.

Semantic Domain/Dimensions	Adjectives	Total Freq.
Taste	spicy (3), flavourless (3), tasteless (2), bland (1), flavoursome (1), metallic (1), overpowering (1), salty (1), tasty (1), unseasoned (1)	36
General negative descriptors	disappointing (8), awful (5), unpleasant (2), inedible (2), poor (2), sloppy (2), strange (2), weird (2), bad (1), disgusting (1), dreadful (1), freaky (1), gross (1), horrible (1), nauseous (1), negative (1), underwhelming (1), uneatable (1)	35
Texture	mushy (4), dry (3), soft (3), rubbery (2), slimy (2), tough (2), chewy (1), crumbly (1), moist (1), pappy (1), solid (1), heavy (1)	25
General positive descriptors	good (5), amazing (1), pleasant (1), great (1)	8
Temperature of processing	hot (1), cold (1)	2
Health	healthy (1)	1
Manner of processing	undercooked (1)	1
People	fussy [eater] (1)	1
Colour	green (1)	1

However, the experience proved to be disappointing; the extracts below exemplify this trend in the negative online reviews:

(1) "Thought I'd give these meatballs a try as they seemed a bit different. But I found them disappointing because they fell apart when cooking and not very tasty".
(2) "Tried twice to check, still awful & flavourless".
(3) "Trying to find healthy food for kids. Unfortunately I had to throw away because the texture was so unpleasant, slimy and rubbery".
(4) "OK because nice to know some veg in there but my husband and I werent (sic) keen on taste".

The analysis of key terms confirms the dominant-negative experience of the taste and texture of the products in the negative online reviews (Table 6). Yet, because of the small sample under consideration (only 40 reviews), it is difficult to generalise from these results. Overall, we can say that both the positive and negative reviews on taste highlight the paramount importance of this sensory attribute.

Table 6. Key terms in negative reviews.

Semantic Domain	Key Terms	Total Freq.
Taste/texture	poor texture (2), metallic aftertaste (1), unseasoned chicken (1), lacked flavour (1), real flavour (1), recognisable flavour (1), awful smell (1), overpowering taste (1), freaky texture (1), slimy texture (1), strange texture (1), pappy thing (1), undercooked veg (1), dry side (1), sloppy mix (1)	16
General positive terms	good sauce (1), natural shape (1), plus side (1)	3
General negative terms	bad news (1), very poor (1)	2
People	single person (1), toddler wouldn't (1)	2
Health	healthy food (1)	1

It is worth pointing out the limitations of this study. The linguistic analyses were carried out on a limited sample of reviews, and further analysis should be conducted on a larger sample. The negative reviews were lower in number compared to the positive one. This could be because genuinely, most of the consumers who tried the products had a positive view on them, however, it is possible that consumers with negative views did not report them online. In addition, many who purchased the hybrid meat products might have decided not to leave a review at all, therefore, the online product reviews we have gathered cannot be used on their own to make specific recommendations on these products.

4. Conclusion and Future Work

This study was the first investigation into commercially available UK hybrid meat products, and the first preliminary exploration into consumers' attitudes towards hybrid meat products using online product reviews. The messages adopted by retailers to promote hybrid meat products have changed greatly since these products were first introduced to the market. The linguistic analysis showed that the most important themes related to hybrid meats were taste, followed by healthiness and convenience. There are still several gaps in the literature that need to be investigated. Future research should include both qualitative and quantitate studies developing the topic of hybrid meat products and further the understanding of its potential. It would be valuable to compare different socio-demographic characteristics and different nations to gather social and cross-country insights on this topic. For hybrid meat products not to be a fad, product development needs to be codesigned with consumers and carried out by teams of multidisciplinary professionals investigating recipe reformulations, sensory aspects and consumer attitudes simultaneously, in a more holistic approach.

Author Contributions: Conceptualisation, S.G. and S.J.; data curation, S.G. and S.J.; formal analysis, S.J.; investigation, S.G. and S.J.; methodology, S.G. and S.J.; project administration, S.G.; resources, S.J.; software, S.G. and S.J.; validation, S.J.; writing—original draft, S.G. and S.J.; writing—review and editing, S.G. and S.J. All authors have read and agreed to the published version of the manuscript.

Funding: This research received no external funding.

Conflicts of Interest: The authors declare no conflict of interest.

References

1. Apostolidis, C.; McLeay, F. Should we stop meating like this? Reducing meat consumption through substitution. *Food Policy* **2016**, *65*, 74–89. [CrossRef]
2. Hoek, A.C.; Elzerman, J.E.; Hageman, R.; Kok, F.J.; Luning, P.A.; de Graaf, C. Are meat substitutes liked better over time? A repeated in-home use test with meat substitutes or meat in meals. *Food Qual. Prefer.* **2013**, *28*, 253–263. [CrossRef]
3. Ryan, R.M.; Deci, E.L. Self-determination theory and the facilitation of intrinsic motivation, social development, and well-being. *Am. Psychol.* **2000**, *55*, 68. [CrossRef] [PubMed]
4. Grunert, K.G. Current issues in the understanding of consumer food choice. *Trends Food Sci. Technol.* **2002**, *13*, 275–285. [CrossRef]
5. Schösler, H.; De Boer, J.; Boersema, J.J. Can we cut out the meat of the dish? Constructing consumer-oriented pathways towards meat substitution. *Appetite* **2012**, *58*, 39–47. [CrossRef]
6. Spencer, M.; Cienfuegos, C.; Guinard, J.-X. The Flexitarian Flip™ in university dining venues: Student and adult consumer acceptance of mixed dishes in which animal protein has been partially replaced with plant protein. *Food Qual. Prefer.* **2018**, *68*, 50–63. [CrossRef]
7. Asher, K.; Green, C.; Gutbrod, H.; Jewell, M.; Hale, G.; Bastian, B. *Study of Current and Former Vegetarians and Vegans: Initial Findings*; Faunalytics: Olympia, WA, USA, 2014.
8. Grasso, S. Hybrid meat. *Food Sci. Technol.* **2020**, *34*, 48–51. [CrossRef]
9. Mintel. Executive summary. In *Meat Free Foods, UK, September 2018*; Mintel: London, UK, 2018.
10. Neville, M.; Tarrega, A.; Hewson, L.; Foster, T. Consumer-orientated development of hybrid beef burger and sausage analogues. *Food Sci. Nutr.* **2017**, *5*, 852–864. [CrossRef]
11. Grasso, S.; Smith, G.; Bowers, S.; Ajayi, O.M.; Swainson, M. Effect of texturised soy protein and yeast on the instrumental and sensory quality of hybrid beef meatballs. *J. Food Sci. Technol.* **2019**, *56*, 1–10. [CrossRef] [PubMed]
12. Mintel. *Processed Poultry and Read Meat Main Meal Components UK*; Mintel: London, UK, 2019.
13. Hicks, T.M.; Knowles, S.O.; Farouk, M.M. Global provisioning of red meat for flexitarian diets. *Front. Nutr.* **2018**, *5*, 11. [CrossRef]

14. England Regulations Statutory Instrument 3001. The Products Containing Meat etc. (England) Regulations. 2014. Available online: https://www.legislation.gov.uk/uksi/2014/3001/contents/made (accessed on 17 December 2020).
15. Petracci, M.; Bianchi, M.; Mudalal, S.; Cavani, C. Functional ingredients for poultry meat products. *Trends Food Sci. Technol.* **2013**, *33*, 27–39. [CrossRef]
16. Singh, P.; Kumar, R.; Sabapathy, S.; Bawa, A. Functional and edible uses of soy protein products. *Compr. Rev. Food Sci. Food Saf.* **2008**, *7*, 14–28. [CrossRef]
17. Asgar, M.A.; Fazilah, A.; Huda, N.; Bhat, R.; Karim, A.A. Nonmeat protein alternatives as meat extenders and meat analogs. *Compr. Rev. Food Sci. Food Saf.* **2010**, *9*, 513–529. [CrossRef]
18. Wong, K.M.; Corradini, M.G.; Autio, W.; Kinchla, A.J. Sodium reduction strategies through use of meat extenders (white button mushrooms vs. textured soy) in beef patties. *Food Sci. Nutr.* **2019**, *7*, 506–518. [CrossRef] [PubMed]
19. Deliza, R.; Saldivar, S.S.; Germani, R.; Benassi, V.; Cabral, L. The effects of colored textured soybean protein (TSP) on sensory and physical attributes of ground beef patties. *J. Sens. Stud.* **2002**, *17*, 121–132. [CrossRef]
20. De Boer, J.; Schösler, H.; Boersema, J.J. Motivational differences in food orientation and the choice of snacks made from lentils, locusts, seaweed or "hybrid" meat. *Food Qual. Prefer.* **2013**, *28*, 32–35. [CrossRef]
21. Biber, D.; Johansson, S.; Leech, G.; Conrad, S.; Finegan, E.; Quirk, R. *Longman Grammar of Spoken and Written English*; Longman: London, UK, 1999; Volume 2.
22. The Grocer. Aldi Launches 'Flexitarian' Full of Beans Chilled Mince. Available online: https://www.thegrocer.co.uk/new-product-development/aldi-launches-flexitarian-full-of-beans-chilled-mince/547310.article (accessed on 17 December 2020).
23. Metro News. Aldi Gets Slammed by Vegan for Launching 'Flexitarian' Burgers. Available online: https://metro.co.uk/2019/01/17/aldi-gets-slammed-vegan-launching-flexitarian-burgers-8355380/?ito=cbshare (accessed on 17 December 2020).
24. Waitrose. Pork and Lentil Sausages? *Just One New Way to Good Health*. Available online: https://waitrose.pressarea.com/pressrelease/details/78/product%20news_12/9090 (accessed on 17 December 2020).
25. Tesco. Tesco Launches New 'Meat & Veg' Range to Help Health Conscious Customers. Available online: https://www.tescoplc.com/news/2019/tesco-launches-new-meat-veg-range-to-help-health-conscious-customers (accessed on 17 December 2020).
26. Marks and Spencer. Hidden Veggies Range. Available online: https://www.facebook.com/MarksandSpencer/posts/our-new-hidden-veggies-range-makes-family-meals-delicious-and-nutritious-and-the/10157423606858612/ (accessed on 17 December 2020).
27. Just-Food. ABP Food Group Expands Debbie & Andrew's Range in the UK. Available online: https://www.just-food.com/news/abp-food-group-expands-debbie-andrews-range-in-uk_id135395.aspx (accessed on 17 December 2020).
28. Food Manufacture. New MOR Sausage Range Lanuches at Tesco. Available online: https://www.foodmanufacture.co.uk/Article/2017/03/22/New-MOR-sausage-range-launches-at-Tesco (accessed on 17 December 2020).
29. Independent. Byron Launches New Flexitarian Beef Burger. Available online: https://www.independent.co.uk/life-style/food-and-drink/flexitarian-beef-burger-byron-launch-sustainable-ethical-food-vegetarian-flex-uk-a8310016.html (accessed on 17 December 2020).
30. YouGov. BrewDog's New Hybrid Flexitarian Burger Isn't That Left Field. Available online: https://yougov.co.uk/topics/consumer/articles-reports/2019/10/09/brewdogs-new-hybrid-flexitarian-burger-isnt-left-f (accessed on 17 December 2020).
31. Grunert, K.G.; Verbeke, W.; Kügler, J.O.; Saeed, F.; Scholderer, J. Use of consumer insight in the new product development process in the meat sector. *Meat Sci.* **2011**, *89*, 251–258. [CrossRef] [PubMed]
32. Dermiki, M.; Mounayar, R.; Suwankanit, C.; Scott, J.; Kennedy, O.B.; Mottram, D.S.; Gosney, M.A.; Blumenthal, H.; Methven, L. Maximising umami taste in meat using natural ingredients: Effects on chemistry, sensory perception and hedonic liking in young and old consumers. *J. Sci. Food Agric.* **2013**, *93*, 3312–3321. [CrossRef]
33. Reed, D.R.; Mainland, J.D.; Arayata, C.J. Sensory nutrition: The role of taste in the reviews of commercial food products. *Physiol. Behav.* **2019**, *209*, 112579. [CrossRef]
34. Shan, L.C.; Henchion, M.; De Brún, A.; Murrin, C.; Wall, P.G.; Monahan, F.J. Factors that predict consumer acceptance of enriched processed meats. *Meat Sci.* **2017**, *133*, 185–193. [CrossRef]

35. Grunert, K.G. European consumers' acceptance of functional foods. *Ann. N. Y. Acad. Sci.* **2010**, *1190*, 166–173. [CrossRef] [PubMed]
36. Bäckström, A.; Pirttilä-Backman, A.M.; Tuorila, H. Willingness to try new foods as predicted by social representations and attitude and trait scales. *Appetite* **2004**, *43*, 75–83. [CrossRef] [PubMed]
37. Verbeke, W. Functional foods: Consumer willingness to compromise on taste for health? *Food Qual. Prefer.* **2006**, *17*, 126–131. [CrossRef]

Publisher's Note: MDPI stays neutral with regard to jurisdictional claims in published maps and institutional affiliations.

 © 2020 by the authors. Licensee MDPI, Basel, Switzerland. This article is an open access article distributed under the terms and conditions of the Creative Commons Attribution (CC BY) license (http://creativecommons.org/licenses/by/4.0/).

Article

Analysis of Association between the Consumer Food Quality Perception and Acceptance of Enhanced Meat Products and Novel Packaging in a Population-Based Sample of Polish Consumers

Dominika Guzek [1,*], **Dominika Głąbska** [2], **Marta Sajdakowska** [1] and **Krystyna Gutkowska** [1]

1. Department of Food Market and Consumer Research, Institute of Human Nutrition Sciences, Warsaw University of Life Sciences (WULS-SGGW), 159C Nowoursynowska Street, 02-776 Warsaw, Poland; marta_sajdakowska@sggw.edu.pl (M.S.); krystyna_gutkowska@sggw.edu.pl (K.G.)
2. Department of Dietetics, Institute of Human Nutrition Sciences, Warsaw University of Life Sciences (WULS-SGGW), 159C Nowoursynowska Street, 02-776 Warsaw, Poland; dominika_glabska@sggw.edu.pl
* Correspondence: dominika_guzek@sggw.edu.pl; Tel.: +48-22-593-71-34

Received: 21 July 2020; Accepted: 21 October 2020; Published: 23 October 2020

Abstract: The consumer acceptance of novel enhanced-quality products and their willingness to buy such products may be a crucial topic in the field of marketing. The aim of this study was to analyze the association between consumers' perceptions of food quality and their acceptance of enhanced meat products and novel packaging. The study was conducted using the Computer-Assisted Personal Interview (CAPI) method in a random group of 1009 respondents, who were recruited as a representative sample based on data from the Polish National Identification Number database. The participants were asked about the most important quality determinants of food products of animal origin and about quality improvement methods and their acceptance of those methods. The quality determinants of animal-based food products were indicated as follows: origin, production technology, manufacturer, components and nutritional value, visual and sensory characteristics, expiry date, and cost. The quality improvement methods were clustered into groups that were associated with product enhancement and application of novel packaging, and the acceptance of those methods was also verified. Indicating specific quality determinants of animal-derived food products affects the consumer acceptance of product enhancement ($p = 0.0264$) and novel packaging as quality improvement methods ($p = 0.0314$). The understanding that enhancement is applied for the purpose of quality improvement did not influence the acceptance of products ($p = 0.1582$), whereas the knowledge that novel packaging is applied influenced the acceptance ($p = 0.0044$). The obtained results suggested that in the case of application of novel packaging, a higher level of knowledge may be a reason for consumer's rejection of the resulting products, but the appearance and taste of products may contribute to the higher acceptance of novel packaging. Educating consumers may improve their acceptance of product enhancement, as concerns about the addition of food preservatives may lead them to reject enhanced products.

Keywords: consumer; preference; food acceptability; knowledge management; quality

---

1. Introduction

Meat, as well as meat products, is an important component of a typical western diet and is consumed almost every day by some people. Although they provide essential amino acids, vitamins (mainly B_{12}), and minerals (mainly iron and zinc) needed for the body, meat products, especially red and processed ones, may increase the risk of chronic diseases typical in developed countries (e.g., cardiovascular diseases, diabetes, and some cancers), as well as general mortality, when consumed in excessive amounts [1].

The role of meat and meat products in the consumer lifestyle is influenced by various factors, such as socioeconomic determinants, traditions, perceived ethics, and religious beliefs [2]. Many studies have discussed consumers' attitudes toward meat and meat products [3,4], with increasing attention paid to the unsustainable consumption of meat and meat-derived saturated fats [5].

Nevertheless, the necessity of a transition toward more sustainable meat consumption is emphasized by nutritionists and dietitians for promoting health-related values [6]. Provision of moderate amounts of protein, from moderate proportions of meat, eggs, and dairy products, is indicated as one of the characteristics of a well-balanced diet [7]. Taking into account that Western diets often include food products of animal origin [8], reducing the intake of red and processed meat can be beneficial to both health and the environment [9].

On the one hand, plant-based products are considered as substitutes for meat, but on the other hand, less-but-better meat products are more suitable for consumers [10], as vegetarian diets may not be preferred by a majority of the population [11] and may be treated as too radical [5]. Therefore, adding more health-promoting components to meat-based food products and improving their nutritional value can be an effective strategy for reducing the intake of other meat products. This is not only due to the fact that health benefits are given more importance by consumers than environmental concerns [12] but also because a sustainable and moderate consumption of meat products is necessary.

Nowadays, consumers choose food products that satisfy their hunger as well as prevent diet-related diseases. The food market, especially the meat market, is currently undergoing many changes, and as a result, new meat products with health-promoting components appear, which are acceptable [13]. Functional foods, including meat products, providing additional health benefits [14] have been studied by many researchers, but consumer attitudes and behaviors toward such products remain a question [15–17].

There are three main types of factors affecting consumer behaviors—individual factors (psychological aspects), product-specific factors, and marketing factors [2]. Among the product-specific sensory factors, the most important ones are quality and sensory features, such as visual appearance, texture, flavor, and taste [18].

To obtain novel functional meat and meat products that have health-promoting properties, one of the two main approaches are used: (1) enhancing or (2) reducing the share of some components. The health-promoting components incorporated into food products of animal origin may be added either to animal fodder or directly to the product during the production process, while reducing the share of other components. However, consumers have a common assumption that all the additives added to the products are "bad" [19]. This belief is prevalent in spite of the fact that labeling food products with information about the additive, including its name and function, is mandatory, as well as that only EU-approved additives may be used in food products according to legal regulations [20]. Although labeling of food products does not change their sensory features, it could be valuable in terms of influencing consumers' opinion about the health-promoting attributes of such products [21].

Therefore, the so-called "feed-to-food" approach (adding bioactive compounds to fodder, rather than to the product) seems to be a promising direction for improving the quality of animal-based food products to increase consumer acceptability [22,23]. However, adding bioactive components directly to products may be easier for producers. Taking this into account, an important task is to keep the initial features of a product intact [24] and sustain them during storage [25].

For maintaining the nutritional value and sensory features of animal-based food products, some advances are made in the form of novel packaging systems, which are divided into two main categories—active and intelligent packaging [26]. Although these types of packaging have existed on the market for quite some time, some consumers do not know and are still worried that innovative packaging might mislead them to buy spoiled food [27]. However, packaging may influence the consumers' color perception of meat. As color is a major determinant of meat quality [28], various packaging systems may allow the meeting of the real market demands, providing products that may be accepted by consumers.

The present study aimed to analyze the association between consumers' perception of food quality and their acceptance of enhanced meat products and novel packaging.

2. Materials and Methods

2.1. Study Participants

The study was conducted using the Computer-Assisted Personal Interview (CAPI) method. A total of 1009 adult Polish inhabitants were randomly recruited (inclusion criteria: age ≥ 18 years; willing to provide informed consent to participate in the study; involved in food purchasing in the household—either own or cooperative purchasing), based on data from the National Identification Number database, which is the universal electronic system for population registration. The included participants were verified to be representatives of the general population of Poland. The characteristics of the study participants are provided in Table 1.

Table 1. The characteristics of the included participants of the study ($n = 1009$).

Characteristic		N	%
Gender	men	496	49.2
	women	513	50.8
Age (years)	19–29	121	12.0
	30–39	220	21.8
	40–49	211	20.9
	50–59	204	20.2
	60–70	181	17.9
	≥70	72	7.1
Educational background *	primary	120	12.1
	vocational	342	34.5
	secondary	377	38.0
	higher	152	15.3
Per capita net income *	<1500 PLN (<350 EUR)	108	18.5
	1500–2500 PLN (~350–600 EUR)	153	26.2
	2500–4000 PLN (~600–1000 EUR)	205	35.0
	>4000 PLN (>1000 EUR)	119	20.3
Village/city (number of inhabitants) *	village	352	35.1
	city < 20,000	133	13.3
	city 20,000–100,000	203	20.3
	city 100,000–500,000	201	20.1
	city > 500,000	113	11.3
Household size (number of persons) *	1	123	13.0
	2	265	27.9
	3	241	25.4
	4	209	22.0
	≥5	111	11.7

* Missing data in category: educational background $n = 18$; per capita net income $n = 424$; town of residence $n = 7$; household size $n = 60$; PLN—Polish Zloty (Polish currency).

In order to verify the representativeness of the studied sample, the distribution was compared with the characteristics of the general Polish population, as described by Statistics Poland [29]. It was verified if the proportions of gender groups were adequate when compared with the general population of adults and the studied population was stated to be representative for gender ($p = 0.3684$, Chi Square Test). Afterwards, it was verified if the proportions of age groups were adequate when compared with the general population of adults and for this assessment, it was assumed that for the youngest and oldest age group (19–29 years and ≥70 years), the proportion should be reduced, as in previous studies [16], to obtain an adequate number of respondents participating in grocery shopping. The general population of adults with indicated age groups reduced to 50% was compared with the studied population and it was stated to be representative for age ($p = 0.5222$, Chi Square Test).

2.2. Assessment of Consumers' Acceptance of Enhanced Animal-Derived Food Products and Novel Packaging

The questionnaire used in the study was developed for assessing the consumers' perception of the quality determinants of animal-based food products, quality improvement methods, and acceptance of those methods. It included questions related to the discussed issue, as presented in a previous study [17] (Table 2).

Table 2. The questionnaire applied in the study.

Question That Was Asked	Characteristics of Question	Interpretation of Answers
(1) Please define the quality determinant of food products of animal origin which is in your opinion the most important one	Open-ended question	If respondent defined more than one determinant, the first one listed was interpreted as the most important.
(2) Do you know any quality improvement methods applied for food products of animal origin?	Yes/no question	If respondent stated that he knows any quality improvement method, he was asked to describe the method which he knows.
(3) Please describe the quality improvement method applied for food products of animal origin which you know.	Open-ended question	Respondents were allowed to describe an unlimited number of methods, if they wanted to.
(4) What is your level of acceptance of the quality improvement/assurance methods applied for food products of animal origin, for the following groups of actions? - Actions associated with origin (e.g., method of farming); - Actions associated with production technology (e.g., certification of production); - Actions associated with manufacturer (e.g., trademark); - Actions associated with components and nutritional value (e.g., food additives); - Actions associated with visual and sensory characteristics (e.g., flavor); - Actions associated with expiry date (e.g., shelf life); - Actions associated with cost of production (e.g., price).	Close-ended questions (7-point scale): 1: Definitely do not accept; 2: Moderately do not accept; 3: Slightly do not accept; 4: Undecided; 5: Slightly accept; 6: Moderately accept; 7: Definitely accept.	1–3: Negative answers; 4: Neutral answer; 5–7: Positive answers.

For the question on the quality determinants of food products of animal origin (Question 1), the researchers interpreted the obtained responses by clustering them in common groups associated with the same area as follows: origin (indicated by respondents as: country of origin, farming, organic farm, etc.), production technology (indicated as: certified production, process of production, technology, etc.), manufacturer (indicated as: known company, trademark, producer, etc.), components and nutritional value (indicated as: vitamins, minerals, food additives, etc.), visual and sensory characteristics (indicated as: appearance, flavor, taste, etc.), expiry date (indicated as: shelf life, fresh product, production data, etc.), and cost (indicated as: price). An additional cluster contained the answers of respondents who declared that they were not able to define the quality determinants of food products of animal origin.

Similarly, for the question about the quality improvement methods applied for the food products of animal origin (Question 3), the researchers interpreted the obtained responses by clustering them in common groups associated with the same area, which allowed analyzing of the quality improvement methods associated with product enhancement and application of novel packaging. In the cluster of methods associated with product enhancement were those responses associated with descriptions such as additives, technology of production, and so on. On the other hand, in the cluster of methods associated with the application of novel packaging were those responses associated with descriptions such as dedicated packaging, proper packaging, or specific packaging described. An additional cluster contained the respondents' answers that were not related to the methods applied by producers (e.g., storage, not consuming products that are past the expiration date).

The structure of the interview and the order of questions allowed extending of the analysis from the most general issue associated with concept definition (what does quality mean for respondents), to market situation (what actions do they observe that can improve the quality), and finally, to the subjectively perceived acceptance (do they accept the observed actions). Thus, the data obtained from the interview were analyzed to verify the following potential associations:

- The association between the perceived quality determinants of food products of animal origin and the level of acceptance of the quality improvement methods (hypothesis: perceived quality determinants can influence the acceptance of the quality improvement methods);
- The association between the known methods of quality improvement applied for the products of animal origin and the level of acceptance of these methods (hypothesis: the known quality improvement methods are more accepted).

2.3. Statistical Analysis

The frequency of specific answers provided in the subgroups was compared using the Chi Square Test. The accepted level of significance was set at $p \leq 0.05$. Statistical analysis was conducted using Statgraphics Plus for Windows 4.0 (Statistical Graphics Corporation, Rockville, MD, USA).

3. Results

3.1. Acceptance of Enhancement of Animal-Derived Food Products as a Quality Improvement Method

The acceptance of animal-derived food products' enhancement as the quality improvement method for respondents declaring diverse quality determinants is presented in Table 3. The additional analysis conducted for the subgroups stratified by income, place of living, and educational background is presented in the Supplementary Materials (Supplementary Tables S1–S6).

It was stated that declaring diverse quality determinants influenced acceptance of the enhancement of animal-derived food products as the quality improvement method ($p = 0.0264$). In the studied group, the highest number of respondents accepted food products' enhancement in the sub-group of respondents not able to define quality determinants (42.9%), while a lower number of respondents accepted it in the sub-groups of respondents declaring as quality determinants: origin (33.6%), production technology (26.6%), manufacturer (35.4%), components and nutritional value (32.8%), visual and sensory characteristics (26.1%), expiry date (29.7%), and cost (35.0%). At the same time, the highest number of respondents did not accept food products' enhancement in the sub-group of respondents declaring expiry date as a quality determinant (41.5%), while a lower number of respondents did not accept it in the sub-groups of respondents declaring as quality determinants: origin (37.7%), production technology (34.4%), manufacturer (23.1%), components and nutritional value (31.2%), visual and sensory characteristics (37.0%), and cost (32.3%), or in the sub-group of respondents not being able to define quality determinants (19.8%).

Table 3. Level of acceptance of enhanced animal-derived food products for respondents stratified by perceived quality determinants.

Perceived Quality Determinants	Acceptance of Animal-Derived Food Products' Enhancement as the Quality Improvement Method								p
	1 *	2	3	4	5	6	7	0	
Origin	10 (15.6%)	3 (4.7%)	4 (6.3%)	13 (20.3%)	11 (17.2%)	7 (10.9%)	10 (15.6%)	6 (9.4%)	
Production technology	25 (17.2%)	12 (8.2%)	12 (8.2%)	24 (16.4%)	30 (20.5%)	10 (6.8%)	25 (17.2%)	8 (5.5%)	
Manufacturer	9 (19.6%)	2 (4.3%)	3 (6.5%)	11 (23.9%)	7 (15.2%)	2 (4.3%)	7 (15.3%)	5 (10.9%)	
Components and nutritional value	32 (14.2%)	17 (7.5%)	14 (6.2%)	49 (21.6%)	41 (18.1%)	32 (14.2%)	25 (11.1%)	16 (7.1%)	0.0264
Visual and sensory characteristics	20 (16.0%)	9 (7.2%)	9 (7.2%)	18 (14.4%)	23 (18.4%)	15 (12.0%)	19 (15.2%)	12 (9.6%)	
Expiry date	52 (21.1%)	14 (5.7%)	18 (7.3%)	46 (18.7%)	32 (13.1%)	16 (6.5%)	37 (15.0%)	31 (12.6%)	
Cost	7 (10.8%)	0 (0.0%)	2 (3.1%)	21 (32.3%)	13 (20.0%)	9 (13.8%)	6 (9.2%)	7 (10.8%)	
Not able to define quality determinants	9 (9.9%)	2 (2.2%)	5 (5.5%)	22 (24.2%)	13 (14.2%)	10 (11.0%)	16 (17.6%)	14 (15.4%)	

* 1–3 negative answers (1—definitely; 2—moderately; 3—slightly do not accept); 4—neutral answer; 5–7 positive answers (5—slightly; 6—moderately; 7—definitely accept); 0—do not observe such method for food products of animal origin in Poland.

The acceptance of animal-derived food products' enhancement as a quality improvement method, for respondents declaring diverse known methods of quality improvement, is presented in Table 4.

Table 4. Level of acceptance of enhanced animal-derived food products for respondents stratified by known methods of quality improvement.

Known Methods of Quality Improvement	Acceptance of Animal-Derived Food Products' Enhancement as the Quality Improvement Method								p
	1 *	2	3	4	5	6	7	0	
Indicates product enhancement	122 (14.8%)	45 (5.5%)	58 (7.0%)	170 (20.6%)	145 (17.6%)	76 (9.2%)	119 (14.4%)	90 (10.9%)	
Indicates method other than product enhancement	37 (25.2%)	13 (8.8%)	7 (4.8%)	26 (17.7%)	20 (13.6%)	17 (11.6%)	20 (13.6%)	7 (4.8%)	0.0783
Does not know any method to improve quality	5 (13.5%)	1 (2.7%)	2 (5.4%)	8 (21.6%)	5 (13.5%)	8 (21.6%)	6 (16.2%)	2 (5.4%)	

* 1–3 negative answers (1—definitely; 2—moderately; 3—slightly do not accept); 4—neutral answer; 5–7 positive answers (5—slightly; 6—moderately; 7—definitely accept); 0—do not observe such method for food products of animal origin in Poland.

It was stated that indicating product enhancement as a method of quality improvement did not influence acceptance of animal-derived food products' enhancement as the quality improvement method ($p = 0.0783$). The acceptance was comparable in sub-groups of respondents indicating product enhancement, methods other than product enhancement, and those who did not know any method to improve quality.

3.2. Acceptance of the Application of Novel Packaging for Animal-Derived Food Products as the Quality Improvement Method

The acceptance of the application of novel packaging for animal-derived food products as the quality improvement method, for respondents declaring diverse quality determinants, is presented in Table 5. The additional analysis conducted for the sub-groups stratified by income, place of living, and educational background is presented in the Supplementary Materials (Supplementary Tables S7–S12).

Table 5. Level of acceptance of novel packaging in animal-derived food products for respondents stratified by perceived quality determinants.

Perceived Quality Determinants	Acceptance of the Application of Novel Packaging for Animal-Derived Food Products as the Quality Improvement Method								p
	1 *	2	3	4	5	6	7	0	
Origin	1 (1.6%)	2 (3.1%)	2 (3.1%)	18 (28.1%)	10 (15.6%)	7 (10.9%)	22 (34.5%)	2 (3.1%)	
Production technology	10 (6.8%)	6 (4.1%)	8 (5.5%)	20 (13.7%)	35 (24.0%)	20 (13.7%)	44 (30.1%)	3 (2.1%)	
Manufacturer	2 (4.3%)	0 (0.0%)	0 (0.0%)	10 (21.7%)	9 (19.6%)	2 (4.3%)	20 (43.6%)	3 (6.5%)	
Components and nutritional value	23 (10.2%)	6 (2.7%)	12 (5.3%)	34 (15.0%)	45 (19.9%)	42 (18.6%)	60 (26.5%)	4 (1.8%)	0.0314
Visual and sensory characteristics	4 (3.2%)	0 (0.0%)	7 (5.6%)	31 (24.8%)	21 (16.8%)	18 (14.4%)	41 (32.8%)	3 (2.4%)	
Expiry date	29 (11.8%)	6 (2.4%)	10 (4.1%)	40 (16.3%)	44 (17.9%)	38 (15.4%)	72 (29.3%)	7 (2.8%)	
Cost	3 (4.6%)	1 (1.5%)	2 (3.1%)	9 (13.8%)	17 (26.2%)	13 (20.0%)	15 (23.1%)	5 (7.7%)	
Does not know what determines food quality	4 (4.4%)	2 (2.2%)	1 (1.1%)	16 (17.6%)	21 (23.1%)	18 (19.8%)	24 (26.3%)	5 (5.5%)	

* 1–3 negative answers (1—definitely; 2—moderately; 3—slightly do not accept); 4—neutral answer; 5–7 positive answers (5—slightly; 6—moderately; 7—definitely accept); 0—do not observe such method for food products of animal origin in Poland.

It was stated that declaring diverse quality determinants influenced the acceptance of the application of novel packaging for animal-derived food products as the quality improvement method ($p = 0.0314$). In the studied group, the highest number of respondents accepted the application of novel packaging in the sub-group of respondents not able to define quality determinants (69.2%) and declaring manufacturer as a quality determinant (69.2%), while a lower number of respondents accepted it in the sub-groups of respondents declaring as quality determinants: origin (67.8%), production technology (60.9%), components and nutritional value (64.0%), visual and sensory characteristics (67.4%), expiry date (62.6%), and cost (65.0%). At the same time, the lowest number of respondents did not accept the application of novel packaging in the sub-group of respondents declaring visual and sensory characteristics (4.3%), while a higher number of respondents did not accept it in the sub-groups of respondents declaring as quality determinants: origin (16.4%), production technology (7.8%), manufacturer (9.2%), components and nutritional value (8.8%), expiry date (18.3%) and cost (18.1%), or in the sub-group of respondents not able to define quality determinants (7.7%).

The acceptance of the application of novel packaging for animal-derived food products as the quality improvement method, for respondents declaring diverse known methods of quality improvement, is presented in Table 6.

It was stated that indicating the application of novel packaging as a method of quality improvement influenced acceptance of novel packaging application for animal-derived food products as the quality improvement method ($p = 0.0044$). The acceptance was comparable in sub-groups of respondents indicating product enhancement, methods other than product enhancement, and those who did not know any method to improve quality. In the case of respondents indicating the application of novel packaging as a known method of quality improvement, the lowest number of respondents accepted the application of novel packaging (33.3%) than for the sub-group indicating methods other than novel packaging application (61.9%) and those who did not know any method to improve quality (66.1%).

Table 6. Level of acceptance of novel packaging for animal-derived food products for respondents stratified by known methods of quality improvement.

Known Methods of Quality Improvement	Acceptance of the Application of Novel Packaging for Animal-Derived Food Products as the Quality Improvement Method								p
	1 *	2	3	4	5	6	7	8	
Indicates application of novel packaging	58 (7.0%)	20 (2.4%)	35 (4.2%)	140 (17.0%)	176 (21.3%)	132 (16.0%)	237 (28.7%)	27 (3.3%)	0.0044
Indicates other method than the application of novel packaging	18 (9.9%)	2 (1.1%)	6 (3.3%)	38 (21.0%)	26 (14.4%)	25 (13.8%)	61 (33.7%)	5 (2.8%)	
Does not know any method to improve quality	0 (0.0%)	1 (33.3%)	1 (33.3%)	0 (0.0%)	0 (0.0%)	1 (33.3%)	0 (0.0%)	0 (0.0%)	

* 1–3 negative answers (1—definitely; 2—moderately; 3—slightly do not accept); 4—neutral answer; 5–7 positive answers (5—slightly; 6—moderately; 7—definitely accept); 0—do not observe such method for food products of animal origin in Poland.

4. Discussion

The main observation from the obtained results indicated that those individuals who do not accept the analyzed methods of product enhancement may not accept such methods probably because they equate them with the addition of food preservatives (respondents who perceived expiration date as a major determinant of quality). Meat products are traditional foods in many countries, so in order to implement dietary behavior changes in those populations, culinary knowledge and understanding of the needed changes are required, as well as providing time to adapt and accept those changes [30]. Moreover, it should be recognized that specific factors determine the perceived quality, such as the cut used and method of thermal treatment applied [31].

The low level of acceptance in this group may be a serious problem for producers and distributors. However, in the study conducted by Bearth et al. [32] in a large sample of Swiss-German households, analyzing consumers' perception of artificial food additives, the factors that influenced the perceived level of risks and benefits were identified. The study concluded that acceptance of food additives is associated with perceived risks and benefits, which are influenced by the knowledge of legal regulations, trust in legal regulators, and a general preference for natural products [32]. Educating consumers may not only increase their knowledge of regulations but also improve their perception of food additives and minimize their worries, as education is indicated as one of the elements of intervention used for treating adult food neophobia [33].

The knowledge that product enhancement is applied does not relieve the stress resulting from the concerns of artificial additives. It is indicated that the names of food additives are sometimes even difficult to pronounce for consumers, and thus, could create an impression of unfamiliarity [22]. In spite of the fact that consumers can be provided reasonable information about risk assessment applied for food additives, it is difficult in practice [34]. Moreover, it should be emphasized that designing functional food products dedicated to specific groups of consumers with more concerns about product enhancement may be a promising strategy, but these products must be accompanied by information about health issues. Consumers with more modern health concerns (associated with the risk of diet-related diseases) tend to accept functional food products that are developed either to reduce the risk of the disease or to help in its cure [35].

The most important observation of the study was that respondents who perceived appearance and taste as major determinants of food quality accepted novel packaging methods. It must be indicated that for some animal-derived food products, appearance may be the main determinant, and therefore, applying suitable packaging is crucial. The study by Grebitus et al. [36] showed that in the case of packaged meat, consumers had a higher preference for meat having a brighter red color (aerobic packaging or carbon monoxide atmosphere packaging applied) and were willing to pay more if such a color was warranted. Nevertheless, it must be stated that consumers have higher preferences for products that they know and like [37]; therefore, the color of products must be characteristic, not changed by packaging.

The study by Giles et al. [38] indicated that improvement in taste, which was perceived as a beneficial modification of food products, especially when accompanied by a lower price, may be accepted as a reason for applying nanotechnology in food production. However, it was emphasized that nanotechnological components were accepted if they were applied only for food packaging, but not when they were integrated into food products [38].

The obtained results are in agreement with other studies conducted on carbon monoxide atmosphere packaging, which is promising for meat production [39]. In the study by Grebitus et al. [34], it was stated that a higher level of knowledge, gained through sources such as media, led to the lower acceptance of carbon monoxide atmosphere packaging.

The study by Chen et al. [40] measured the resistance to the use of new technology in food production, with an example of vacuum packaging used for beef in Canada, which is not uncommon in Europe. The information that vacuum packaging was applied influenced consumers' choices and increased their acceptability. The authors also highlighted that, in general, there was a problem with a low share of respondents being well-informed about this technology. This suggests that, in some cases, consumers' knowledge about the risks and benefits of the packaging system could dispel their doubts and have positive effects on market demands.

In addition, the microbiological safety of meat, which is another aspect associated with the risks and benefits of packaging, must be considered in the context of consumers' preferences and acceptances. The perspective of consumers should be regarded irrespective of the fact that novel packaging methods could prevent quality deterioration of enhanced products, as well as reduce product spoilage, as a part of sustainable development. The study of Dastile et al. [41] emphasized that visual assessment, rather than knowledge, may be a crucial determinant of sustainable meat consumption.

In the study of Akehurst et al. [42], the gap between purchase intentions and behaviors was stated to be less visible among respondents who had a higher level of consciousness. This may be due to the fact that respondents generally underestimated the impact of the meat industry and meat consumption on the environment [43]; therefore, proper knowledge could help in transmitting attitudes into purchase decisions.

Although the study presents some novel observations in a population-based sample, it has certain limitations that must be addressed. It should be mentioned that the study analyzed only declarative acceptance of the quality improvement methods applied for food products of animal origin, and such assessment is associated with self-reporting bias [44]. This is because conscious and unconscious behaviors differ, and even if the respondents do not intend to report false answers, their responses may be different from their unconscious preferences [45].

5. Conclusions

The results obtained in the study indicated that for the application of novel packaging, a higher level of knowledge may be a reason for consumers' rejection of the resulting products, but the appearance and taste of products may contribute to the higher acceptance of novel packaging. Educating consumers may improve their acceptance of product enhancement, as concerns about the addition of food preservatives, due to insufficient knowledge, may lead them to reject enhanced products.

Supplementary Materials: The following are available online at http://www.mdpi.com/2304-8158/9/11/1526/s1. Supplementary Table S1. Acceptance of the enhancement of animal-derived food products as the quality improvement method, for low income respondents who declared diverse quality determinants ($n = 261$). Supplementary Table S2. Acceptance of the enhancement of animal-derived food products as the quality improvement method, for high income respondents who declared diverse quality determinants ($n = 324$). Supplementary Table S3. Acceptance of the enhancement of animal-derived food products as the quality improvement method, for respondents of primary and vocational education level who declared diverse quality determinants (n = 462). Supplementary Table S4. Acceptance of the enhancement of animal-derived food products as the quality improvement method, for respondents of secondary and higher education level respondents who declared diverse quality determinants (n = 529). Supplementary Table S5. Acceptance of the enhancement of animal-derived food products as the quality improvement method, for respondents living in cities and villages of less than 100,000 inhabitants who declared diverse quality determinants (n = 688). Supplementary Table S6. Acceptance of the enhancement of animal-derived food products as the quality improvement method, for respondents living in cites of more than 100,000 inhabitants who declared diverse quality determinants (n = 314). Supplementary Table S7. Acceptance of the application of novel packaging for animal-derived food products as the quality improvement method, for low income respondents who declared diverse quality determinants (n = 261). Supplementary Table S8. Acceptance of the application of novel packaging for animal-derived food products as the quality improvement method, for high income respondents who declared diverse quality determinants (n = 324). Supplementary Table S9. Acceptance of the application of novel packaging for animal-derived food products as the quality improvement method, for respondents of primary and vocational education level declaring diverse quality determinants (n = 462). Supplementary Table S10. Acceptance of the application of novel packaging for animal-derived food products as the quality improvement method, for respondents of secondary and higher education level who declared diverse quality determinants (n = 529). Supplementary Table S11. Acceptance of the application of novel packaging for animal-derived food products as the quality improvement method, for respondents living in cities and villages of less than 100,000 inhabitants who declared diverse quality determinants (n = 688). Supplementary Table S12. Acceptance of the application of novel packaging for animal-derived food products as the quality improvement method, for respondents living in cities of more than 100,000 inhabitants who declared diverse quality determinants (n = 314).

Author Contributions: D.G. (Dominika Guzek) and D.G. (Dominika Głąbska) made study conception and design; D.G. (Dominika Guzek), D.G. (Dominika Głąbska), M.S., and K.G. performed the research; D.G. (Dominika Guzek) and D.G. (Dominika Głąbska) analyzed the data; D.G. (Dominika Guzek) and D.G. (Dominika Głąbska) interpreted the data; D.G. (Dominika Guzek), D.G. (Dominika Głąbska), M.S., and K.G. wrote the paper. All authors have read and agreed to the published version of the manuscript.

Funding: The presented study was conducted within the project "BIOFOOD–innovative, functional products of animal origin" no. POIG.01.01.02-014-090, co-financed by the European Union from the European Regional Development Fund within the Innovative Economy Operational Programme. The analysis was co-financed by the Polish Ministry of Science and Higher Education within funds of Institute of Human Nutrition Sciences, Warsaw University of Life Sciences (WULS), for scientific research.

Conflicts of Interest: The authors declare no conflict of interest.

References

1. Wolk, A. Potential health hazards of eating red meat. *J. Intern. Med.* **2017**, *281*, 106–122. [CrossRef] [PubMed]
2. Font-I-Furnols, M.; Guerrero, L. Consumer preference, behavior and perception about meat and meat products: An overview. *Meat Sci.* **2014**, *98*, 361–371. [CrossRef]
3. Grunert, K.G. Food quality and safety: Consumer perception and demand. *Eur. Rev. Agric. Econ.* **2005**, *32*, 369–391. [CrossRef]
4. Grunert, K.G. Future trends and consumer lifestyles with regard to meat consumption. *Meat Sci.* **2006**, *74*, 149–160. [CrossRef] [PubMed]
5. De Groeve, B.; Bleys, B. Less Meat Initiatives at Ghent University: Assessing the Support among Students and How to Increase It. *Sustainability* **2017**, *9*, 1550. [CrossRef]
6. De Bakker, E.; Dagevos, H. Reducing Meat Consumption in Today's Consumer Society: Questioning the Citizen-Consumer Gap. *J. Agric. Environ. Ethics* **2011**, *25*, 877–894. [CrossRef]
7. WHO; World Health Organization–Food and Agriculture Organization of the United Nations. *Diet, Nutrition, and the Prevention of Chronic Diseases: Report of a Joint WHO/FAO Expert Consultation*; WHO Technical Report Series no. 916; World Health Organization: Geneva, Switzerland, 2003.
8. Westhoek, H.; Lesschen, J.P.; Rood, T.; Wagner, S.; De Marco, A.; Murphy-Bokern, D.; Leip, A.; Van Grinsven, H.; Sutton, M.A.; Oenema, O. Food choices, health and environment: Effects of cutting Europe's meat and dairy intake. *Glob. Environ. Chang.* **2014**, *26*, 196–205. [CrossRef]
9. Aston, L.M.; Smith, J.N.; Powles, J.W. Impact of a reduced red and processed meat dietary pattern on disease risks and greenhouse gas emissions in the UK: A modelling study. *BMJ Open* **2012**, *2*, e001072. [CrossRef]
10. De Boer, J.; Schösler, H.; Aiking, H. "Meatless days" or "less but better"? Exploring strategies to adapt Western meat consumption to health and sustainability challenges. *Appetite* **2014**, *76*, 120–128. [CrossRef]

11. MacDiarmid, J.I.; Douglas, F.; Campbell, J. Eating like there's no tomorrow: Public awareness of the environmental impact of food and reluctance to eat less meat as part of a sustainable diet. *Appetite* **2016**, *96*, 487–493. [CrossRef]
12. Aiking, H. Protein production: Planet, profit, plus people? *Am. J. Clin. Nutr.* **2014**, *100*, 483S–489S. [CrossRef] [PubMed]
13. Decker, E.A.; Park, Y. Healthier meat products as functional foods. *Meat Sci.* **2010**, *86*, 49–55. [CrossRef] [PubMed]
14. Roberfroid, M.B. An European consensus of scientific concepts of functional foods. *Nutrition* **2000**, *16*, 689–691. [CrossRef]
15. Hathwar, S.C.; Rai, A.K.; Modi, V.K.; Narayan, B. Characteristics and consumer acceptance of healthier meat and meat product formulations—A review. *J. Food Sci. Technol.* **2011**, *49*, 653–664. [CrossRef] [PubMed]
16. Olewnik-Mikołajewska, A.; Guzek, D.; Głąbska, D.; Gutkowska, K. Consumer Behaviors Toward Novel Functional and Convenient Meat Products in Poland. *J. Sens. Stud.* **2016**, *31*, 193–205. [CrossRef]
17. Olewnik-Mikołajewska, A.; Guzek, D.; Głąbska, D.; Sajdakowska, M.; Gutkowska, K. Fodder enrichment and sustaining animal well-being as methods of improving quality of animal-derived food products, in the aspect of consumer perception and acceptance. *Anim. Sci. Pap. Rep.* **2016**, *4*, 361–372.
18. Glitsh, K. Consumer perceptions of fresh meat quality: Cross-national comparison. *Br. Food J.* **2000**, *102*, 177–194. [CrossRef]
19. Lee, H.K. Consumers' awareness of food additives. *Safe Food* **2012**, *7*, 21–25.
20. EU Commission. Regulation (EC) No 1331/2008 of the European Parliament and of the Council of 16 December 2008 establishing a common authorisation procedure for food additives, food enzymes and food flavourings. *Off. J. Eur. Communities L* **2008**, *354*, 1–6.
21. Grunert, K.G.; Skytte, H.; Esbjerg, L.; Poulsen, C.S.; Hviid, M. *Dokumenteret Kødkvalitet*; MAPP Project Paper No. 2-02; Aarhus School of Business: Aarhus, Denmark, 2002.
22. Kaptan, B.; Kayisoglu, S. Consumers' attitude towards food additives. *Am. J. Food Sci. Nutr. Res.* **2015**, *2*, 21–25.
23. Guzek, D.; Głąbska, D.; Głąbski, K.; Wierzbicka, A. Influence of Duroc breed inclusion into Polish Landrace maternal line on pork meat quality traits. *An. Acad. Bras. Cienc.* **2016**, *88*, 1079–1088. [CrossRef] [PubMed]
24. Guzek, D.; Głąbska, D.; Sakowska, A.; Wierzbicka, A. Colour of pork loin produced of meat of animals fed with bioactive compounds forage. *Pesq. Agropec. Bras.* **2012**, *47*, 1504–1510. [CrossRef]
25. Brodowska, M.; Guzek, D.; Kołota, A.; Głąbska, D.; Górska-Horczyczak, E.; Wojtasik-Kalinowska, I.; Wierzbicka, A. The effect of diet on oxidation and profile of volatile compounds of pork after freezing storage. *J. Food Nutr. Res.* **2016**, *55*, 40–47.
26. O'Grady, M.; Kerry, J.P. Smart packaging technologies and their application in conventional met packaging systems. In *Meat Biotechnology*; Toldra, F., Ed.; Springer Science and Business Media: New York, NY, USA, 2008; pp. 425–451.
27. Aday, M.S.; Yener, U. Assessing consumers' adoption of active and intelligent packaging. *Br. Food J.* **2015**, *117*, 157–177. [CrossRef]
28. Chulayo, A.Y.; Bradley, G.; Muchenje, V. Effects of transport distance, lairage time and stunning efficiency on cortisol, glucose, HSPA1A and how they relate with meat quality in cattle. *Meat Sci.* **2016**, *117*, 89–96. [CrossRef]
29. The Statistics Poland December. Available online: https://stat.gov.pl/obszary-tematyczne/roczniki-statystyczne/roczniki-statystyczne/rocznik-statystyczny-rzeczypospolitej-polskiej-2016,2,16.html (accessed on 19 September 2020).
30. Meyer, N.L.; Reguant-Closa, A. "Eat as If You Could Save the Planet and Win!" Sustainability Integration into Nutrition for Exercise and Sport. *Nutrients* **2017**, *9*, 412. [CrossRef]
31. Guzek, D.; Głąbska, D.; Gutkowska, K.; Woźniak, A.; Wierzbicki, J.; Wierzbicka, A. Influence of cut and thermal treatment on consumer perception of beef in Polish trials. *Pak. J. Agric. Sci.* **2015**, *78*, 533–538.
32. Bearth, A.; Cousin, M.-E.; Siegrist, M. The consumer's perception of artificial food additives: Influences on acceptance, risk and benefit perceptions. *Food Qual. Prefer.* **2014**, *38*, 14–23. [CrossRef]
33. Marcontell, D.K.; Laster, A.E.; Johnson, J. Cognitive-behavioral treatment of food neophobia in adults. *J. Anxiety Disord.* **2003**, *17*, 341–349. [CrossRef]

34. Bearth, A.; Cousin, M.-E.; Siegrist, M. "The Dose Makes the Poison": Informing Consumers About the Scientific Risk Assessment of Food Additives. *Risk Anal.* **2016**, *36*, 130–144. [CrossRef]
35. Devcich, D.A.; Pedersen, I.K.; Petrie, K.J. You eat what you are: Modern health worries and the acceptance of natural and synthetic additives in functional foods. *Appetite* **2007**, *48*, 333–337. [CrossRef]
36. Grebitus, C.; Jensen, H.H.; Roosen, J.; Sebranek, J.G. Fresh Meat Packaging: Consumer Acceptance of Modified Atmosphere Packaging including Carbon Monoxide. *J. Food Prot.* **2013**, *76*, 99–107. [CrossRef]
37. Gray, Z.D.; Haefner, J.E.; Rosenbloom, A. The role of global brand familiarity, trust and liking in predicting global brand purchase intent: A Hungarianâ American comparison. *Int. J. Bus. Emerg. Mark.* **2012**, *4*, 4. [CrossRef]
38. Giles, E.L.; Kuznesof, S.; Clark, B.; Hubbard, C.; Frewer, L.J. Consumer acceptance of and willingness to pay for food nanotechnology: A systematic review. *J. Nanopart. Res.* **2015**, *17*, 1–26. [CrossRef] [PubMed]
39. Sakowska, A.; Guzek, D.; Głąbska, D.; Wierzbicka, A. Carbon monoxide concentration and exposure time effects on the depth of CO penetration and surface color of raw and cooked beef longissimus lumborum steaks. *Meat Sci.* **2016**, *121*, 182–188. [CrossRef]
40. Chen, Q.; Anders, S.; An, H. Measuring consumer resistance to a new food technology: A choice experiment in meat packaging. *Food Qual. Prefer.* **2013**, *28*, 419–428. [CrossRef]
41. Dastile, L.S.; Francis, J.; Muchenje, V. Consumers' Social Representations of Meat Safety in Two Selected Restaurants of Raymond Mhlaba Municipality in the Eastern Cape, South Africa. *Sustainability* **2017**, *9*, 1651. [CrossRef]
42. Akehurst, G.; Afonso, C.; Gonçalves, H.M. Re-examining green purchase behaviour and the green consumer profile: New evidences. *Manag. Decis.* **2012**, *50*, 972–988. [CrossRef]
43. Hartmann, C.; Siegrist, M. Consumer perception and behaviour regarding sustainable protein consumption: A systematic review. *Trends Food Sci. Technol.* **2017**, *61*, 11–25. [CrossRef]
44. Bell, L.; Vogt, J.; Willemse, C.; Routledge, T.; Butler, L.T.; Sakaki, M. Beyond Self-Report: A Review of Physiological and Neuroscientific Methods to Investigate Consumer Behavior. *Front. Psychol.* **2018**, *7*, 1655. [CrossRef]
45. Morsella, E.; Poehlman, T.A. The inevitable contrast: Conscious vs. unconscious processes in action control. *Front. Psychol.* **2013**, *10*, 590. [CrossRef]

Publisher's Note: MDPI stays neutral with regard to jurisdictional claims in published maps and institutional affiliations.

© 2020 by the authors. Licensee MDPI, Basel, Switzerland. This article is an open access article distributed under the terms and conditions of the Creative Commons Attribution (CC BY) license (http://creativecommons.org/licenses/by/4.0/).

Article

Sensory Analysis in Assessing the Possibility of Using Ethanol Extracts of Spices to Develop New Meat Products

Krystyna Szymandera-Buszka, Katarzyna Waszkowiak *, Anna Jędrusek-Golińska and Marzanna Hęś

Department of Gastronomy Science and Functional Foods, Faculty of Food Science and Nutrition, Poznan University of Life Sciences, Wojska Polskiego 31, 60-624 Poznan, Poland; krystyna.szymandera_buszka@up.poznan.pl (K.S.-B.); anna.jedrusek-golinska@up.poznan.pl (A.J.-G.); marzanna.hes@up.poznan.pl (M.H.)
* Correspondence: katarzyna.waszkowiak@up.poznan.pl; Tel.: +48-061-848-7379

Received: 17 November 2019; Accepted: 13 February 2020; Published: 18 February 2020

Abstract: The food industry has endeavoured to move toward the direction of clean labelling. Therefore, replacing synthetic preservatives with natural plant extracts has gained significant importance. It is necessary to determine whether products enriched with such extracts are still accepted by consumers. In this study, consumer tests (n = 246) and sensory profiling were used to assess the impact of ethanol extracts of spices (lovage, marjoram, thyme, oregano, rosemary, and basil; concentration 0.05%) on the sensory quality of pork meatballs and hamburgers. The desirability of meat products with spice extracts to consumers depended on the added extract. The highest scores were for products with lovage extract, whose sensory profile was the most similar to the control sample without the addition of an extract (with higher intensity of broth taste compared with the others). Products with rosemary and thyme extracts were characterised by lower desirability than the control. This was related to the high intensity of spicy and essential oil tastes, as well as the bitter taste in the case of products with thyme. The studied extracts of spices allow for the creation of meat products (meatballs and hamburgers) with high consumer desirability, however, the high intensity of essential oil and spicy tastes might be a limitation.

Keywords: new product development; consumer research; sensory profiling; meat products; ethanol extracts of spices

---

1. Introduction

Lipid oxidation is one of the main factors resulting in undesirable changes in food. The oxidation of meat lipids, taking place during processing and storage, leads to a considerable deterioration of processed meat quality [1]. As a result of fatty acid oxidation (especially polyenic acids), various products are formed, both volatile and non-volatile [2]. These oxidation products deteriorate physico-chemical properties and affect sensory attributes and the health quality of meat products [3,4]. For this reason, the food industry is interested in the development of novel food products enriched with natural antioxidants [5]. The development of food additives, which are extracts of natural compounds exhibiting antioxidant and antimicrobial activity, is a crucial step toward the production of food with health benefits [6,7]. A source of those antioxidant compounds can be spice plants, such as rosemary (*Rosmarinus officinalis* L.), thyme (*Thymus vulgaris* L.), marjoram (*Origanum majorana* L.), oregano (*Origanum vulgare* L.), basil (*Ocimum basilicum* L.), and lovage (*Levisticum officinale*) [8–10], which are more and more frequently used in the food industry. The results of earlier studies indicated that phenolic compounds from the spices (their ethanol extracts) show strong antioxidant activities both in model systems and food [10–14]. These ethanol extracts also showed antibacterial activity [15–17]. In the previous studies, their effectiveness was assessed in various differently processed meat products. Naveena et al. [18] investigated the effect of oil soluble and water dispersible carnosic acid extract from dried rosemary leaves at two different concentrations (0.0023% and 0.013%) on the inhibition of oxidation in raw and cooked ground buffalo meat patties. It was reported that carnosic acid extract reduced the TBARS (thiobarbituric acid reactive substances) content by 39%–47% at a lower concentration and by 86%–96% at a higher concentration in cooked buffalo meat patties compared to controls. The extracts also inhibited peroxide value and free fatty acid formation in the products. Moreover, the extract added at the higher concentrations stabilised the colour of raw buffalo meat. Sebranek et al. [19] tested the commercial rosemary extract at concentrations of 0.15% and 0.25% in frozen and precooked-frozen pork sausage, and between 0.05% and 0.3% in refrigerated fresh pork sausage. They reported that the rosemary extract added to refrigerated sausage at 0.25% showed similar antioxidant activity to synthetic antioxidants (BHA/BHT). Bilska et al. [20] added 0.05% of rosemary extract to enhance the oxidative stability of pork liver pâté with fat replaced by 20% of flaxseed oil and showed that the addition significantly slowed down lipid oxidation during the storage of the pâté. Oregano extract (0.02%) has been shown to inhibit lipid oxidation in cooked ground beef and pork (30% fat, fresh meat basis) [21,22] and in lean raw beef [23]. Rojas and Brewer [22] studied the effects of water-soluble oregano extract in cooked beef and pork. They found that the extract added at 0.02% was effective at reducing lipid oxidation in vacuum-packaged cooked beef samples stored at −18 °C for 4 months. Trindade et al. [24] studied the effect of the addition of rosemary and oregano extracts (0.04%) on the sensory qualities of irradiated (7 kGy) beef burger during frozen storage, and it was concluded that these natural antioxidants could prevent lipid oxidation without affecting the sensory scores of treated meat samples. The similar effect was observed by Manhani et al. [25] in the case of the addition of rosemary and oregano commercial extracts (0.04%) to precooked beef hamburgers. Gramatina et al. [17] investigate the potential use of ethanol extracts from selected herbals to maintain the pork meat quality during refrigerated storage and they found that oregano and lovage extracts inhibited microbial growth after soaking in the extracts, thus extending the shelf-life of pork meat. Thyme, marjoram, and basil were also studied as antioxidants in meat products. El-Alim et al. [26] investigated the use of ground spices and spice extracts (including marjoram, basil and thyme) as antioxidants in raw ground chicken and ground pork. Authors reported that TBARS formation was inhibited in refrigerated and frozen samples of ground chicken which were treated with 10 g dried spices/kg. They also examined the ethanol extracts of basil, sage, thyme, and ginger (prepared in the laboratory) as antioxidants in ground pork (ground pork treated with the extracts at a concentration of 1 mL/10 g) [26], and found that sage, thyme, and basil extracts were more effective at inhibiting TBARS values in ground pork than ginger extract. Hęś et al. [12] investigated the effect of rosemary, green tea, and thyme extracts (0.05%) on lipid stability and the protein nutritional value of frozen-stored fried balls from ground pork and found that the additions limited lipid oxidation, as well as reduced changes in methionine and lysine content, and protein digestibility compared to the control. Sarıçoban and Yilmaz [27] showed that the addition of thyme essential oils at 0.05% decreased oxidation and microbial grow, extending the shelf-life of chicken pâté.

Adding the extracts into product formulations may allow developing new processed meats and culinary products with a longer shelf-life and high nutritional value [3,12,28,29]. Additionally, some of the health effects attributed to the antioxidant, antimicrobial, and anti-inflammatory effects of phenolic compounds present in the extracts indicate their potential protection against cardiovascular disease, neurodegeneration, type 2 diabetes, and cancer [30–33].

In the process of food product development, there are two aspects that are necessary to determine, apart from studies on antioxidant activity of plant extracts during production and storage of meat products. These aspects are whether products enriched with such extracts still show good sensory quality and whether they are accepted by consumers. In technological research, a new product is first created and then its nutritional value, technological features, shelf life, etc., are examined and, finally, its sensory quality is determined. On the contrary, in the sensory approach to designing new products, the sensory quality of the designed products is checked in detail first to know which product attributes are worth strengthening or masking. In this case, it is vital to combine the results of consumer assessment (in which consumer preferences are assessed and thus their subjective feelings) with the results of objective methods (e.g., Quantitative Descriptive Analysis) [34]. By comparing the results of both methods, it is possible to obtain full information concerning sensory quality and to determine which product attributes and their intensities are accepted by consumers [34,35]. Only tasty food has a chance of being bought and consumed.

Therefore, the aim of this study was the application of consumer tests and the sensory profiling method to assess the impact of ethanol extracts of selected spices (rosemary, thyme, marjoram, oregano, basil, and lovage) on the sensory quality of pork meatballs and hamburgers.

2. Materials and Methods

The pork meatballs and hamburgers enriched with ethanol extracts from the selected spice (rosemary, thyme, marjoram, oregano, basil, and lovage) were the research material of this study.

The rosemary extract (Oxy'Less U; powder form, dry extract: 95 ± 3%) was purchased from Naturex (Avignon, France). It was an unrefined rosemary extract obtained by ethanol extraction of rosemary leaves (the commercial characteristic based on the producer information). The other extracts were prepared according to the procedure described by Hęś and Gramza-Michałowska [12]. Briefly, dried leaves of thyme (*Thymus vulgaris* L.), marjoram (*Origanum majorana* L.), oregano (*Origanum vulgare* L.), basil (*Ocimum basilicum* L.), and lovage (*Levisticum officinale*) were purchased from a local herb shop, which guaranteed the origin and the freshness of raw materials. Dried materials (100 g) were macerated with 250 mL of 80% ethanol overnight at room temperature. The suspension was filtered, the residue was mixed with another portion of ethanol, and the procedure was repeated three times. The extracts were collected, and ethanol was removed at 50 °C using a rotary vacuum evaporator (Buchi, Flawil, Switzerland). Then, the extracts were freeze-dried (Alpha 1–4 LSC Freeze dryer, Christ, Osterode am Harz, Germany). The dried extracts were stored at 4 °C in a dark place. The average total phenolic contents of the ethanol extracts (expressed as mg gallic acid per g of dry mass) are as follows: 161.3 mg rosemary extract, 229.6 mg thyme extract, 105.5 mg oregano extract, 36.4 mg lovage extract, 113.2 mg marjoram extract, and 129.6 mg basil extract [12,36] and unpublished results.

For pork meatball and hamburger preparation, pork (from the best end of the neck) was minced (mesh size of 3mm) and mixed thoroughly (approx. 10 min) with the other ingredients using a homogenizer (Foss, Hilleroed, Denmark), producing batter containing 70% meat, 15.8% water, 8% breadcrumbs, 5% eggs, 1% salt, and 0.2% pepper. The batter was divided into seven portions. One was the control sample, and ethanol extracts were added to the others: rosemary, thyme, marjoram, oregano, basil, and lovage (0.05% of meat batter). The addition of the extracts was selected based on our previous study concerning the protective effect of the extracts against oxidation [12,20,28] and EU Regulations [37]. For heat-treated meat products, the EU Regulations No 723/2013 [37] set the maximum addition of rosemary extract at 15 mg/kg. Our previously published studies reported [12,20] that 0.05% additives of rosemary and thyme extracts significantly slowed down lipid oxidation during storage of meat products. This observation was supported by the studies of other scientists who have shown the protective effect of rosemary [19,24,25], thyme [27], and oregano [21–25] extracts added at a similar level to meat products (that is 0.02%–0.05%). Therefore, the 0.05% addition was used in the case of all tested extracts to compare the impact of the extracts on the sensory quality of meatballs and hamburgers.

Round samples of a similar weight (50 g ± 1 g) were formed. Next, the samples were roasted (hot air at 160 °C for 8 min) to prepare the hamburgers, or steamed (100 °C for 16 min) to prepare the meatballs, in a convection oven (Rational, Landsberg am Lech, Germany). The products were prepared (on a laboratory scale) in the technology laboratory located at the Faculty of Food Science and Nutrition, Poznań University of Life Sciences (Poland).

Sensory tests were conducted in a sensory analysis laboratory equipped with individual booths (at a controlled temperature of 21 ± 1 °C and combined natural/artificial light) designed according to ISO standards [38]. The laboratory was located at the Faculty of Food Science and Nutrition, Poznań University of Life Sciences (Poland). The tests were carried out in 2018 between 9 a.m. and 3 p.m. The samples of the products were evaluated when warm, on the same day they were made, and were served on odourless white plates. The samples were coded with random three-digit numbers, and the serving order of samples was random (program Analsens was used for coding and arrangement of serving order). Water was used to cleanse the mouth between samples. No ethical approval was required for this study. Participants were informed about the study's aim and that their participation was entirely voluntary, so that they could stop the analysis at any point and the responses would be anonymous.

For detailed sensory characteristics of hamburgers and meatballs with spice ethanol extracts, the quantitative descriptive analysis was applied, i.e., sensory profiling. The analysis was conducted by an 8-member trained panel [39]. A total of 13 descriptors, elementary attributes of aroma (meat, peppery, fried, spicy, bitter, and essential oil aromas) and taste (salty, broth, fried, spicy, bitter, and essential oil tastes) of tested meat products were evaluated. These descriptors of aroma and taste were selected in preliminary tests. The intensity of each score was determined using a 10-cm linear scale with appropriate margin descriptions. For attributes of aroma and taste, uniform margin denotations were applied: "undetectable"–"very intensive". Four samples were served during a session to the panellists (three samples with spice extracts and control sample in one session); two sessions were carried out in one day. All samples were assessed in two independent replications.

Simultaneously with the profile analysis of meats products, the consumer test was conducted among a group of 246 people who consume meat products at least once every week, aged 20–50. They were neither students, professors, nor administrative employees of Poznań University of Life Sciences; 53% of the participants were women. Consumers evaluated the aroma, taste, and overall desirability of the meat products. For the consumer test, a 10 cm hedonic graphic scale was applied with the following margin denotations: "undesirable"–"highly desirable". All consumers evaluated both meatballs and hamburgers in one session (order of serving: seven samples, a break of 0.5 h, and seven samples).

Statistical analyses were conducted using Statistica (v.13.1, StatSoft Tulsa, OK, USA). The effects of spice ethanol extracts (L = 7; six selected spice extracts and control sample with no extract) were analysed. The results of sensory tests were subjected to the analysis of variance (ANOVA), and then post hoc Tukey's test was applied at a significance level of $p < 0.05$ to compare the means. Hierarchical Cluster Analysis (HCA) was performed to identify similar groups of meat products based on consumer test results (i.e., aroma, taste, and overall consumer desirability). Ward's method was used. According to the method, the means for all variables are calculated for each cluster. For each case, the squared Euclidean distance to the cluster means is calculated; the two clusters that merge are those that result in the smallest increase in the overall sum of the squared within-cluster distances [40]. Principal Component Analysis (PCA) was applied to the data sets from the sensory profiling of products to assess differences and similarities in sensory profiles based on their aroma and taste descriptors.

3. Results

3.1. Consumer Test Results

The results of the consumer evaluation showed that pork meatballs and hamburgers with ethanol extracts of spices were characterised by varied overall desirability (Table 1, one-way ANOVA, $p < 0.05$). Both meat products with lovage extract were characterised by the significantly higher overall consumer desirability compared to the products with rosemary extract and thyme extract. The same was observed in the case of these products' aroma and taste assessment. Moreover, the consumer desirability of aroma and taste of the hamburgers and meatballs with lovage extract were significantly higher than those with basil extract and oregano extract.

Table 1. Mean scores (n = 246) ± standard deviation of the desirability of pork meatballs and hamburgers with ethanol extracts of the spices (rosemary, thyme, lovage, marjoram, basil, oregano; 0.05%) and the control sample (without any extract) to the consumers.

Consumer Desirability	Addition of Spice Ethanol Extracts						
	Rosemary	Thyme	Lovage	Marjoram	Basil	Oregano	Control
Meatballs							
aroma	4.8 ± 0.98 [a]	5.1 ± 1.00 [a]	7.6 ± 0.98 [b]	6.9 ± 0.98 [ab]	6.4 ± 0.92 [ab]	6.3 ± 0.92 [a]	6.6 ± 0.88 [ab]
taste	4.9 ± 0.94 [a]	5.1 ± 0.95 [a]	7.7 ± 0.96 [b]	7.0 ± 0.96 [ab]	5.7 ± 0.88 [a]	5.8 ± 0.98 [a]	6.5 ± 0.91 [ab]
overall	4.9 ± 0.94 [a]	5.1 ± 0.95 [a]	7.7 ± 1.02 [b]	7.0 ± 1.02 [ab]	5.9 ± 0.94 [ab]	5.8 ± 1.06 [ab]	6.5 ± 0.91 [ab]
Hamburgers							
aroma	5.2 ± 0.95 [a]	5.1 ± 0.98 [a]	7.6 ± 0.97 [b]	6.9 ± 1.00 [ab]	6.0 ± 0.89 [ab]	5.6 ± 0.88 [ab]	6.8 ± 1.00 [ab]
taste	5.4 ± 0.94 [a]	5.1 ± 0.95 [a]	7.6 ± 1.00 [b]	7.0 ± 0.98 [ab]	6.0 ± 0.92 [a]	5.7 ± 1.10 [a]	6.9 ± 1.04 [ab]
overall	4.9 ± 0.96 [a]	5.1 ± 0.95 [a]	7.7 ± 1.03 [b]	7.0 ± 1.07 [ab]	5.7 ± 0.97 [ab]	5.8 ± 1.15 [ab]	6.5 ± 1.01 [ab]

Different letters in the same raw for particular products denote a significant difference at $p < 0.05$ (one-way ANOVA, and post hoc Tukey test).

The hierarchical cluster analysis (HCA) grouped the meat products (meatballs and hamburgers) into clusters based on the similarity of overall consumer desirability and desirability of aroma and taste. HCA results are presented as dendrograms (Figures 1 and 2).

The results of statistical analysis showed that the addition of spice extract influenced the aroma and taste desirability of both meat products (Table 1; one-way ANOVA, $p < 0.05$). In the case of meatballs and hamburgers, HCA analysis showed (Figures 1a and 2a) that consumers distinguished the aroma of the samples with thyme extract and rosemary extract from the other samples (they formed a separate cluster). A similarity was found between meat products with marjoram extract and lovage extract (they were arranged in one cluster based on their aroma desirability). Based on the results of taste desirability, HCA analysis showed a different arrangement of the tested products in clusters when compared to aroma desirability (Figures 1b and 2b). The taste desirability of samples with lovage extract and samples with marjoram extract was the closest (they formed clusters according to HCA). On the other hand, samples with rosemary extract and thyme extract showed similarity to those with oregano extract and basil extract. For both meatballs and hamburgers, the taste desirability of the control samples showed the highest similarity to those with marjoram extract and lovage extract.

The consumers also assessed the overall consumer desirability of both meat products. For meatballs (Figure 1c), a high similarity was observed between samples with lovage extract and marjoram extract, as they were arranged in one cluster. Meatballs with oregano extract and basil extract formed the second cluster of products with the overall desirability to the consumers closest to the overall desirability of the control sample. Samples with thyme extract and rosemary extract formed a separate cluster, which had the weakest association with the others. Similar clusters were arranged in the case of overall desirability of hamburgers (Figure 2c) to the consumers. The strongest similarity was found between the samples with marjoram extract and lovage extract, and the weakest similarity was between the control sample and all other spice extracts. The HCA analysis showed that the spice extract additions influenced the overall consumer desirability of the tested meat products, and this effect was stronger for the fried product (hamburgers) than the steamed one (meatballs).

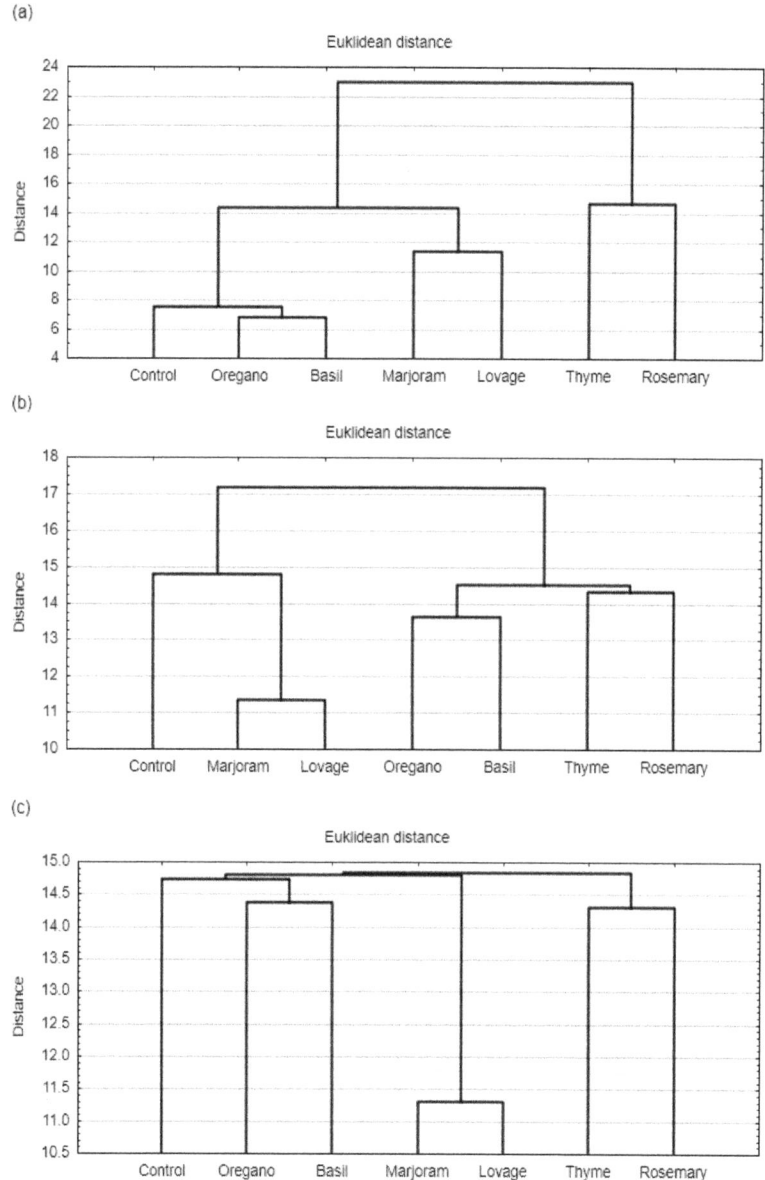

Figure 1. Dendrograms of hierarchical cluster analysis (HCA) for (**a**) the desirability of the aroma, (**b**) the desirability of the taste, and (**c**) the overall desirability of the pork meatballs with ethanol extracts of the spices (rosemary, thyme, lovage, marjoram, basil, oregano; 0.05%) and the control sample (without any extract) to the consumers.

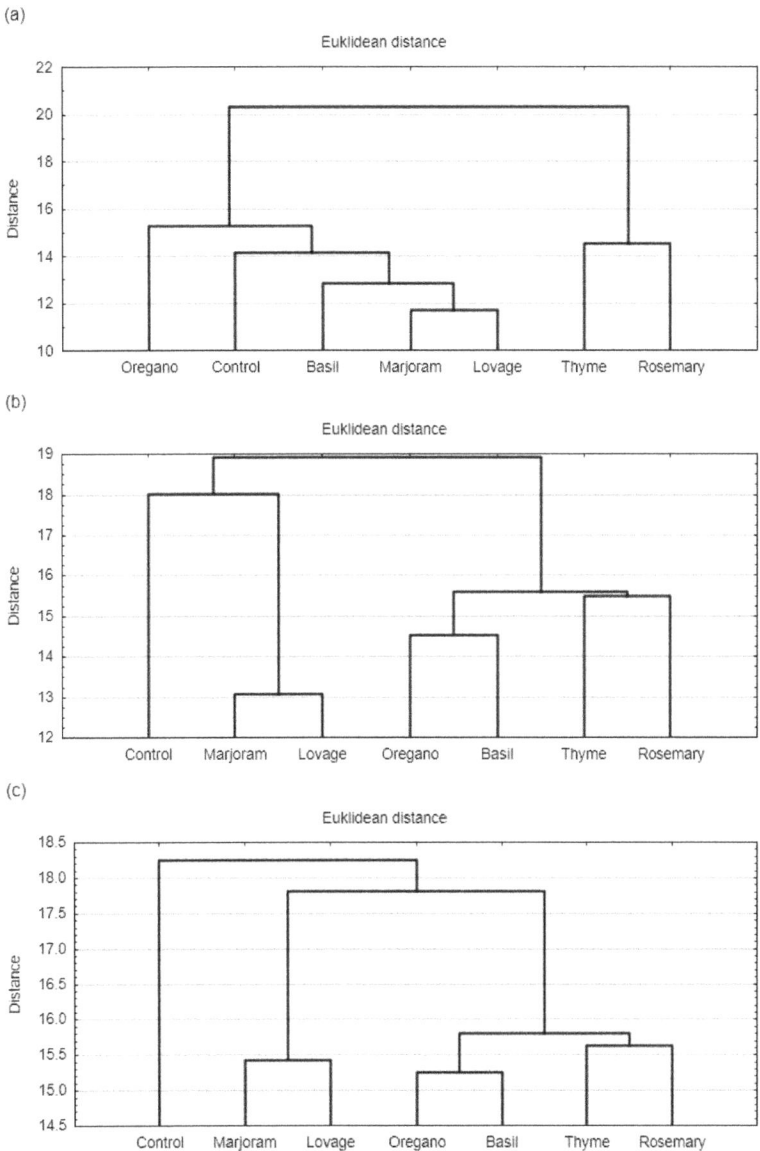

Figure 2. Dendrograms of hierarchical cluster analysis (HCA) for (**a**) the desirability of the aroma, (**b**) the desirability of the taste, and (**c**) the overall desirability of the pork hamburgers with ethanol extracts of the spices (rosemary, thyme, lovage, marjoram, basil, oregano; 0.05%) and the control sample (without any extract) to the consumers.

3.2. Sensory Profiling Results

In the sensory profiling, the perception of the following descriptors was defined and determined: aroma (meat, peppery, fried, spicy, essential oil, and bitter) and taste (meat, salty, broth, fried, spicy, essential oil, and bitter). The results of the sensory profiling of the meatballs and hamburgers with spice ethanol extracts and the control sample (without any extract) are presented in Table 2.

Table 2. Mean scores (n = 8) of sensory taste and aroma profiling of pork meatballs and hamburgers with ethanol extracts of the spices (rosemary, thyme, lovage, marjoram, basil, oregano; 0.05%) and the control sample (without any extract).

Sensory Attributes	Addition of Spice Ethanol Extracts						
	Rosemary	Thyme	Lovage	Marjoram	Basil	Oregano	Control
Meatballs							
Aroma							
meat	5.0 ± 0.3 [a]	5.6 ± 0.2 [ba]	5.9 ± 0.3 [b]	5.1 ± 0.4 [a]	5.0 ± 0.4 [a]	5.0 ± 0.4 [a]	5.6 ± 0.4 [ba]
peppery	0.8 ± 0.2 [a]	0.6 ± 0.2 [a]	0.7 ± 0.3 [a]	0.8 ± 0.2 [a]	0.8 ± 0.2 [a]	1.0 ± 0.2 [a]	1.0 ± 0.2 [a]
fried	0.4 ± 0.3 [a]	0.2 ± 0.2 [a]	0.2 ± 0.2 [a]	0.2 ± 0.1 [a]	0.0 ± 0.0 [a]	0.2 ± 0.1 [a]	0.2 ± 0.2 [a]
spicy	1.9 ± 0.2 [bc]	0.8 ± 0.2 [ab]	0.4 ± 0.2 [a]	2.0 ± 0.4 [bc]	1.7 ± 0.4 [bc]	2.2 ± 0.3 [c]	0.2 ± 0.1 [a]
essential oil	2.8 ± 0.2 [c]	0.2 ± 0.2 [a]	0.2 ± 0.2 [a]	0.2 ± 0.1 [a]	0.7 ± 0.2 [ab]	1.4 ± 0.2 [bc]	0.0 ± 0.0 [a]
bitter	0.4 ± 0.2 [ab]	1.20 ± 0.3 [b]	0.0 ± 0.0 [a]	0.0 ± 0.0 [a]	0.2 ± 0.2 [a]	0.2 ± 0.2 [a]	0.2 ± 0.1 [a]
Taste							
meat	4.9 ± 0.3 [a]	5.2 ± 0.3 [a]	6.4 ± 0.3 [b]	5.4 ± 0.3 [a]	5.2 ± 0.3 [a]	5.0 ± 0.3 [a]	5.6 ± 0.3 [ab]
salty	2.0 ± 0.2 [a]	2.0 ± 0.2 [a]	3.0 ± 0.3 [a]	2.4 ± 0.3 [a]	2.0 ± 0.2 [a]	2.4 ± 0.2 [a]	2.3 ± 0.3 [a]
broth	1.9 ± 0.3 [a]	1.8 ± 0.3 [a]	3.5 ± 0.3 [b]	2.0 ± 0.3 [a]	2.3 ± 0.3 [a]	1.9 ± 0.3 [a]	2.8 ± 0.4 [ab]
fried	0.1 ± 0.2 [a]	0.2 ± 0.2 [a]	0.0 ± 0.0 [a]	0.2 ± 0.2 [a]	0.2 ± 0.2 [a]	0.1 ± 0.1 [a]	0.2 ± 0.2 [a]
spicy	4.9 ± 0.3 [c]	5.0 ± 0.3 [c]	2.8 ± 0.2 [ab]	4.0 ± 0.2 [bc]	3.8 ± 0.3 [abc]	4.0 ± 0.3 [bc]	2.0 ± 0.3 [a]
essential oil	1.5 ± 0.3 [b]	1.5 ± 0.2 [b]	0.2 ± 0.2 [a]	1.0 ± 0.3 [ab]	0.6 ± 0.2 [a]	1.2 ± 0.4 [b]	0.0 ± 0.1 [a]
bitter	0.4 ± 0.2 [a]	1.7 ± 0.3 [b]	0.2 ± 0.2 [a]	1.0 ± 0.2 [ab]	1.8 ± 0.2 [b]	1.2 ± 0.3 [b]	0.4 ± 0.2 [a]
Hamburgers							
Aroma							
meat	5.7 ± 0.3 [a]	6.1 ± 0.4 [a]	6.9 ± 0.3 [a]	5.8 ± 0.4 [a]	6.2 ± 0.3 [a]	5.7 ± 0.3 [a]	6.7 ± 0.3 [a]
peppery	3.4 ± 0.3 [a]	3.4 ± 0.3 [a]	3.6 ± 0.3 [a]	3.3 ± 0.3 [a]	3.2 ± 0.2 [a]	3.0 ± 0.3 [a]	3.4 ± 0.3 [a]
fried	2.6 ± 0.3 [a]	2.7 ± 0.3 [a]	2.9 ± 0.3 [a]	2.7 ± 0.3 [a]	2.9 ± 0.2 [a]	2.5 ± 0.2 [a]	2.7 ± 0.2 [a]
spicy	1.7 ± 0.3 [b]	0.2 ± 0.2 [a]	0.4 ± 0.2 [a]	1.9 ± 0.2 [b]	1.8 ± 0.2 [b]	1.8 ± 0.2 [b]	0.2 ± 0.2 [a]
essential oil	2.2 ± 0.4 [c]	0.4 ± 0.2 [a]	0.2 ± 0.2 [a]	0.2 ± 0.2 [a]	0.5 ± 0.2 [a]	1.2 ± 0.2 [b]	0.0 ± 0.0 [a]
bitter	0.5 ± 0.3 [ab]	1.7 ± 0.3 [b]	0.0 ± 0.0 [a]	0.2 ± 0.2 [a]	0.6 ± 0.3 [ab]	0.2 ± 0.2 [a]	0.0 ± 0.0 [a]
Taste							
meat	5.7 ± 0.2 [a]	6.1 ± 0.3 [ab]	7.4 ± 0.3 [b]	6.5 ± 0.3 [a]	6.0 ± 0.2 [ab]	5.8 ± 0.3 [a]	6.5 ± 0.3 [ab]
salty	3.1 ± 0.2 [ab]	2.9 ± 0.2 [a]	4.2 ± 0.2 [b]	3.8 ± 0.3 [ab]	2.9 ± 0.2 [a]	3.1 ± 0.2 [ab]	3.3 ± 0.3 [ab]
broth	4.9 ± 0.3 [a]	5.3 ± 0.3 [a]	7.3 ± 0.3 [b]	5.2 ± 0.3 [a]	4.9 ± 0.2 [a]	4.9 ± 0.3 [a]	7.0 ± 0.2 [b]
fried	0.5 ± 0.2 [a]	0.2 ± 0.2 [a]	0.9 ± 0.2 [a]	0.5 ± 0.3 [a]	0.5 ± 0.2 [a]	0.5 ± 0.2 [a]	0.2 ± 0.2 [a]
spicy	4.7 ± 0.3 [c]	4.9 ± 0.2 [c]	2.9 ± 0.3 [a]	3.6 ± 0.3 [ab]	3.4 ± 0.3 [ab]	4.2 ± 0.2 [ab]	2.9 ± 0.3 [a]
essential oil	1.2 ± 0.2 [bc]	1.7 ± 0.2 [c]	0.1 ± 0.1 [a]	0.8 ± 0.2 [b]	0.8 ± 0.2 [b]	1.1 ± 0.3 [bc]	0.1 ± 0.1 [a]
bitter	0.2 ± 0.2 [a]	1.4 ± 0.3 [bc]	0.2 ± 0.2 [a]	0.7 ± 0.2 [ab]	1.5 ± 0.3 [c]	1.2 ± 0.3 [b]	0.2 ± 0.1 [a]

Different letters in the same raw for particular products denote a significant difference at $p < 0.05$ (one-way ANOVA, and *post hoc* Tukey test).

Principal component analysis (PCA) was used to study the relations between the aroma and/or taste attributes characteristic sensory profiles of meat products (variables) and to derive factors according to which these variables can be classified. PCA was also used to map the variants tested in our experiment (i.e., samples with selected spice extracts) into these factors. The PCA showed that the first two factors (F1 and F2) were the most important elements explaining variation in the data. For meatballs, they explained approximately 71% of the total variance for aroma and 87% for taste. For hamburgers, those factors (F1 and F2) explained approximately 74% of the total variance for aroma and 82% for taste. Therefore, they were selected for data interpretation. It is worth highlighting that F1 dominated in the explanation of taste: it explained approximately 70% of the taste variance.

The absolute values of the factor coordinates of the variables show the relationship between the factors and the sensory attributes (Table 3). For the aroma attributes of both meat products, the first factor (F1) was most strongly related to meat, spicy and essential oil aromas, the second factor (F2) to the bitter aroma. For taste attributes of meatballs and hamburgers, F1, the most strongly related to the meat, salty, broth, essential oil, and spicy taste, and F2 to the fried taste.

Table 3. The factor loadings for the aroma and taste attributes of pork meatballs and hamburgers with ethanol extracts of spices (rosemary, thyme, lovage, marjoram, basil, oregano; 0.05%) and the control sample (without any extract).

Sensory Attributes	Hamburgers		Meatballs	
	F1	F2	F1	F2
Aroma				
meat	0.948 *	−0.185	0.918 *	−0.082
peppery	−0.770 *	0.198	−0.475	0.560
fried	0.759	−0.164	−0.300	−0.665
spicy	−0.891 *	−0.408	−0.892 *	−0.001
essential oil	−0.738 *	0.159	−0.804 *	−0.467
bitter	−0.066	0.937 *	0.291	−0.744 *
Taste				
meat	0.917 *	−0.078	0.921 *	0.078
salty	0.884 *	−0.347	0.871 *	0.347
broth	0.873 *	−0.336	0.967 *	−0.091
fried	0.585	0.760 *	−0.560	−0.767 *
spicy	−0.845 *	0.255	−0.835 *	0.482
essential oil	−0.913 *	0.258	−0.839 *	0.525
bitter	−0.688	0.110	−0.704	−0.349

Two factors (F1 and F2) were extracted by applying Principal Component Analysis (PCA) on the mean values of descriptive sensory scores. Sensory attributes with numbers marked * are believed to be most important.

In the case of meatball and hamburger aroma profiles (case-factor coordinate plots—Figures 3a and 4a, Table 2), the control samples and the samples with lovage extract (which were characterised by a low intensity of spicy, essential oil, and bitter aromas) were plotted close together to the right side of the F1 axis. The samples with thyme, whose aroma profile had a low intensity of spice and essential oil aroma but a higher intensity of bitter aroma compared with the control sample, was also placed to this side of the F1 axis but the opposite side of the F2 axis. On the left side of the F1 axis were placed samples with rosemary extract (with a higher intensity of the essential oil aroma and spicy aroma compared with the control sample), samples with basil extract and marjoram extract (with a higher intensity of the spicy aroma than the control sample and low essential oil aroma), and those with oregano extract (with high intensity of spicy aroma).

For taste descriptors, the projection of the meatball variants on the factor-plane F1 × F2 (Figure 3b) shows that the control samples and samples with lovage extract were plotted to the right side of the F1 axis (i.e., they have positive coordinate values for F1). Meatballs with lovage, which were characterised by a low intensity of the spicy, essential oil and bitter tastes and a high intensity of broth taste, had a taste profile similar to the control sample (the differences between mean intensity of the descriptors for the sample with lovage and control were insignificant; Table 2). Samples with other spice extracts were plotted to the left side (negative coordinate values for F1). Their taste profiles were characterised by a higher intensity of the spicy and essential oil tastes (samples with rosemary extract and thyme extract) and/or the bitter taste (those with thyme, basil, and oregano extract) than the control sample. For hamburgers, the mapping of the samples based on their taste profiles (Figure 4b) shows that the control samples and the samples with lovage and marjoram extract were placed on the right side of the F1. In contrast, the other samples were placed on the left side—the taste profile of those samples was characterised by a higher intensity of the essential oil and spicy tastes (samples with rosemary and thyme extracts) and the bitter taste (for thyme, basil and oregano extracts), and a lower intensity of the broth tastes compared with the control sample (Table 2).

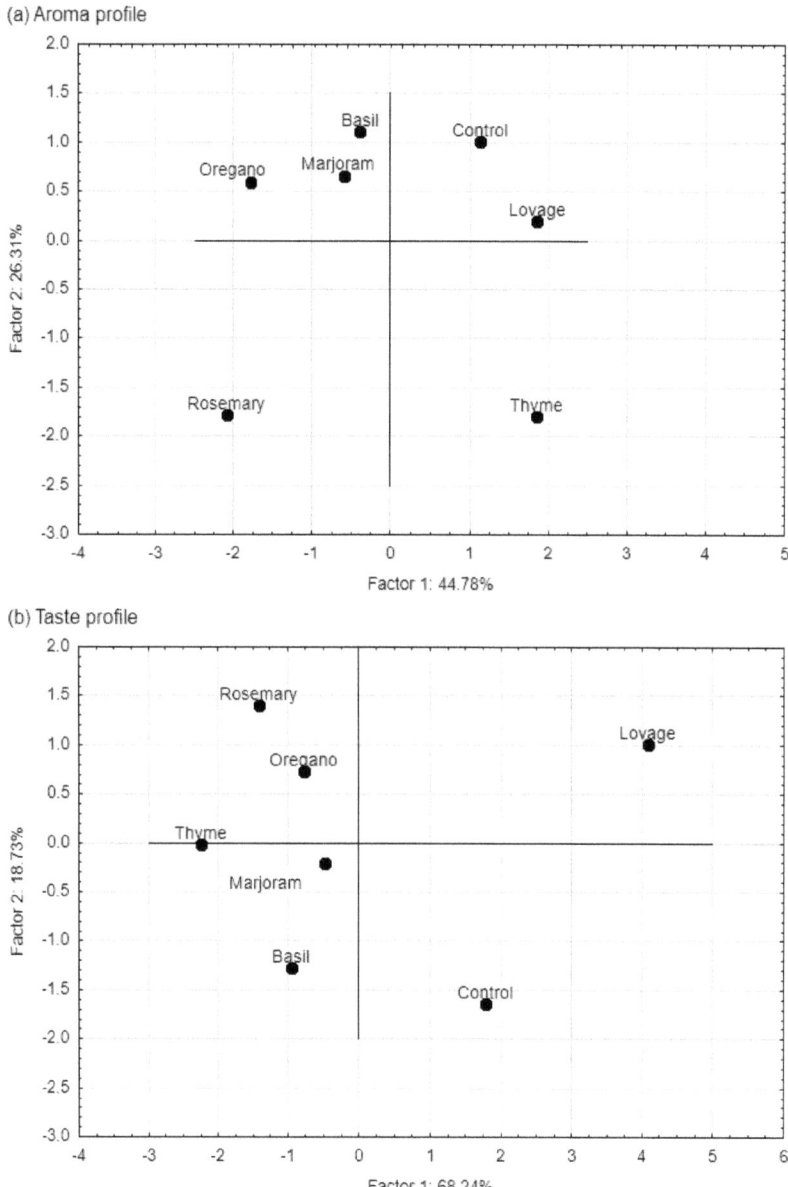

Figure 3. Map of the variants of pork meatballs with the addition of ethanol extracts of the spices (rosemary, thyme, lovage, marjoram, basil, oregano; 0.05%) and the control sample (without any extract) into factors (F1 × F2). Case–factor coordinate plots based on the attributes of (**a**) aroma profiles and (**b**) taste profiles (PCA analysis).

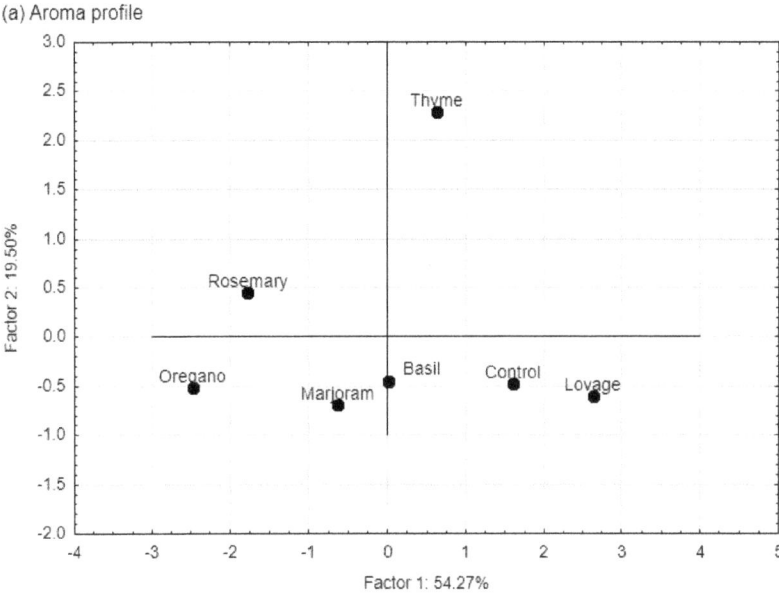

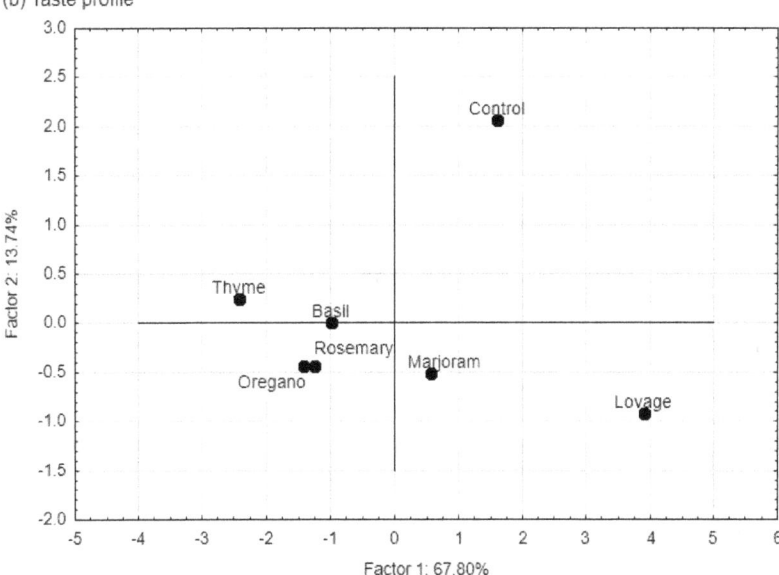

Figure 4. Map of the variants of pork hamburgers with the addition of ethanol extracts of the spices (rosemary, thyme, lovage, marjoram, basil, oregano; 0.05%) and the control sample (without any extract) into factors (F1 × F2). Case–factor coordinate plots based on the attributes of (**a**) aroma profiles and (**b**) taste profiles (PCA analysis).

4. Discussion

Due to fast changes in consumer expectations and demands, developing new products is a necessity for producers to survive on the competitive food market [41]. The participation of consumers in the initial phase of the new food product development is one of the vital factors determining success on the market [42]. According to Lord [43], the unsatisfied needs of consumers help to create opportunities for the development of new products. Consumer satisfaction, achieved through fulfilling sensory expectations, among others, increases the likelihood of a product's marketability [34,44].

Various active compounds of spice origin and their extracts were investigated for antimicrobial and antioxidant properties [36,45] as well as health benefits [8]. Some studies on the quality and stability of meat products with spice extract (particularly rosemary extract) also investigated their overall sensory quality [25,46–48]. Sutha et al. [48] evaluated the effect of rosemary essential oil addition (0.05%, 0.10%, and 0.25%) on the physico-chemical (pH, emulsion stability, product yield, and shear force value) and sensory qualities of chicken nuggets. They found that the additions were accepted by the panellists. Manhani et al. [25] compared the antioxidant potential of rosemary and oregano deodorized commercial extracts (0.04% addition) in precooked beef hamburgers by assessing the changes in lipid oxidation (TBRS values) and sensory analysis (colour, taste, odour evaluation using a 9-point hedonic scale) during 30 days of frozen storage. They showed that the hamburgers with those extracts were characterized by higher oxidation stability and satisfactory sensory quality compared to the control at the end of storage time. They also reported that the sensory analysis did not allow establishing a correlation between the scores of sensory attributes and the chemical changes. However, there is not much research regarding the effect of these additives on the sensory profile of food meat products and its acceptance by consumers.

In our research, both consumer tests and sensory profiling concerning meat products with the addition of selected spice extracts were carried out. The results of sensory profiling (Table 2) helped to explain the desirability of meat products with the addition of the selected spice extracts to consumers. These results can be an indication of which of the attributes of taste and smell, introduced into the product together with spice extracts, satisfy consumers, and which should be levelled.

According to the HCA analyses concerning the results of consumer tests, the meat products with the addition of rosemary extract and thyme extract formed one group (Figures 1 and 2). These products are characterised by a higher intensity of both the spice and essential oil tastes than the control sample, which influenced consumer acceptance of the products. Products with thyme extract were also characterised by a higher intensity of the bitter taste, but it was not the case in the profile of products with rosemary extracts. The higher intensity of the bitter taste in the products with thyme extract compared to the products with rosemary extract may be a result of the higher content of phenolic compounds in the thyme extract. The previous studies reported the higher total phenolic content in the case of thyme extract than rosemary extract [12]. Spicy tastes and aromas usually do not come from a single compound but result from the presence of various compounds, including terpenes, essential oils, and aldehydes in plant products and foods. Many of these substances are volatile, which gives them a potent fragrance [49]. Therefore, they have great potential in creating the smell and taste of products. The essential oil taste and smell probably come from essential oils present in a particular spice (e.g., thymol and carvacrol in thyme [50], and carnosol, rosmanol, and geraniol in rosemary [8]). Our research shows that the combination of these high-intensity attributes in the product profile diminished the consumer acceptance of the meat product.

Products with basil extract and oregano extract formed the second separate group based on the consumer test results. They were characterised by higher bitter tastes and spice aromas compared with the control sample. Phenolic compounds are responsible for the bitterness and astringency of many foods and beverages [51], and so are spices. The bitter taste can decrease the consumer acceptance of food products [52,53], particularly those for which this taste is not characteristic. Our study shows that some of the spice ethanol extracts are carriers of the bitter taste, and this can decrease the overall desirability of pork meatballs and hamburgers. Consumers can rate overall palatability without being consciously aware of all food ingredients, especially those present at near-threshold levels [54,55] which are characteristic for individuals [56]. For this reason, the masking of bitterness can be crucial in the development of new products with spice addition. On the other hand, not all consumers are averse to bitter tastes. The findings of the previous study reveal that females and respondents with higher education levels demonstrate a higher acceptance of bitter tastes, as well as specific individual characteristics, such as high compensatory health beliefs [57]. Perhaps these groups of consumers could be encouraged to buy meatballs and hamburgers with the addition of extracts from basil and oregano if the label highlighted their positive impact, e.g., on the oxidative stability of the product.

According to overall consumer desirability, the third group was formed by products with lovage extract and marjoram extract. The intensity of the essential oil and bitter aroma and bitter taste of the samples were not statistically different from the control sample (Table 2). The aroma and taste profile of meat products with lovage extract was the most similar to the control sample. It may result from a relatively low content of total phenolics in lovage extracts compared with other spice extracts, which was reported in the previous studies [36]. However, it is worth highlighting that the lovage extract was also a carrier of broth taste and had a positive effect on its intensity in the meat products compared with the other extracts. It seems to be important for explaining the consumer acceptance of products with lovage extract. Further studies are necessary to find a relationship between a particular compound/compounds of lovage extract and the broth taste intensity in meat products with its addition. A high consumer acceptance of meat products with marjoram extract, despite the high intensity of the spicy aroma, spicy taste, and essential oil taste in their sensory profile, may explain the popularity of this spice. Consumers who often use marjoram for seasoning dishes may get used to its characteristic aroma and flavour.

5. Conclusions

The sensory quality of foods is one of the main factors in the acceptance of a product on the market. For this reason, studying factors that influence consumer desirability is crucial in food product development. The study shows that ethanol extracts of spices allow for the creation of new meat products with high consumer desirability. However, the high intensity of essential oil and spicy tastes of the spice extracts might be a limitation on the application of the extracts.

Author Contributions: Conceptualization, K.S.-B. and M.H. Formal analysis, K.S.-B. and A.J.-G. Investigation, K.S.-B. Methodology, K.S.-B. and A.J.-G. Software, K.S.-B., K.W. and A.J.-G. Validation, K.S.-B., K.W. and A.J.-G. Writing-original draft, K.S.-B., K.W and A.J.-G. Writing-review & editing, K.S.-B., K.W., A.J.-G. and M.H. All authors have read and agreed to the published version of the manuscript.

Funding: This research was funded by the Ministry of Scientific Research and Information Technology in Poland (project number N312 025 31/2049). The publication was co-financed within the framework of the Ministry of Science and Higher Education programme as "Regional Initiative Excellence" in years 2019–2022, project number 005/RID/2018/19.

Conflicts of Interest: The authors declare no conflict of interest.

References

1. Falowo, A.B.; Fayemi, P.O.; Muchenje, V. Natural antioxidants against lipid-protein oxidative deterioration in meat and meat products: A review. *Food Res. Int.* **2014**, *64*, 171–181. [CrossRef] [PubMed]
2. Wąsowicz, E.; Gramza, A.; Hęś, M.; Jelen, H.H.; Korczak, J.; Małecka, M.; Mildner-Szkudlarz, S.; Rudzinska, M.; Samotyja, U.; Zawirska-Wojtasiak, R. Oxidation of lipids in food. *Pol. J. Food Nutr. Sci.* **2004**, *13*, 87–100.
3. Hęś, M. Protein-lipid interactions in different meat systems in the presence of natural antioxidants—A review. *Pol. J. Food Nutr. Sci.* **2017**, *67*, 5–17. [CrossRef]
4. Kanner, J. Oxidative processes in meat and meat products: Quality implications. *Meat Sci.* **1994**, *36*, 169–189. [CrossRef]
5. Tomovic, V.; Jokanovic, M.; Sojic, B.; Ivic, M. Plants as natural antioxidants for meat products. In *Proceedings of the IOP Conference Series: Earth and Environmental Science*; IOP Publishing: Bristol, UK, 2017; Volume 85.
6. Gonzalez-Gallego, J.; Sanchez-Campos, S.; Tunon, M.J. Anti-inflammatory properties of dietary flavonoids. *Nutr. Hosp.* **2007**, *22*, 287–293.
7. Andre, C.M.; Larondelle, Y.; Evers, D. Dietary antioxidants and oxidative stress from a human and plant perspective: A review. *Curr. Nutr. Food Sci.* **2010**, *6*, 2–12. [CrossRef]

8. Yashin, A.; Yashin, Y.; Xia, X.; Nemzer, B. Antioxidant activity of spices and their impact on human health: A review. *Antioxidants* **2017**, *6*, 70. [CrossRef]
9. Nakatani, N. Phenolic antioxidants from herbs and spices. *BioFactors* **2000**, *13*, 141–146. [CrossRef]
10. Nour, V.; Trandafir, I.; Cosmulescu, S. Bioactive compounds, antioxidant activity and nutritional quality of different culinary aromatic herbs. *Not. Bot. Horti Agrobot.* **2017**, *45*, 179–184. [CrossRef]
11. Stanciu, G.; Cristache, N.; Lupsor, S.; Dobrinas, S. Evaluation of antioxidant activity and total phenols content in selected spices. *Rev. Chim.* **2017**, *68*, 1429–1434.
12. Heś, M.; Gramza-Michałowska, A. Effect of plant extracts on lipid oxidation and changes in nutritive value of protein in frozen-stored meat products. *J. Food Process. Preserv.* **2017**, *41*, 1–9. [CrossRef]
13. Waszkowiak, K.; Dolata, W. The application of collagen preparations as carriers of rosemary extract in the production of processed meat. *Meat Sci.* **2007**, *75*, 178–183. [CrossRef] [PubMed]
14. Szymandera-Buszka, K.; Heś, M.; Waszkowiak, K.; Jędrusek-Golińska, A. Thiamine losses during storage of pasteurised and sterilized model systems of minced chicken meat with addition of fresh and oxidized fat, and antioxidants. *Acta Sci. Pol. Technol. Aliment.* **2014**, *13*, 393–401. [CrossRef] [PubMed]
15. Jałosińska, M.; Wilczak, J. Influence of plant extracts on the microbiological shelf life of meat products. *Polish J. Food Nutr. Sci.* **2009**, *59*, 303–308.
16. Nugboon, K.; Intarapichet, K. Antioxidant and antibacterial activities of Thai culinary herb and spice extracts and application in pork meatballs. *Int. Food Res. J.* **2015**, *22*, 1788–1800.
17. Gramatiņa, I.; Sazonova, S.; Kruma, Z.; Skudra, L.; Priecina, L. Herbal extracts for ensuring pork meat quality during cold storage. *Proc. Latv. Acad. Sci.* **2017**, *71*, 453–460. [CrossRef]
18. Naveena, B.M.; Vaithiyanathan, S.; Muthukumar, M.; Sen, A.R.; Kumar, Y.P.; Kiran, M.; Shaju, V.A.; Chandran, K.R. Relationship between the solubility, dosage and antioxidant capacity of carnosic acid in raw and cooked ground buffalo meat patties and chicken patties. *Meat Sci.* **2013**, *95*, 195–202. [CrossRef]
19. Sebranek, J.G.; Sewalt, V.J.H.; Robbins, K.L.; Houser, T.A. Comparison of a natural rosemary extract and BHA/BHT for relative antioxidant effectiveness in pork sausage. *Meat Sci.* **2005**, *69*, 289–296. [CrossRef]
20. Bilska, A.; Waszkowiak, K.; Błaszyk, M.; Rudzińska, M.; Kowalski, R. Effect of liver pâté enrichment with flaxseed oil and flaxseed extract on lipid composition and stability. *J. Sci. Food Agric.* **2018**, *98*, 4112–4120. [CrossRef]
21. Rojas, M.C.; Brewer, M.S. Effect of natural antioxidants on oxidative stability of cooked, refrigerated beef and pork. *J. Food Sci.* **2007**, *72*, 282–288. [CrossRef]
22. Rojas, M.C.; Brewer, M.S. Effect of natural antioxidants on oxidative stability of frozen, vacuum-packaged beef and pork. *J. Food Qual.* **2008**, *31*, 173–188. [CrossRef]
23. Sánchez-Escalante, A.; Djenane, D.; Torrescano, G.; Beltrán, J.A.; Roncales, P. Antioxidant action of borage, rosemary, oregano, and ascorbic acid in beef patties packaged in modified atmosphere. *J. Food Sci.* **2003**, *68*, 339–344. [CrossRef]
24. Trindade, R.A.; Lima, A.; Andrade-Wartha, E.R.; Oliveira e Silva, A.M.; Mancini-Filho, J.; Villavicencio, A.L.C.H. Consumer's evaluation of the effects of gamma irradiation and natural antioxidants on general acceptance of frozen beef burger. *Radiat. Phys. Chem.* **2009**, *78*, 293–300. [CrossRef]
25. Manhani, M.R.; Nicoletti, M.A.; Da Silva Barretto, C.A.; De Jesus, G.R.; Munhoz, C.; De Abreu, G.R.; Zaccarelli-Magalhães, J.; Fukushima, A.R. Antioxidant action of rosemary and oregano extract in pre-cooked meat hamburger. *Food Nutr. Sci.* **2018**, *9*, 806–817. [CrossRef]
26. El-Alim, S.S.L.A.; Lugasi, A.; Hóvári, J.; Dworschák, E. Culinary herbs inhibit lipid oxidation in raw and cooked minced meat patties during storage. *J. Sci. Food Agric.* **1999**, *79*, 277–285. [CrossRef]
27. Sariçoban, C.; Yilmaz, M.T. Effect of thyme/cumin essential oils and butylated hydroxyl anisole/butylated hydroxyl toluene on physicochemical properties and oxidative/microbial stability of chicken patties. *Poult. Sci.* **2014**, *93*, 456–463. [CrossRef]
28. Heś, M.; Gliszczyńska-Świgło, A.; Gramza-Michałowska, A. The effect of antioxidants on quantitative changes of lysine and methionine in linoleic acid emulsions at different pH conditions. *Acta Sci. Pol. Technol. Aliment.* **2017**, *16*, 1–15.
29. Jongberg, S.; Torgngren, M.A.; Gunvig, A.; Skibsted, L.H.; Lund, M.N. Effect of green tea or rosemary extract on protein oxidation in Bologna type sausages prepared from oxidatively stressed pork. *Meat Sci.* **2013**, *93*, 538–546. [CrossRef]

30. Opara, E.I. Culinary herbs and spices: What can human studies tell us about their role in the prevention of chronic non-communicable diseases? *J. Sci. Food Agric.* **2019**, *99*, 4511–4517. [CrossRef]
31. Vázquez-Fresno, R.; Rosana, A.R.R.; Sajed, T.; Onookome-Okome, T.; Wishart, N.A.; Wishart, D.S. Herbs and spices—Biomarkers of intake based on human intervention studies—A systematic review. *Genes Nutr.* **2019**, *18*, 1–27. [CrossRef]
32. Hathwar, S.C.; Rai, A.K.; Modi, V.K.; Narayan, B. Characteristics and consumer acceptance of healthier meat and meat product formulations—A review. *J. Food Sci. Technol.* **2012**, *49*, 653–664. [CrossRef] [PubMed]
33. Haugaard, P.; Hansen, F.; Jensen, M.; Grunert, K.G. Consumer attitudes toward new technique for preserving organic meat using herbs and berries. *Meat Sci.* **2014**, *96*, 126–135. [CrossRef] [PubMed]
34. Halagarda, M.; Suwała, G. Sensory optimisation in new food product development: A case study of polish apple juice. *Ital. J. Food Sci.* **2018**, *30*, 317.
35. Baryłko-Pikielna, N.; Matuszewska, I. *Sensory Food Studies*; Scientific Publishing House PTTŻ: Krakow, Poland, 2013; pp. 1–375.
36. Ulewicz-Magulska, B.; Wesolowski, M. Total phenolic contents and antioxidant potential of herbs used for medical and culinary purposes. *Plant Foods Hum. Nutr.* **2019**, *74*, 61–67. [CrossRef]
37. Commission Regulation (EU) No 723/2013 of 26 July 2013 amending Annex II to Regulation (EC) No 1333/2008 of the European Parliament and of the Council as regards the use of extracts of rosemary (E 392) in certain low fat meat and fish products. 2013. Available online: https://eur-lex.europa.eu/legal-content/EN/TXT/?uri=celex%3A32013R0723 (accessed on 17 November 2019).
38. *ISO 8589: 2009 Sensory Analysis—General Guidance for the Design of Test Rooms*; ISO: Geneva, Switzerland, 2009.
39. *ISO 6564: 1985 Sensory Analysis—Methodology—Flavour Profile Methods*; ISO: Geneva, Switzerland, 1985.
40. Tekień, A.; Gutkowska, K.; Żakowska-Biemans, S.; Jóźwik, A.; Krotki, M. Using cluster analysis and choice-based conjoint in research on consumers preferences towards animal origin food products. Theoretical review, results and recommendations. *Anim. Sci. Pap. Rep.* **2018**, *36*, 171–184.
41. Halagarda, M. Decomposition analysis and consumer research as essential elements of the new food product development process. *Brit. Food J.* **2017**, *119*, 511–526. [CrossRef]
42. Mattsson, J.; Helmersson, H. Food product development. A consumer-led text analytic approach to generate preference structures. *Brit. Food J.* **2007**, *3*, 249–259.
43. Lord, J.B. New product failure and success. In *Developing New Food Products for a Changing Marketplace*; Brody, A.L., Lord, J.B., Eds.; CRC Press: Boca Raton, FL, USA, 2008; pp. 47–74.
44. Grunert, K.G. Current issues in the understanding of consumer food choice. *Trends Food Sci. Tech.* **2002**, *13*, 275–285. [CrossRef]
45. Karre, L.; Lopez, K.; Getty, K.J.K. Natural antioxidants in meat and poultry products. *Meat Sci.* **2013**, *94*, 220–227. [CrossRef]
46. Essid, I.; Smeti, S.; Atti, N. Effect of rosemary powder on the quality of dry ewe sausages. *Ital. J. Food Sci.* **2018**, *30*, 641–649.
47. Schilling, M.W.; Pham, A.J.; Dhowlaghar, N.; Campbell, Y.L.; Williams, J.B.; Xiong, Y.L.; Perez, S.M.; Kin, S. Effects of rosemary (*Rosmarinus Officinalis* L.) and green tea (*Camellia Sinensis* L.) extracts on sensory properties and shelf-life of fresh pork sausage during long-term frozen storage and subsequent retail display. *Meat Muscle Biol.* **2018**, *2*, 375–390. [CrossRef]
48. Sutha, M.; Selvaraj, R.; Kulkarni, V.V.; Chandirasekaran, V.; Edwin, S.C.; Malmarugan, S. Effect of rosemary essential oil on the quality characteristics of chicken nuggets. *Int. J. Curr. Microbiol. Appl. Sci.* **2018**, *7*, 4686–4693. [CrossRef]
49. Roger, W.W. The flavors of herbs and foods—Commonly overlooked, but crucial to your health. *Herb. Rev.* **2019**, *1*, 1–4.
50. Gedikoğlu, A.; Sökmen, M.; Çivit, A. Evaluation of *Thymus vulgaris* and *Thymbra spicata* essential oils and plant extracts for chemical composition, antioxidant, and antimicrobial properties. *Food Sci. Nutr.* **2019**, *7*, 1704–1714. [CrossRef]
51. Drewnowski, A.; Gomez-Carneros, C. Bitter taste, phytonutrients, and the consumer: A review. *Am. J. Clin. Nutr.* **2000**, *72*, 1424–1435. [CrossRef]
52. Drewnowski, A. Taste preferences and food intake. *Annu. Rev. Nutr.* **1997**, *17*, 237–253. [CrossRef]

53. Feng, Y.; Tapia, M.A.; Okada, K.; Lazo, N.B.; Chapman-Novakofski, K.; Phillips, C.; Lee, S.-Y. Consumer acceptance comparison between seasoned and unseasoned vegetables. *J. Food Sci.* **2018**, *83*, 446–453. [CrossRef]
54. Drewnowski, A. The science and complexity of bitter taste. *Nutr. Rev.* **2001**, *59*, 163–169. [CrossRef]
55. Makita, Y.; Tomoko, I.; Kobayashi, N.; Fujio, M.; Fujimoto, K.; Moritomo, R.; Fujita, J.; Fujiwara, S. Evaluation of the bitterness-masking effect of powdered roasted soybeans. *Foods* **2016**, *5*, 44. [CrossRef]
56. Szymandera-Buszka, K.; Jędrusek-Golińska, A.; Waszkowiak, K.; Hęś, M. Sensory sensitivity to sour and bitter taste among people with Crohn's disease and folic acid supplementation. *J. Sens. Stud.* **2020**, *35*, e12550. [CrossRef]
57. Vecchio, R.; Cavallo, C.; Cicia, G.; Del Giudice, T. Are (all) consumers averse to bitter taste? *Nutrients* **2019**, *11*, 323. [CrossRef] [PubMed]

© 2020 by the authors. Licensee MDPI, Basel, Switzerland. This article is an open access article distributed under the terms and conditions of the Creative Commons Attribution (CC BY) license (http://creativecommons.org/licenses/by/4.0/).

Article

Adding Value to Bycatch Fish Species Captured in the Portuguese Coast—Development of New Food Products

Frederica Silva [1,2], Ana M. Duarte [1], Susana Mendes [3], Patrícia Borges [3], Elisabete Magalhães [2], Filipa R. Pinto [1], Sónia Barroso [1], Ana Neves [2], Vera Sequeira [2,4], Ana Rita Vieira [2,4], Maria Filomena Magalhães [4,5], Rui Rebelo [4,5], Carlos Assis [2,4], Leonel Serrano Gordo [2,4] and Maria Manuel Gil [3,*]

1. MARE—Marine and Environmental Sciences Centre, Polytechnic of Leiria, Cetemares, 2520-620 Peniche, Portugal; frederica.g.silva@ipleiria.pt (F.S.); ana.c.duarte@ipleiria.pt (A.M.D.); filipa.gomes@ipleiria.pt (F.R.P.); sonia.barroso@ipleiria.pt (S.B.)
2. MARE—Marine and Environmental Sciences Centre, Faculdade de Ciências, Universidade de Lisboa, 1749-016 Lisbon, Portugal; eli_magalhaes_silva@hotmail.com (E.M.); amneves@fc.ul.pt (A.N.); vlsequeira@fc.ul.pt (V.S.); arivieira@fc.ul.pt (A.R.V.); caassis@fc.ul.pt (C.A.); lsgordo@fc.ul.pt (L.S.G.)
3. MARE—Marine and Environmental Sciences Centre, ESTM, Polytechnic of Leiria, Cetemares, 2520-620 Peniche, Portugal; susana.mendes@ipleiria.pt (S.M.); patricia.borges@ipleiria.pt (P.B.)
4. Departamento de Biologia Animal, Faculdade de Ciências, Universidade de Lisboa, 1749-016 Lisbon, Portugal; mfmagalhaes@fc.ul.pt (M.F.M.); rmrebelo@fc.ul.pt (R.R.)
5. CE3C—Centre for Ecology, Evolution and Environmental Changes, Faculdade de Ciências, Universidade de Lisboa, 1749-016 Lisboa, Portugal
* Correspondence: maria.m.gil@ipleiria.pt

Citation: Silva, F.; Duarte, A.M.; Mendes, S.; Borges, P.; Magalhães, E.; Pinto, F.R.; Barroso, S.; Neves, A.; Sequeira, V.; Vieira, A.R.; et al. Adding Value to Bycatch Fish Species Captured in the Portuguese Coast—Development of New Food Products. *Foods* **2021**, *10*, 68. https://doi.org/10.3390/foods10010068

Received: 27 November 2020
Accepted: 25 December 2020
Published: 31 December 2020

Publisher's Note: MDPI stays neutral with regard to jurisdictional claims in published maps and institutional affiliations.

Copyright: © 2020 by the authors. Licensee MDPI, Basel, Switzerland. This article is an open access article distributed under the terms and conditions of the Creative Commons Attribution (CC BY) license (https://creativecommons.org/licenses/by/4.0/).

Abstract: We live in a world of limited biological resources and ecosystems, which are essential to feed people. Consequently, diversifying target species and considering full exploitation are essential for fishery sustainability. The present study focuses on the valorization of three low commercial value fish species (blue jack mackerel, *Trachurus picturatus*; black seabream, *Spondyliosoma cantharus*; and piper gurnard, *Trigla lyra*) and of two unexploited species (comber, *Serranus cabrilla* and boarfish, *Capros aper*) through the development of marine-based food products with added value. A preliminary inquiry with 155 consumers from Região de Lisboa e Vale do Tejo (Center of Portugal) was conducted to assess fish consumption, the applicability of fish product innovation, and the importance of valorizing discarded fish. Five products (black seabream *ceviche*, smoked blue jack mackerel pâté, dehydrated piper gurnard, fried boarfish, and comber pastries) were developed and investigated for their sensory characteristics and consumer liking by hedonic tests to 90 consumers. The most important descriptors were identified for each product (texture, flavor, color, and appearance). Comber pastries had the highest purchase intention (88%), followed by black seabream *ceviche* (85%) and blue jack mackerel pâté (76%). Sensory evaluations showed a clear tendency of consumers to accept reformulated products, with the introduction of the low-value and unexploited species under study.

Keywords: fish valorization; consumer; unexploited; low commercial value; hedonic tests; sustainability; product reformulation

1. Introduction

The growing world population and, consequently, the increasing need for food resources are demanding a more sustainable and circular bio-economy. In fact, the planet's sustainability is one of the most discussed topics worldwide, with particular emphasis on marine resources. Despite the ocean's richness in different fish species, consumers and industries only know a few of them, which are the ones with the highest commercial value. However, underexploited and underutilized fish species may offer added benefits for consumer health, given that their nutritional content as well as sensory attributes are equally pleasant as more commonly consumed fish [1].

Fish species that are not targeted by fisheries are called bycatches, which means (a) species with low or no commercial value that are discarded to gain room on the vessel for commercial species caught later; (b) discarded species with no commercial value due to restrictive measures in terms of landing (quota, size below the minimum conservation reference size); (c) deteriorated species; or (d) protected species that cannot be caught, which in most cases do not survive [2,3]. Fish disposal results in marine ecosystem changes, which can be avoided with the exploitation of waste from bio-based industrial processes, including by-products, co-products, and discards (from fisheries), contributing to resource-efficient process improvement [4].

In Portugal, the most consumed fish species are cod (*Gadus morhua*), hake (*Merluccius merluccius*) [1], and canned tuna (made with some species from the Scombridae family) [5,6]. On the other hand, in Europe the top five consumed species, which accounted for 44% of the total volume in 2017, were tuna, cod, salmon, Alaska pollock, and shrimp [7]. This consumption preference can be explained by the lack of consumer awareness of other species, which discourages their purchase. However, this may not be the only explanation, due to the complexity of food choice, which is influenced by many interrelated factors, including social, environmental, political, economic, and cultural aspects [8]. According to Coralo et al. [8] consumers' attitude towards food can be categorized as (a) the Individualist: food choices based on their personal interests (e.g., economic convenience, personal mood); (b) the Foodie: food choices based on the sensory aspects related to food (e.g., better taste, right price–quality ratio); (c) the Environmentalist: food choices based on environmental sustainability issues (e.g., the integrity of the farmer, food origin); and (d) the Health Enthusiast: food choices based on contemporary diets (e.g., low-calorie diet) and analyzing label contents, claims, and health effects. Therefore, with consumer surveys, it is possible to quantify and measure consumption options and frequencies, which allows a pattern of eating behavior to be traced. Regardless of the population's consumption habits, one of the ways to make a new fish species known is through its use in traditional products or in products that are familiar to the consumer, replacing the more commercial species. After acknowledging population consumption habits and expectations regarding new food products, this information is applied to product development and, subsequently, sensory tests are carried out to assess acceptability and purchase intention. Sensory evaluation allows for the establishment of a target market, the identification of the most important product features the avoidance of wasted effort during product development, quality issues to be dealt with; a comparison between brands, and an attempt at ensuring long shelf-life [9]. Therefore, sensory evaluation is a result of human decision and, consequently, it is an outcome of complex interactions conditioned by personal history, environmental variables, subjective covariates, and object characteristics that also interact with the modality of the survey [9]. In fact, sensory evaluation is essential for both new food product development and food reformulation of existing products.

Food reformulation is considered one of the contributors to achieving population nutrient goals (e.g., energy and salt reduction, clean labels) [10,11]. Therefore, food reformulation can be a way to valorize fish species with low or no commercial value by using them to replace the commercial ones. This both increases the value of underexploited and underutilized fish species and reduces the final price of the reformulated product, with the added advantage of avoiding overexploitation of species with significant commercial value.

When considering development of new fish-based food products, attention should be given to seasonality to achieve greater nutritional (e.g., fat and protein content) and sensorial (e.g., appearance, odor, flavor, and texture) features. Therefore, it is important to identify the best time of the year to capture fish considering the product formulation and, consequently, consumer acceptance, taking into consideration their life history, including their reproductive cycle and growth characteristics. Considering that there is a high probability of market failure when introducing new food products, it is crucial to know in advance the consumer habits and preferences to create or redesign and reformulate a product that satisfies their necessities [12]. One of the approaches to gain this knowledge is to perform a market survey about consumers' diet and the reason for their food choices (e.g., nutritionally rich, easy to prepare). Consumer characterization (e.g., gender, age, occupation) is also important to identify a potential target market segment.

Despite world fisheries and aquaculture production increasing in the last 30 years from 100 million tonnes in 1986 to 178 million tonnes in 2018 [13], fisheries showed a stable value around 90 million tonnes between 1996 and 2016 and a slight increase to 96 million tonnes in 2018 [13]. However, regardless of this stability in catches, the number of unsustainable (overfished) stocks has increased from 10% in 1974 to 34.2% in 2017. In addition, within the sustainable stocks, the number of underfished stocks decreased continuously from 1974 (around 38%) to 2017 (less than 8%), with the majority of stocks in the "maximally sustainably fished" category [13]. In Portugal, as in other countries from temperate waters, most fisheries are multi-specific, meaning that besides the target species (usually species with high commercial value) other species are also caught. Among these, some are discarded at sea and others are landed but normally reach a low commercial value. In terms of stock status, many commercial species of more global (European) interest are studied and their stocks are assessed and managed by international organizations (International Council for the Exploration of the Sea—ICES). The other species, the great majority of them, namely those included in this paper, are not assessed and managed and the biological information is scarce and very often totally lacking. The minimization of discards and the valorization of non-commercial or low commercial species is therefore very important to present alternatives to the currently most exploited species. The inclusion of new alternative species in the market would allow the increase of first auction prices and then the interest of fishermen to shift their effort to these species, as occurred in the 1990s with the black scabbardfish fishery in Portuguese waters.

Therefore, the present study aims to value bycatch species with low commercial and non-commercial value. Among the former, three low commercial fish species were chosen: blue jack mackerel (*Trachurus picturatus*), black seabream (*Spondyliosoma cantharus*) and piper gurnard (*Trigla lyra*). Why these species? These species are particularly relevant either because of their high landing values or their first auction prices. Among species of non-commercial value, the comber (*Serranus cabrilla*) and the boarfish (*Capros aper*) are particularly abundant in Portuguese waters, with *S. cabrilla* being one of the most important species in terms of discards of several fisheries, namely gillnets and trawls, while *C. aper* is among the 10 most important species in terms of abundance, being caught as bycatch in both crustacean and fish trawl fishing [14,15]. The valorization of these species is aimed at through the development of five innovative and differentiating marine-based products in relation to what currently exists on the market. A market survey about fish consumption habits, the applicability of fish product innovation, and the importance of valuing discarded fish was also performed.

2. Materials and Methods

2.1. Market Survey—Fish Consumption, Applicability of Fish Products Innovation, and Importance of Fish Valorisation with Low or Absent Commercial Value

A total of 155 consumers from Região de Lisboa e Vale do Tejo (Center of Portugal), answered to an anonymous market survey with 13 closed-ended questions divided into three groups: (1) consumer baseline characteristics, (2) fish consumption habits and their characterization, and (3) individual consumer preference and consumption patterns of processed fish products. In the first group, the consumer's personal features (gender, age, level of education, and occupation) were identified. The second group was conceived to assess individual importance of fish consumption (rated on a scale of 5—Very important, 4—"Important", 3—"Indifferent", 2—"Less important," and 1—"Not important"), the main attributes that are valorized in fish consumption (e.g., low fat content, protein, flavor), consumption frequency characterization (rated on a scale of 5—"Occasionally," 4—"More than once a week," 3—"Every week," 2—"Twice a month," and 1—"< once a month"), and to perceive the consumption of some fish species with low or no commercial value. The third group allowed for the appraisal of the willingness to consume new products derived from fish species with low or no commercial value, the main motivations for its consumption (e.g., low-price products, fast and practical products, nutritionally rich), how to consume them (e.g., to prepare in the oven, ready-to-eat products, or frozen products), an evaluation of the importance given to their valorization (rated on a scale of 5—"Very important," 4—"Important," 3—"Indifferent," 2—"Less important," and 1—"Not important"), and the importance of creating new fish products from them (rated on a scale of 5—"Very important," 4—"Important," 3—"Indifferent," 2—"Less important," and 1—"Not important"). This data collection, through a market survey, was done personally and ran from September to November 2019.

2.2. Fish-Based Food Product Production and Preparation

The fish species used in this work were blue jack mackerel, black seabream, piper gurnard, comber, and boarfish. The fish (captured on the Portuguese coast) were acquired at Peniche fish auction or purchased from local fishermen. The initial fish preparation included the removal of scales, internal organs, and head.

The fish products were developed by a Portuguese chef, considering the consumers' answers to the survey, and the need to be sensorially pleasing and preferably ready-to-eat or with few preparation requirements. In addition, no food additives were added to any fish product. Given the consumers' lack of knowledge about the fish species under study, it was decided to develop reformulations of products familiar to consumers. In addition, new consumption trends were also considered.

The black seabream used in ceviche formulation was initially filleted, the skin was removed, and then sliced into small cubes. The black seabream cubes were then mixed with the remaining ingredients, which included lime, lemon, chives, vinegar, ginger, mango, and red pepper.

For the preparation of the smoked blue jack mackerel pâté, the fish fillets were smoked with bay leaves, the skin was removed, they were minced, and then added to the remaining ingredients that included chili, cream cheese, pickles, onion, and mustard.

Piper gurnard skinless fillets were first deep-fried with vegetable oil and, after excess oil removal, the fish was dehydrated.

For the product formulation, the boarfish was first scaled, leaving the caudal fin, and then coated with flour and fried in vegetable oil.

For the preparation of the comber pastries, the fish was smoked with bay leaves, minced, and added to the other ingredients: mashed potato, onion, parsley, and egg. After mixing all the ingredients the pastries were molded with the help of spoons and deep-fried.

2.3. Hedonic Tests

Each of the fish products developed was evaluated by 90 consumers of both genders aged between 18 and 75 years old with a hedonic sensorial test at the "European night of researchers" event in Lisbon, Portugal. The aim was to measure the overall hedonic perception of a product by consumers. This test had 9 closed-response questions and was divided into three groups: (1) consumer characterization (gender and age), (2) product sensorial evaluation (general appearance, color, odor, flavor, texture, and global appreciation), and (3) purchase intention. The sensorial evaluation was performed considering a scale of 9 ("Extremely pleasant"), 8 ("Very pleasant"), 7 ("Pleasant"), 6 ("Not very pleasant"), 5 ("Neither pleasant nor unpleasant"), 4 ("Not very unpleasant"), 3 ("Unpleasant"), 2 ("Very unpleasant"), and 1 ("Extremely unpleasant"), and the purchase intention from 5 ("I'm sure I would buy"), 4 ("I would probably buy"), 3 ("I don't know if I would buy"), 2 ("I probably wouldn't buy") and 1 ("I'm sure I wouldn't buy").

2.4. Statistical Analysis

Descriptive statistics and exploratory data analysis were performed on 155 consumers sampled in the same region to characterize the fish consumer profile, the applicability of fish product innovation, and the importance of valorizing discarded fish. An exploratory factorial analysis (EFA) was performed to achieve a parsimonious representation of the associations among measured descriptors (that is, appearance, color, odor, flavor, texture, and global appreciation). Prior to running the EFA, the uniformity of the sample was tested by the Kaiser–Meyer–Olkin (KMO) measure of sampling adequacy. A KMO value of more than 0.50 was considered acceptable. In addition, the measure of sampling adequacy of the anti-image correlation matrix was all greater than 0.5. The presence of correlations between measured descriptors was tested using the Bartlett test of sphericity and was accepted when it was significant at p-value < 0.05. The correlation matrix was also analyzed and values of more than 0.3 were considered suitable [16]. As regards the sample size, there are several guiding rules cited in the literature [16–18] and the lack of agreement is noted in all of them. In this work, it was an option to follow the recommendation denoted as a ratio of N:p, where N refers to the number of participants and p to the number of variables. Thus, the ratio of respondents to variables should be at least 10:1 [19], which was satisfied by the data set under study. Principal component analysis (PCA) was used for factor extraction and orthogonal rotation (varimax option) to minimize the number of indicators that have high loading on one factor [20]. The number of factors was retained based on the eigenvalue >1.0 criterion (Kaiser criterion) and a plot of the eigenvalues, which shows the total variance associated with each other [21]. For all analyses, this resulted in the extraction of the first two factors.

Spearman's rank–order correlation coefficient was used to assess the co-variation of the descriptor scores. The Kruskal–Wallis non-parametric test was performed to assess the statistical differences between descriptor scores and purchase intention when comparing age groups. When significant differences were achieved, the Games–Howell or Bonferroni multiple comparison tests were executed. The use of the Kruskal–Wallis test proved to be appropriate, since it allows for the comparison of distributions of two or more ordinal variables less observed in two or more independent samples. Finally, the Mann–Whitney non-parametric test was performed to assess the statistical differences between descriptors and purchase intention when comparing consumer gender.

All analyses were performed for the five fish-based products that were developed. The significance level was set to p-value < 0.05. Analyses were carried out using IBM SPSS Statistics 26 software.

3. Results and Discussion

3.1. Market Survey

The respondents from the market survey were mostly employed (57%), doctoral (33%), women (69%), and aged between 31 and 45 (42%) (Table 1). The participants give high importance to fish consumption (72%), mainly due to their fatty acid and omega content (79%) (Table 2). Portugal is among the countries with the highest consumption of fish (about 60 kg per capita per year [13]) and the sample reflected this tendency with fish consumption every week (44%) or more than once a week (43%), i.e., 87% of respondents often consumed fish, some of which with low or absent commercial value (48%) (Table 2). The consumption of processed fish products was present in the consumers' diet (75%) due to their quick and easy preparation (38%). In addition, the consumers were receptive to new products (86%), preferably frozen, ready-to-eat, or to be prepared in the oven (51%) (Table 3). These findings on nutritional fish factors and purchasing convenience were also reported by Samoggia and Castellini's study developed in Italy [22]. In their study, 740 consumers were asked about fish frequency consumption, with a focus on healthy attitude and sociodemographic or family structure characteristics [22]. Their results showed that the consumption frequency was mostly driven by health-orientation and social influences, compared with socio-economic characteristics (including households with children) [22].

Table 1. Results of market survey concerning consumers' gender, age, education level, and occupation (n = 155).

Baseline Characteristics	Survey Respondents (%)
Gender	
Male	31%
Female	69%
Age	
18–30	30%
31–45	42%
46–64	26%
>65	2%
Education Level	
Basic education	4%
High school	11%
Graduation	21%
Master's degree	31%
PhD	33%
Occupation	
Student	17%
Employee	57%
Self-employed	6%
Retired	2%
Unemployed	2%
Research fellow	15%
Teacher	1%

Table 2. Results of the survey concerning the importance and frequency of fish consumption and its main reason, as well as the presence of fish with low or absent commercial value in the consumers diet (n = 155).

Questions	Survey Respondents (%)
1. Do you consider fish consumption to be important?	
Not very important	1%
Indifferent	1%
Important	26%
Very important	72%
1.1. If so, what are the main attributes you valorize in fish consumption?	
Fatty acids/omegas	79%
Protein	7%
Flavor	6%
Low-fat	6%
Low-sugar	1%
High metal content	1%
2. How often do you eat fish?	
<Once a month	1%
Two times a month	9%
Every week	44%
More than once a week	43%
Occasionally	3%

Table 2. *Cont.*

Questions	Survey Respondents (%)
2.1 Do you know of or consume any kind of fish with low or no commercial value?	
Yes	48%
No	22%
Maybe	30%

Table 3. Results of the survey concerning consumers' consumption habits of processed fish products (n = 155), the reason for their consumption (n = 117), their receptivity to these products' innovation (n = 155), what type of innovative products the consumer would like (n = 124), the significance of fish species valorization (n = 155), and the importance of new product formulation with their inclusion and promotion (n = 155).

Questions	Survey Respondents (%)
3. Do you consume processed fish?	
Yes	75%
No	25%
3.1 If so, why do you consume these processed foods?	
Practical and quick meals	38%
Tasty products	20%
Low-price	18%
Nutritionally rich	3%
Does not need to be cooked	10%
Long shelf-life	11%
4. Are you receptive to new processed fish products?	
Yes	86%
No	6%
Maybe	8%
4.1 If you answered yes in the previous answer, please indicate which ones:	
Oven products	40%
Ready-to-eat products	6%
Frozen products	3%
All previous answers	51%
5. Do you think it is important to valorize fish species without commercial value?	
Not important	1%
Little importance	1%
Indifferent	1%
Important	36%
Very important	61%
6. Do you consider it important that fish valorization be done through the formulation of new food products?	
Not important	1%
Little importance	1%
Indifferent	9%
Important	54%
Very important	35%
7. What is the importance of promoting this type of product that aims at the valorization of species without commercial value?	
Not important	1%
Little importance	0%
Indifferent	1%
Important	41%
Very important	57%

The consumers give a major importance to fish valorization (61%), especially through new food product formulation (54%) with proper promotion (57%) (Table 3). Although the number of respondents was not representative of the Portuguese population, this study allowed for a preliminary perception of their receptivity to new fish products and the expectations consumers have of these products.

The importance of ecological sustainability with respect to seafood was also reported [23], where more than half of the interviewees demonstrated good understanding about this topic and were willing to pay more to improve them.

In a systematic review the relation between fish consumption and nutritional benefits, especially omega-3 fatty acids, was also reported [24]. In addition, the authors considered a number of barriers to fish consumption, where the most important was related to sensory disliking of fish, lack of convenience, lack of self-efficacy in selecting and preparing fish, health risk concerns, lack of fish availability, and high prices [24]. These authors also reported that country and origin, production and preserving methods, product innovation, packaging, and eco-labelling were the most relevant fish attributes that affected consumers' choice [24]. Therefore, rapid and effective responses are needed to provide new insights for major actors (fishermen and policymakers) to improve the environmental sustainability of fish consumption [24].

3.2. Fish-Based Developed Food Products

The fish products developed during this study were black seabream ceviche, smoked blue jack mackerel pâté, dehydrated piper gurnard (chip-like treats), fried boarfish, and comber pastries (Figure 1). Black seabream ceviche, smoked blue jack mackerel pâté, and dehydrated piper gurnard are ready-to-eat products, whereas breaded boarfish and comber pastries are both products to be sold frozen and ready-to-fry.

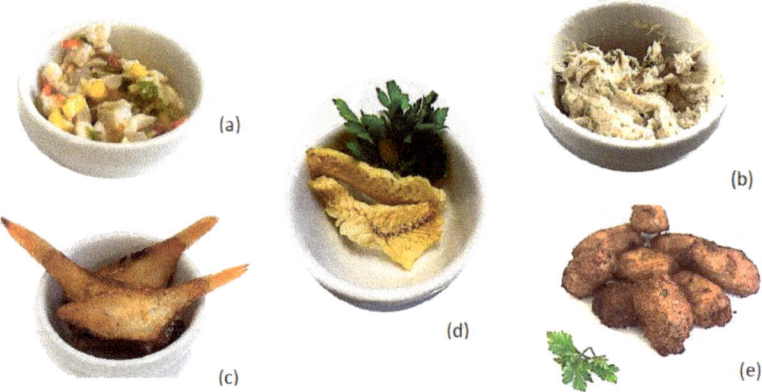

Figure 1. Reformulated fish products developed with each species. Legend: (**a**) black seabream ceviche, (**b**) smoked blue jack mackerel pâté, (**c**) fried boarfish, (**d**) dehydrated piper gurnard, and (**e**) comber pastries. Adapted from [1].

When developing new fish-based products, the season of capture may be an important factor as the fish may have different nutritional composition, many times related with the reproductive cycle, which influences the fat reserves of the fish and thus creates different sensory perceptions when consumed. Hence, for the product formulations, the previous work of Silva et al. [25], related to the sensorial characterization of fresh fish species throughout the year, was taken into consideration [25].

Ceviche development should be carried out with a fish species with cohesion, chewability, and stiffness, which give a favorable texture to the product. In addition, an ivory color is also an advantage to the product appearance, enabling the fish to be highlighted. Thus, considering our previous work [25], the black seabream should be caught during winter, when it is characterized by sea, butter, and seaweed odors, which are also good features for this product.

The blue jack mackerel, used in pâté formulation, should be caught during winter, especially due to the texture, flavor, and odor features related to that season, namely chewability, cohesion, and stiffness, as well as seaweed odor, sea odor, and flavor, which enhance the pâté sensory experience [25].

Piper gurnard should be caught during autumn due do its fat content, despite the absence of descriptors with statistical significance in our previous work [25]. In addition, piper gurnard captured in this season is characterized by a white color, contributing to a better fry product appearance [25].

The boarfish should be caught during winter due do its texture attributes, namely its cohesion, stiffness, and chewability properties, which allow texture maintenance after frying [25]. The comber fish was captured during winter due to its "ideal" stiffness, also being related to the laminar structures and sweet taste at the end of this season, which are essential to enhancing the fish flavor in the final product [25].

Regardless of the sensorial features, other factors should be considered to identify the most favorable season to catch the fish species under study, such as sexual cycle and abundance.

Fish is a food source with different beneficial nutrients for human health, including fatty acids, proteins, vitamins, and mineral elements [26]. However, food processing methods are related to nutrient loss in an extent that depends on, for example, temperature and pressure due to chemical and physical changes [27]. Therefore, when different ingredients are combined, there is a formulation enrichment that is a way of valorizing certain foods such as fish.

The valorization of a food ingredient through the development of new products or by the reformulation of existing ones allows the consumer to be more receptive to buying them, especially when the ingredient to be valorized is a discarded fish species with an uncommon aspect. Therefore, the development of the new fish products described in the present study was carried out considering the market study data, as well as the importance of using new fish species in familiar products. For example, the comber pastries were presented as a reformulation of a typical Portuguese product (cod pastries), adjusted for more traditional consumers. Additionally, the need to increase the snacks, salad ingredients, or side dishes offered (dehydrated piper gurnard) was also considered. Regardless of these needs, the formulation of dehydrated piper gurnard was intended to improve the traditional Portuguese processing of fish drying, in which salt is added and the drying process is performed by sun drying in open air, which is subject to all forms of contamination (e.g., pollution, flies). The need to increase the offer of ready-to-fry breaded products (boarfish) and fish-based pâtés (blue jack mackerel) was also considered.

Pâté is a value-added food that can be developed with meat (e.g., liver from goose or pig) or fish [28]. However, the fish species used in pâté formulation are often those with significant commercial value, promoting their overexploitation, which needs to be avoided [28]. In addition to using low-value fish species, the pâté developed in the present study also offers an advantage for consumers' health—the fish cooking process through smoking. Smoking is a type of preservation method that provides both antimicrobial smoke chemicals (e.g., formaldehydes, phenols) and reduced water activity, with the addition of giving an attractive color and flavor to the fish, improving the final product both nutritionally and sensorially [27].

The black seabream ceviche was developed to maintain the nutritional content in one of the developed fish products to adapt in new diets based on raw foods and new gastronomic trends.

According to the initial survey referred to in this study, the processed fish products are present in consumers' diets due to their practical and quick preparation. Therefore, three of the fish products developed in this study were ready-to-eat, and the boarfish and comber pastries only needed to be fried, reducing the consumer's time spent on their preparation.

3.3. New Fish-Based Food Product Acceptance—Hedonic Tests

3.3.1. Black Seabream Ceviche

Most of consumers who performed the hedonic tests for black seabream ceviche were women (71%), and 52% were aged between 31 and 45 (Figure 2b,c, respectively). Regarding this product, 33% and 32% of the consumers considered its appearance and color very pleasant, respectively, while 34% rated its odor as pleasant (Figure 2a). Regarding flavor, texture, and global appreciation, consumers found these three descriptors very pleasant (46%, 43%, and 47%, respectively) (Figure 2a). The data also showed that most of the consumers would probably (43%) or certainly (42%) buy it (Figure 2d).

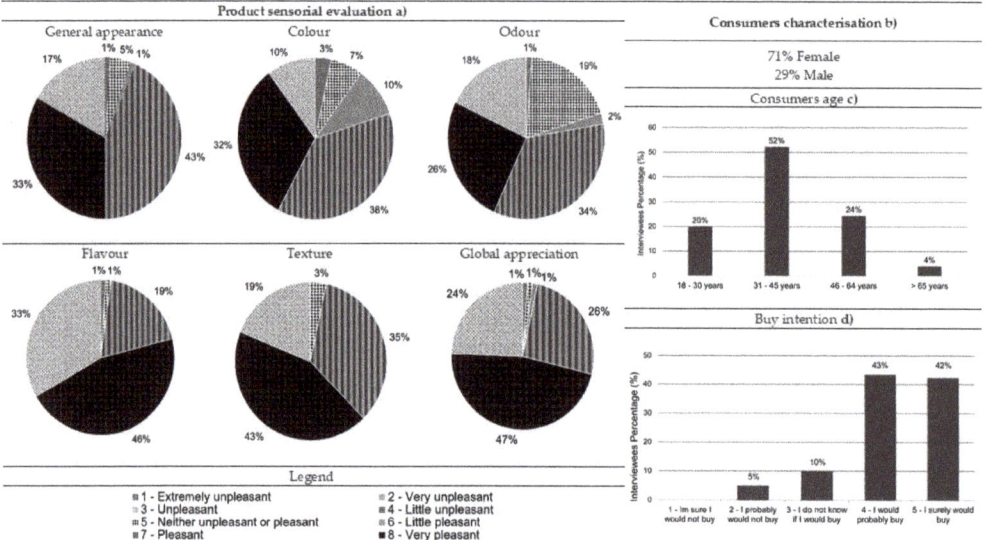

Figure 2. Results from the black seabream ceviche consumer acceptance survey: (**a**) sensorial evaluation (product's general appearance, colour, odour, flavour, texture and global appreciation), (**b**) consumers characterisation (gender percentage), (**c**) consumers age and (**d**) buy intention (n = 90).

After the validation of the sample adequation (KMO = 0.798, Bartlett, p-value < 0.001), it was clear that black seabream ceviche texture, global appreciation, and flavor were related to each other (Figure 3). These descriptors belonged to the first factor, which explained 62.61% of the data variability and was labelled as "touch, taste, and overall sensory experience" (Figure 3, Table 4). The second factor, which explained 16.65% of the total data variability, was represented by color, appearance, and odor, which had less significance for the products' description by consumers and was labelled as "sight and smell" (Figure 3, Table 4). These two groups characterized the product, although they did not have opposite behaviors (Figure 3). Consumers who rated texture, flavor and global appreciation with higher values considered these descriptors as the most important for black seabream ceviche characterization (Figure 3). Those who classified color with higher values also did so for appearance and odor (Figure 3).

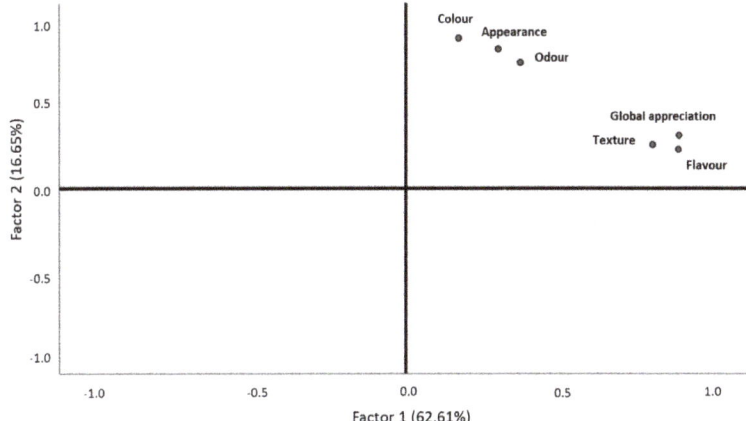

Figure 3. Behavior of black seabream ceviche descriptors, according to consumer classifications.

Table 4. Rotated component matrix for each black seabream ceviche descriptor.

Descriptor	Component	
	1	2
	Touch, Taste, and Overall Sensory Experience	Sight and Smell
Appearance	0.299	0.837
Color	0.172	0.901
Odor	0.370	0.755
Flavor	0.883	0.232
Texture	0.797	0.261
Global Appreciation	0.885	0.316

Extraction method; principal component analysis; rotation method; varimax with Kaiser normalization; rotation converged in three iterations.

Both genders of all ages had the same pattern regarding the descriptors with greater importance, such as texture, flavor, and odor, with these descriptors being decisive for the products' global appreciation by all consumers (Table 5).

In fact, flavor was one of the consumers' favorite attributes, as well as texture, with a "very pleasant" classification, followed by the appearance, color, and odor, classified as "pleasant" (Figure 2a). Black seabream ceviche was more accepted by women, who gave higher scores to flavor and purchase intention (middle rank = 48.83 and 48.82, respectively) than men (middle rank = 37.31 and 37.33, respectively; Mann–Whitney, p-value = 0.041 for both flavor and purchase intention). No statistical differences were reported when comparing age ranges (Kruskal—Wallis, p-value > 0.05).

3.3.2. Dehydrated Piper Gurnard

Most of the consumers who performed the hedonic testes for dehydrated piper gurnard were women (71%), and 52% were aged between 31 and 45 (Figure 4b,c, respectively). The results revealed that all the descriptors were mainly considered "pleasant" by the consumers, with 43% for dehydrated piper gurnard appearance and color, 39% for odor, 42% for flavor, 34% for texture, and 40% for its global appreciation (Figure 4a). Concerning the purchase intention, 47% of consumers would probably or surely buy this product if it were on the market (Figure 4d). The levels of acceptance of this product could be increased if the conservation methods were improved, since they were not optimized during this formulation, which may have led to a loss of "crunchiness" or a gain in humidity that impaired the sensory evaluation of the product.

Table 5. Correlation with Spearman's non-parametric test between all descriptors (appearance, color, odor, flavor, texture, and global appreciation) for each fish product developed ($n = 90$).

Population Group	Descriptor	Appearance	Color	Odor	Flavor	Texture	Global Appreciation
		\multicolumn{6}{c}{Black Seabream Ceviche}					
All	Appearance	-	0.000 *	0.000 *	0.000 *	0.000 *	0.000 *
	Color	0.000 *	-	0.000 *	0.000 *	0.000 *	0.000 *
	Odor	0.000 *	0.000 *	-	0.000 *	0.000 *	0.000 *
	Flavor	0.000 *	0.000 *	0.000 *	-	0.000 *	0.000 *
	Texture	0.000 *	0.000 *	0.000 *	0.000 *	-	0.000 *
	Global Appreciation	0.000 *	0.000 *	0.000 *	0.000 *	0.000 *	-
		Dehydrated Piper Gurnard					
All	Appearance	-	0.000 *	0.000 *	0.000 *	0.000 *	0.000 *
	Color	0.000 *	-	0.000 *	0.002 *	0.000 *	0.000 *
	Odor	0.000 *	0.000 *	-	0.026 *	0.566	0.013 *
	Flavor	0.000 *	0.000 *	0.026 *	-	0.000 *	0.000 *
	Texture	0.000 *	0.000 *	0.566	0.000 *	-	0.000 *
	Global Appreciation	0.000 *	0.000 *	0.013 *	0.000 *	0.000 *	-
		Fried Boarfish					
All	Appearance	-	0.000 *	0.000 *	0.000 *	0.000 *	0.000 *
	Color	0.000 *	-	0.000 *	0.000 *	0.000 *	0.000 *
	Odor	0.000 *	0.000 *	-	0.000 *	0.000 *	0.000 *
	Flavor	0.000 *	0.000 *	0.000 *	-	0.000 *	0.000 *
	Texture	0.000 *	0.000 *	0.000 *	0.000 *	-	0.000 *
	Global Appreciation	0.000 *	0.000 *	0.000 *	0.000 *	0.000 *	-
		Comber Pastries					
All	Appearance	-	0.000 *	0.000 *	0.004 *	0.009 *	0.001 *
	Color	0.000 *	-	0.000 *	0.000 *	0.000 *	0.000 *
	Odor	0.000 *	0.000 *	-	0.010 *	0.044 *	0.041 *
	Flavor	0.004 *	0.000 *	0.010 *	-	0.000 *	0.000 *
	Texture	0.009 *	0.000 *	0.044 *	0.000 *	-	0.000 *
	Global Appreciation	0.001 *	0.000 *	0.041 *	0.000 *	0.000 *	-
		Smoked Blue Jack Mackerel Pâté					
All	Appearance	-	0.000 *	0.000 *	0.000 *	0.000 *	0.000 *
	Color	0.000 *	-	0.000 *	0.000 *	0.000 *	0.000 *
	Odor	0.000 *	0.000 *	-	0.000 *	0.000 *	0.000 *
	Flavor	0.000 *	0.000 *	0.000 *	-	0.000 *	0.000 *
	Texture	0.000 *	0.000 *	0.000 *	0.000 *	-	0.000 *
	Global Appreciation	0.000 *	0.000 *	0.000 *	0.000 *	0.000 *	-

* Statistically significant correlation between descriptors; - not applicable.

After the validation of the sample adequation (KMO = 0.709, Bartlett, p-value < 0.001), the results revealed that dehydrated piper gurnard texture, global appreciation, and flavor were grouped together and belonged to the first factor, which explained most of the data variability (56.93%) and was labelled as "touch, taste, and overall sensory experience" (Figure 5, Table 6). The second factor, formed by odor, color, and appearance, explained 20.86% of the data variability and was labelled as "sight and smell" (Figure 5, Table 6). However, odor clearly differentiated itself from the other descriptors, assuming a differentiation role in this product (Figure 5). Therefore, these two groups were the ones that differentiated this product, although they did not have opposite behaviors (Figure 5). With this data, it is clear that those who classified texture with higher values also did so for overall appreciation and flavor, with these being the descriptors with most importance for consumers (Figure 5). Odor was the second most important descriptor, as it was the one that most contributed to the second axis characterization and so was responsible for another large part of the data variability (Figure 5). Those who classified color with higher values also did so for appearance (Figure 5). Not all descriptors related to each other with strong intensity, such as odor to flavor and global appreciation to odor, which correlated but with weaker intensity when compared to the other correlations (Table 5). The purchase intention was not statistically different among genders (Man–Whitney, p-value > 0.05) and age ranges (Kruskal–Wallis, p-value > 0.05).

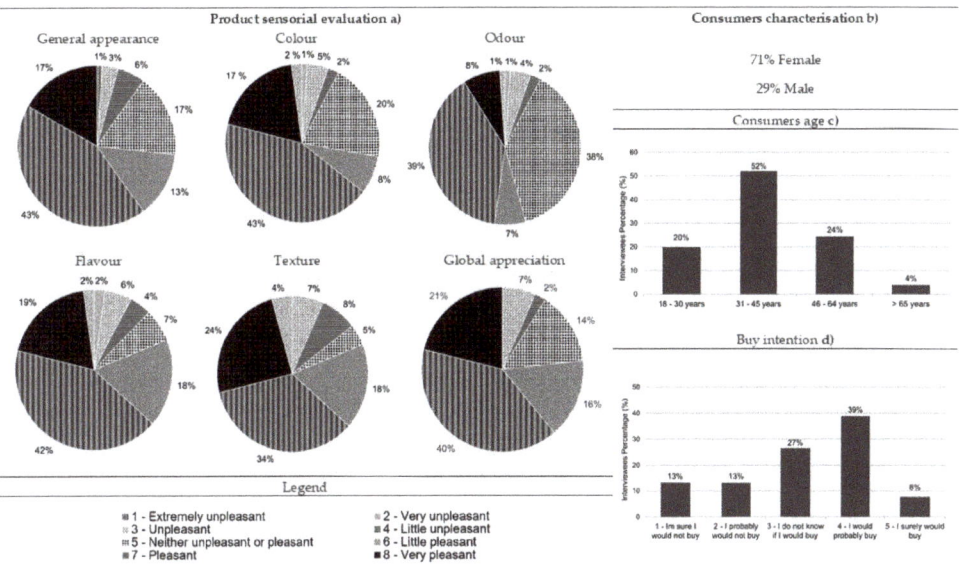

Figure 4. Results from the dehydrated piper gurnard consumer acceptance survey: (**a**) sensorial evaluation (product's general appearance, colour, odour, flavour, texture and global appreciation), (**b**) consumers characterisation (gender percentage), (**c**) consumers age and (**d**) buy intention (n = 90).

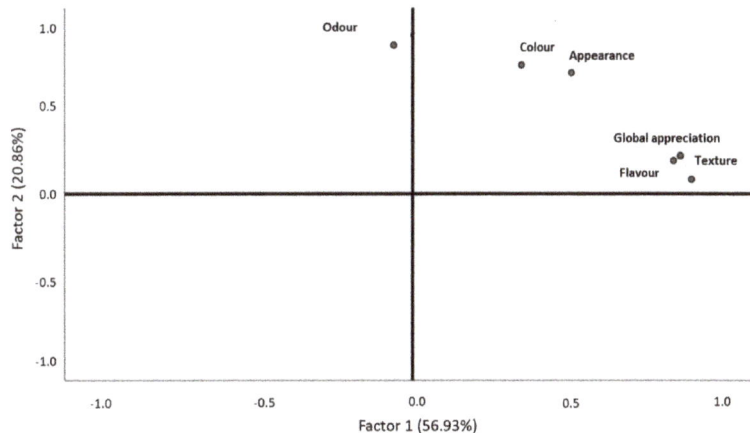

Figure 5. Behavior of dehydrated piper gurnard descriptors, according to consumer classifications.

Table 6. Rotated component matrix for each dehydrated piper gurnard descriptor.

Descriptor	Component	
	1	2
	Touch, Taste, and Overall Sensory Experience	Sight and Smell
Appearance	0.513	0.722
Color	0.349	0.769
Odor	−0.061	0.887
Flavor	0.845	0.200
Texture	0.902	0.086
Global Appreciation	0.867	0.228

Extraction method; principal component analysis; rotation method; varimax with Kaiser normalization; rotation converged in three iterations.

3.3.3. Fried Boarfish

Most of the consumers who performed the hedonic testes for fried boarfish were women (74%), and 66% were aged between 18 and 30 (Figure 6). Fried boarfish is a product that was well accepted by consumers as all the descriptors were considered essentially very nice or pleasant (Figure 6). The fried boarfish's general appearance was considered pleasant (49%), as were its color (56%), odor (53%), and texture (50%) (Figure 6). This product's flavor and global appreciation were mostly considered very pleasant (39% and 43%, respectively) (Figure 6). In addition, the data also showed that 50% of the consumers would probably buy it (Figure 6).

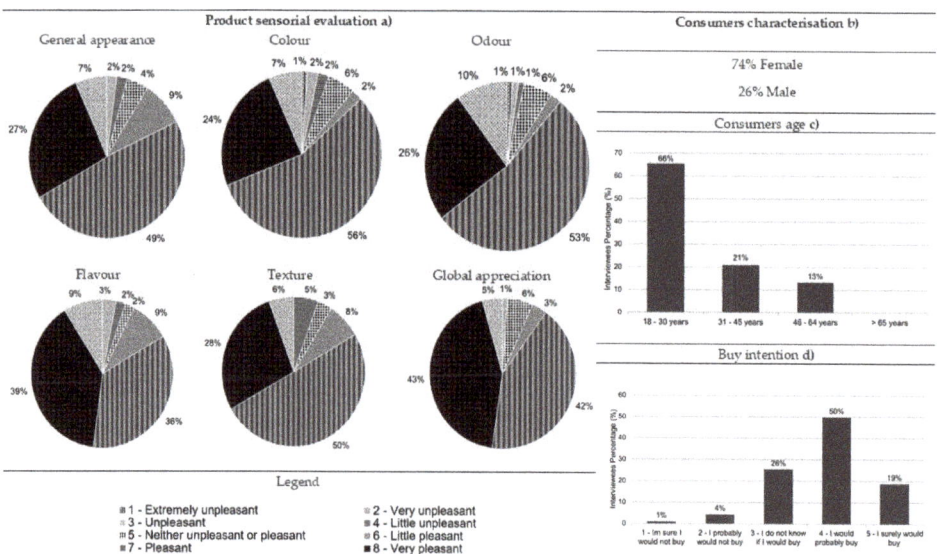

Figure 6. Results from the fried boarfish consumer acceptance survey: (**a**) sensorial evaluation (product's general appearance, colour, odour, flavour, texture and global appreciation), (**b**) consumers characterisation (gender percentage), (**c**) consumers age and (**d**) buy intention (n = 90).

After the validation of the sample adequacy (KMO = 0.859, Bartlett, p-value = 0.000), the results revealed that 62.54% of the data variability was explained by odor, texture, color, flavor, and global appreciation, which belonged to the first factor that was labelled as "all senses and overall sensory experience" (Figure 7, Table 7). The second factor was formed by appearance, which explained 11.08% of data variability and was labelled as "sight" (Figure 7, Table 7). Although the odor and texture were closer to the first factor, they were represented in the middle of the quadrant, revealing that they may not differentiate much either in terms of color, global appreciation, and flavor, or in terms of appearance (Figure 7). These two groups allowed for the differentiation of the fried boarfish, although they did not have opposite behaviors (Figure 7). Consumers who classified texture with higher values also did so for odor (Figure 7). Those who classified flavor with higher values also did so for color and global appreciation, these being the descriptors with most importance to consumers (Figure 7). The product's appearance had almost no association with the color, global appreciation, or flavor of the fried boarfish. Therefore, the sensory evaluation of the product's external aspect did not show any relevance to the evaluation of its color, global appreciation, or flavor (Figure 7). All the descriptors were statistically significant with intense correlation with one another (Table 5). No statistical differences were reported between descriptor classifications and purchase intention when comparing genders (Man–Whitney, p-value > 0.05). The purchase intention was statistically different between age ranges (Kruskal–Wallis, p-value = 0.037), where the older consumers liked the product more than the younger ones (Games–Howell, p-value = 0.005).

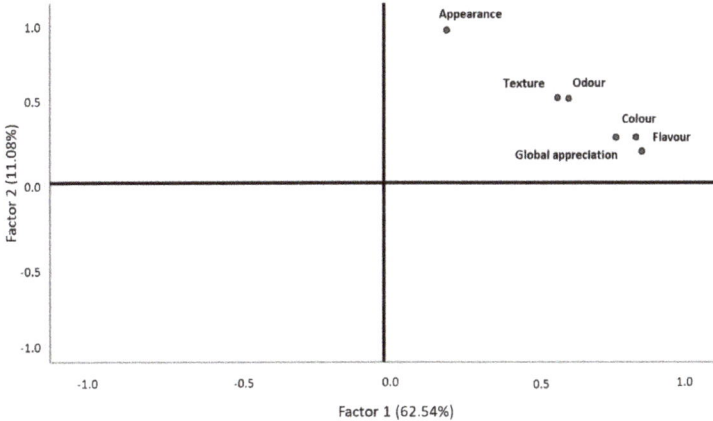

Figure 7. Behavior of fried boarfish descriptors, according to consumer classifications.

Table 7. Rotated component matrix for each fried boarfish descriptor.

Descriptor	Component	
	1	2
	All Senses and Overall Sensory Experience	Sight
Appearance	0.206	0.944
Colour	0.773	0.280
Odour	0.612	0.519
Flavour	0.859	0.194
Texture	0.574	0.525
Global appreciation	0.840	0.280

Extraction method; Principal component analysis; Rotation method; Varimax with Kaiser normalization; Rotation converged in 3 iterations.

3.3.4. Comber Pastries

Most of the consumers who performed the hedonic testes for comber pastries were women (71%), and 70% were aged between 18 and 30 (Figure 8b,c, respectively). In general, comber pastries were considered very pleasant by the consumers in a quantity of 49% for its general appearance, 42% for color, 53% for odor, 41% for flavor, 51% for texture, and

56% for global appreciation (Figure 8a). Regarding the purchase intention of this product, the results revealed that most of the consumers would probably or definitely buy it (50% and 38%, respectively) (Figure 8d).

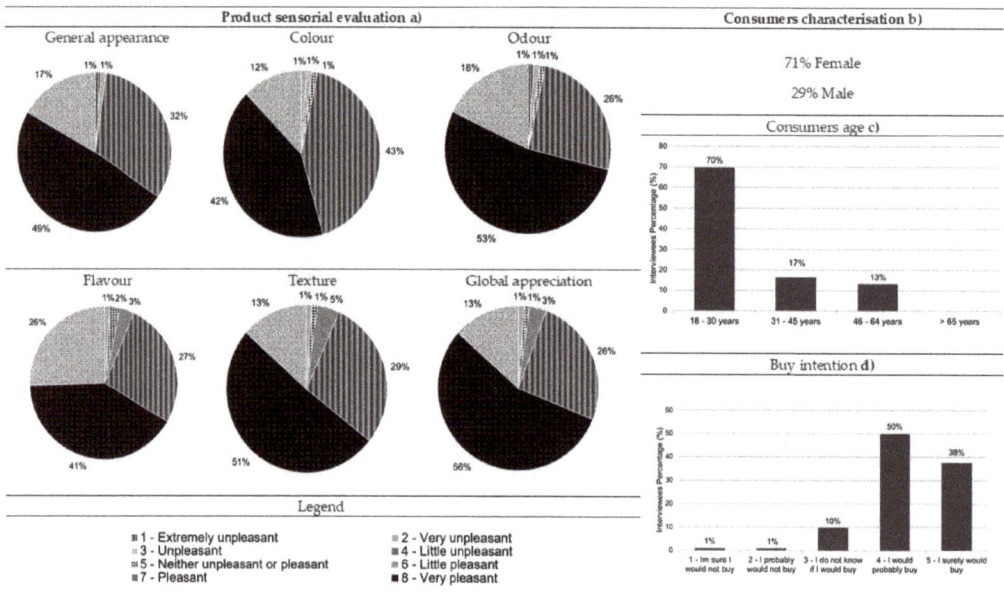

Figure 8. Results from the comber pastry consumer acceptance survey: (**a**) sensorial evaluation (product's general appearance, colour, odour, flavour, texture and global appreciation), (**b**) consumers characterisation (gender percentage), (**c**) consumers age and (**d**) buy intention (n = 90).

After the validation of the sample adequation (KMO = 0.799, Bartlett, p-value = 0.000), it was noticed that comber pastries' color, flavor, texture, and global appreciation were related to each other and helped to explain 62.40% of the data variability, belonging to the first factor that was labelled as "touch, taste, vision, and overall sensory experience" (Figure 9, Table 8). The second factor, which explained 19.47% of the data variability, is characterized by the odor and appearance and was labelled as "sight and smell" (Figure 9, Table 8). The consumers who ranked the appearance with higher values also did so for odor (Figure 9), whereas those who classified flavor with higher values also did so for texture, color, and global appreciation, with these being the descriptors with most importance for consumers (Figure 9). These two groups allowed for the differentiation of the comber pastries, although they did not reveal opposite behavior (Figure 9). The comber pastries' odor and appearance had no association, especially with the global appreciation and texture, revealing that the visual aspect and the olfactory senses were not relevant in the product's general evaluation and texture (Figure 9).

All the descriptors had a statistically significant correlation with each other (Table 5). Only the correlation between texture and odor, as well as between global appreciation and odor, showed an association with a lower intensity (Table 5). The purchase intention was not statistically different among genders (Mann–Whitney, p-value > 0.05) and age ranges (Kruskal–Wallis, p-value > 0.05).

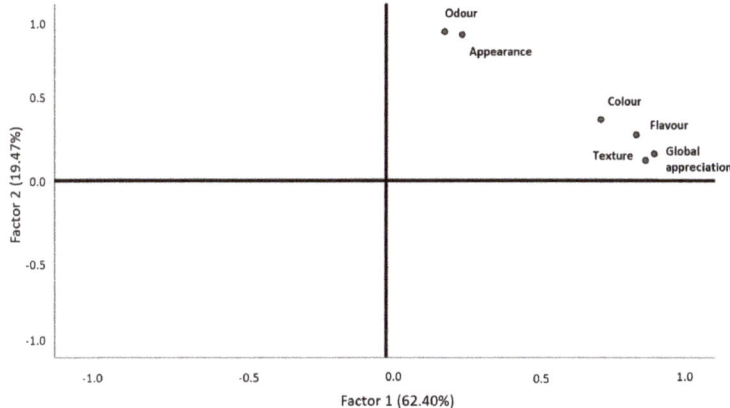

Figure 9. Behavior of comber pastry descriptors, according to consumer classifications.

Table 8. Rotated component matrix for each comber pastry descriptor.

Descriptor	Component	
	1	2
	Touch, Taste, Vision, and Overall Sensory Experience	Sight and Smell
Appearance	0.257	0.917
Color	0.723	0.384
Odor	0.197	0.935
Flavor	0.843	0.286
Texture	0.875	0.126
Global Appreciation	0.905	0.167

Extraction method; principal component analysis; rotation method; varimax with Kaiser normalization; rotation converged in three iterations.

3.3.5. Smoked Blue Jack Mackerel Pâté

Most of the consumers who performed the hedonic testes for smoked blue jack mackerel pâté were women (70%), and 53% were aged between 31 and 45 (Figure 10b,c, respectively). Overall, this product was very well accepted by consumers. Regarding general appearance and odor, these descriptors were mostly classified as "pleasant" (39% and 42%, respectively) (Figure 10a). Color, flavor, texture, and global appreciation were considered by consumers as "very pleasant" attributes for this product (38%, 38%, 38%, and 46%, respectively) (Figure 10a). If the product was on sale in the market, most consumers indicated that they would probably or definitely buy it (47% and 29%, respectively) (Figure 10d).

After the validation of the sample adequation (KMO = 0.795, Bartlett, p-value = 0.000), the results revealed that smoked blue jack mackerel pâté texture, global appreciation, appearance, color, and flavor were correlated with each other, characterizing the first factor that explained 61.23% of the data variability and was labelled as "touch, taste, sight, and overall sensory experience" (Figure 11, Table 9). The second factor, which explained 13.19% of the data variability, was characterized by odor and was labelled as "smell" (Figure 11, Table 9). Odor clearly differentiated itself from the other descriptors, assuming a differentiation role in smoked blue jack mackerel pâté (Figure 11). Therefore, these two groups were the ones that differentiated the product, although they did not have opposite behaviors (Figure 11). Consumers who classified texture with higher values also did so for global appreciation, appearance, color, and flavor, with these being the descriptors with most importance for consumers (Figure 11). Odor followed as the second most important descriptor, as it was the descriptor that most contributed for the second axis characterization and as such was responsible for another large part of the data variability (Figure 11). As mentioned for the remaining products with the exception of black seabream ceviche, odor revealed no association with other descriptors (Figure 11). In the case of the pâté, such a null association was found between its texture, appearance, and color, revealing that its visual aspect and texture did not show any relevance in the evaluation of the olfactory experience that the pâté offered to consumers (Figure 11). All the descriptors related to each other significantly and with strong intensity (Table 5). The purchase intention was not statistically different among genders (Mann–Whitney, p-value > 0.05) or age ranges (Kruskal–Wallis, p-value > 0.05).

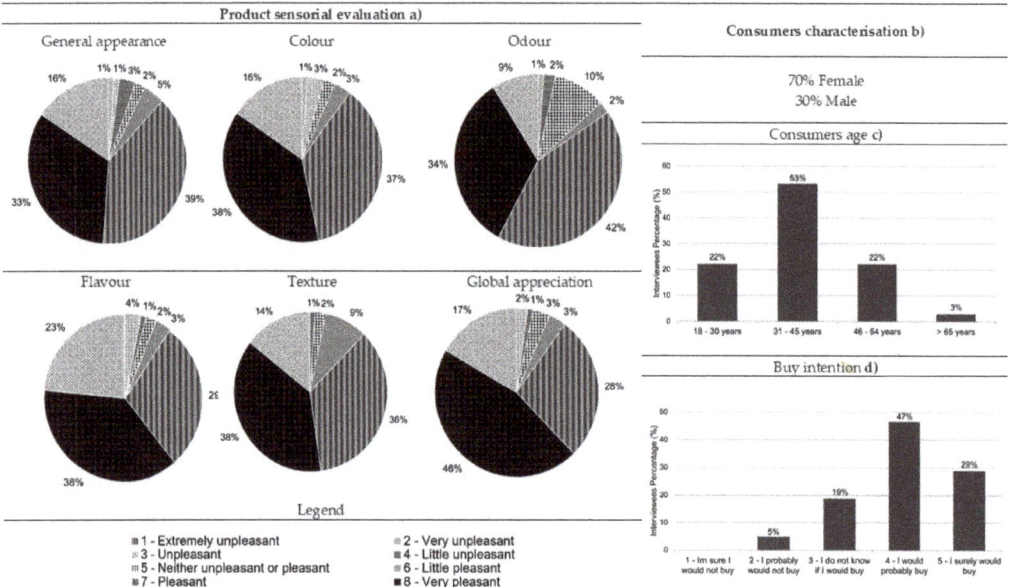

Figure 10. Results from the smoked blue jack mackerel pâté consumer acceptance survey: (**a**) sensorial evaluation (product's general appearance, colour, odour, flavour, texture and global appreciation), (**b**) consumers characterisation (gender percentage), (**c**) consumers age and (**d**) buy intention (n = 90).

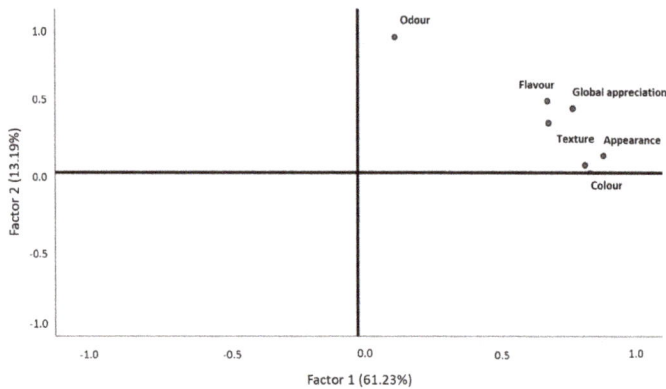

Figure 11. Behavior of smoked blue jack mackerel pâté descriptors, according to consumer classifications.

Table 9. Rotated component matrix for each smoked blue jack mackerel pâté descriptor.

Descriptor	Component	
	1	2
	Touch, Taste, Sight, and Overall Sensory Experience	Smell
Appearance	0.893	0.114
Color	0.826	0.050
Odor	0.135	0.920
Flavor	0.689	0.485
Texture	0.694	0.336
Global Appreciation	0.782	0.435

Extraction method; principal component analysis; rotation method; varimax with Kaiser normalization; rotation converged in three iterations.

3.3.6. Global Analysis

Considering the data as a whole, most of the developed food products showed statistically significant correlations between all the descriptors under study (Table 5). These results were expected considering that the perception and evaluation of food is an inherently multisensory experience, its pleasantness being influenced by food appearance, smell, taste, oral texture, and even by the sound that it makes in the mouth when it is eaten [29]. Therefore, the flavor perception is influenced by different factors including changes in viscosity, color cues, and interaction between oral texture and both olfactory and gustatory cues [29], whereas the color evaluation is an indicator of edibility, flavor identity, and intensity [30]. Therefore, the fish products developed in the present study had high acceptability by the consumers revealing their market potential. However, besides the "food-internal stimuli," which is when flavor and taste are influenced by appearance, texture, and other descriptors, there are also "food-external stimuli" [31]. This stimulus regards health information, societal influence, and availability of certain foods that can affect liking of food products [31]. Thus, considering that consumers increasingly seek healthy food without food additives that are both quick and easy to prepare (as concluded by the initial market survey of this investigation), the products developed may offer new alternatives in addition to those existing in the market. Additionally, fishing is almost an immemorial activity on the Portuguese coast, always linked to the geographical position of Portugal and its contours, where fish has an added importance due to the gastronomic heritage of this country [32]. However, the demand on fish supply has been rising due to the rapid growth of the world's population, favoring an imbalance with the demand of fishery products [33]. Consequently, the overexploitation or depletion of the world's fish stocks increases, revealing the need to search for available fish species underutilized in human food that can be a major source of nutrients, including lipids and fatty acids [33]. Therefore, considering our data from the hedonic evaluation of the fish products developed in the present study, these species can be an alternative to commercialized fish species. In addition, the fish species valorization would increase the fisherman's income, because fish that would be discarded on the high sea would have some commercial value. However, it must be noticed that the consumer dimension under study is not representative of the Portuguese population, and further studies on fish consumption and acceptance of reformulated products are necessary to allow the extrapolation of the data. Nonetheless, the present study provides some insight on this field and can be a valuable contribution for further studies on new fish product development and the characterization of fish consumption in the center of Portugal.

4. Conclusions

The present study allowed us to achieve great findings on interviewees' habits of fish consumption, as well as their marine sustainability conscience, namely on underexploited and underutilized fish valorization and the main attributes for their consumption. According to the initial survey data, it was also possible to understand the receptivity to new fish products, preferably frozen, ready to eat, or to be prepared in the oven, which need to be properly promoted. The knowledge on consumers' interests acquired with the data from the initial survey allowed for the development of new fish products, with the addition or substitution of the fish species under study through the reformulation of existing products that are familiar to the consumer. These products have a clean label, great acceptability, and purchase intention by the consumers. The results from hedonic tests revealed their potential application by companies to increase their revenue due to low or absent commercial value of discarded fish species. In fact, consumer evaluation revealed that they probably would buy all the developed products, with their global appreciation classified mainly as "very pleasant," except for the dehydrated piper gurnard, which was classified as "pleasant." In addition, the consumers favored the texture and flavor of the black seabream ceviche and dehydrated piper gurnard, with the addition of color for the comber pastries. The flavor and color of the fried boarfish were the most important features for consumers, with the addition of texture and appearance in the smoked blue jack mackerel pâté. Therefore, considering all our data and the need to avoid fish overexploitation, the species in this study revealed to be a great alternative to commercialized species commonly used in fish products. However, more research is needed to study Portuguese fish consumption, as well as their opinion about the inclusion of unexploited fish in their diet.

As future perspectives, it is intended to study and extend the shelf-life of the developed products through appropriate and sustainable packaging and maintaining the clean label.

Author Contributions: F.S., A.M.D., S.M., P.B. and M.M.G. contributed to the conceptualization and experimental design of the study; F.S. and A.M.D. undertook the study (investigation); F.S. and S.M. conducted the formal data analysis and presentation of the data; V.S., A.R.V., A.N., L.S.G., and E.M. contributed to sample collection; F.S. and A.M.D. wrote the original manuscript draft; reviews and editing of the completed manuscript were conducted by F.R.P., S.B., A.N., V.S., A.R.V., M.F.M., R.R., C.A., L.S.G., S.M., and M.M.G.; and project administration was the responsibility of M.M.G. All authors have read and agreed to the published version of the manuscript.

Funding: The authors are very grateful to the FCT (Fundação para a Ciência e a Tecnologia) for the financial support of this work through the project UID/MAR/04292/2020, attributed to MARE-Marine and Environmental Sciences Centre, Portugal. A.R.V and V.S. were supported by the FCT through CEECIND/01528/2017 and CEECIND/02705/2017, respectively. This work was also partially funded by the Integrated Programme of SR&TD "SmartBioR" (reference Centro-01-0145-FEDER-000018), co-funded by the Centro 2020 program, Portugal2020, European Union, through the European Regional Development Fund, and by project VALOREJET (16-01-03-FMP-0003), Operational Program Mar 2020, Portugal.

Conflicts of Interest: The authors declare no conflict of interest.

References

1. Barroso, S.; Pinto, F.R.; Silva, A.; Silva, F.; Duarte, A.M.; Gil, M.M. The Circular Economy Solution to Ocean Sustainability: Innovative Approaches for the Blue Economy. In *Mapping, Managing, and Crafting Sustainable Business Strategies for the Circular Economy*; Rodrigues, S.S., Almeida, P.J., Almeida, N.M., Eds.; IGI Global: Hershey, PA, USA, 2020; pp. 139–165.
2. Weissenberger, J. Discarding fish under the Common Fisheries Policy—Towards an end to mandated waste. In *Library Briefing: Library of the European Parliament*; European Parliament EP Library: Brussels, Belgium, 2013; pp. 1–6.
3. Blanco, M.; Domínguez-Timón, F.; Pérez-Martín, R.I.; Fraguas, J.; Ramos-Ariza, P.; Vázquez, J.A.; Borderías, A.J.; Moreno, H.M. Valorization of recurrently discarded fish species in trawler fisheries in North-West Spain. *J. Food Sci. Tech.* **2018**, *55*, 4477–4484. [CrossRef] [PubMed]
4. Burch, M.V.; Rigaud, A.; Binet, T.; Barthélemy, C. *Circular Economy in Fisheries and Aquaculture Areas—Guide #17*; FARNET: Brussels, Belgium, 2019.
5. Guillotreau, P.; Squires, D.; Sun, J.; Compeán, G.A. Local, regional and global markets: What drives the tuna fisheries? *Rev. Fish Biol. Fish.* **2016**, *27*, 909–929. [CrossRef]
6. Gamarro, E.G.; Orawattanamateekul, W.; Sentina, J.; Gopal, T.K.S. *By-Products of Tuna Processing*; GLOBEFISH Research Programme; FAO: Rome, Italy, 2013.

7. European Commission. The EU Fish Market 2019 Edition Is out: Everything You Wanted to Know about the EU Market for Fish and Seafood. Available online: https://ec.europa.eu/fisheries/press/eu-fish-market-2019-edition-out-everything-you-wanted-know-about-eu-market-fish-and-seafood_en (accessed on 24 March 2020).
8. Corallo, A.; Latino, M.E.; Menegoli, M.; Spennato, A. A survey to discover current food choice behaviors. *Sustainability* **2019**, *11*, 5041. [CrossRef]
9. Innario, M.; Manisera, M.; Piccolo, D.; Zuccolotto, P. Sensory analysis in the food industry as a tool for marketing decisions. *Adv. Data Anal. Classif.* **2012**, *6*, 303–321. [CrossRef]
10. Buttriss, J.L. Food reformulation: The challenges to the food industry. *Proc. Nutr. Soc.* **2013**, *72*, 61–69. [CrossRef]
11. Raaij, J.; Hendriksen, M.; Verhagen, H. Potential for improvement of population diet through reformulation of commonly eaten foods. *Public Health Nutr.* **2008**, *12*, 325–330.
12. Horvat, A.; Granato, G.; Fogliano, V.; Luning, P.A. Understanding consumer data use in new product development and the product life cycle in European food firms—An empirical study. *Food Qual. Prefer* **2019**, *76*, 20–32. [CrossRef]
13. FAO. *The State of World Fisheries and Aquaculture. Sustainability in Action*; FAO: Rome, Italy, 2020.
14. Erzini, K.; Gonçalves, J.M.S.; Bentes, L.; Moutopoulos, D.K.; Casal, J.A.H.; Soriguer, M.C.; Puente, E.; Errazkin, L.A.; Stergiou, K.I. Size selectivity of trammel nets in southern European small-scale fisheries. *Fish. Res.* **2006**, *79*, 183–201. [CrossRef]
15. Costa, M.E. Bycatch and Discards of Commercial Trawl Fisheries in the South Coast of Portugal. Ph.D. Thesis, University of Algarve, Faro, Portugal, 2014; 281p.
16. Tabachnick, B.G.; Fidell, L.S. *Using Multivariate Statistics*; Pearson Education Inc: Boston, MA, USA, 2007.
17. Henson, R.K.; Roberts, J.K. Use of Exploratory Factor Analysis in Published Research: Common Errors and Some Comment on Improved Practice. *Educ. Psychol. Meas.* **2006**, *66*, 393–416. [CrossRef]
18. Hogarty, K.; Hines, C.; Kromrey, J.; Ferron, J.; Mumford, K. The Quality of Factor Solutions in Exploratory Factor Analysis: The Influence of Sample Size, Communality, and Overdetermination. *Educ. Psychol. Meas.* **2005**, *65*, 202–226. [CrossRef]
19. Yong, A.G.; Pearce, S. A Beginner's Guide to Factor Analysis: Focusing on Exploratory Factor Analysis. *Tutor Quant Methods Psychol.* **2013**, *9*, 79–94. [CrossRef]
20. Ter Braak, C.; Smilauer, P. *CANOCO Reference Manual and CanoDraw for Windows User's Guide: Software for Canonical Community Ordination (Version 4.5)*; Microcomputer Power: Ithaca, NY, USA, 2002.
21. Bro, R.; Smilde, A.K. Principal component analysis. *Anal. Methods* **2014**, *6*, 2812–2831. [CrossRef]
22. Samoggia, A.; Castellini, A. Health-orientation and socio-demographic characteristics as determinants of fish consumption. *J. Int. Food Agribus. Mark.* **2017**, *30*, 1–16. [CrossRef]
23. McClenachan, L.; Dissanayake, S.T.M.; Chen, X. Fair trade fish: Consumer support for broader seafood sustainability. *Fish Fish.* **2016**, *17*, 1–14. [CrossRef]
24. Carlucci, D.; Nocella, G.; De Devitiis, B.; Viscecchia, R.; Bimbo, F.; Nardone, G. Consumer purchasing behaviour towards fish and seafood products. Patterns and insights from a sample of international studies. *Appetite* **2014**, *84*, 1–16. [CrossRef] [PubMed]
25. Silva, F.; Duarte, A.M.; Mendes, S.; Pinto, F.R.; Barroso, S.; Ganhão, R.; Gil, M.M. CATA vs. FCP for a rapid descriptive analysis in sensory characterization of fish. *J. Sens. Stud.* **2020**, *35*, e12605. [CrossRef]
26. Tilami, S.K.; Sampels, S. Nutritional Value of Fish: Lipids, Proteins, Vitamins, and Minerals. *Rev. Fish. Sci. Aquac.* **2017**, *26*, 1–11.
27. Abraha, B.; Admassu, H.; Mahmud, A.; Tsighe, N.; Shui, X.W.; Fang, Y. Effect of processing methods on nutritional and physico-chemical composition of fish: A review. *MOJ Food Process Technol.* **2018**, *6*, 376–382. [CrossRef]
28. Aquerreta, Y.; Astiasarán, I.; Mohino, A.; Bello, J. Composition of pâtés elaborated with mackerel flesh (*Scomber scombrus*) and tuna liver (*Thunnus thynnus*): Comparison with commercial fish pâtés. *Food Chem.* **2002**, *77*, 147–153. [CrossRef]
29. Zampini, M.; Sanabria, D.; Phillips, N.; Spence, C. The multisensory perception of flavor: Assessing the influence of color cues on flavor discrimination responses. *Food Qual. Prefer.* **2007**, *18*, 975–984. [CrossRef]
30. Shankar, M.U.; Levitan, C.A.; Prescott, J.; Spence, C. The influence of color and label information on flavor perception. *Chemosens. Percept.* **2009**, *2*, 53–58. [CrossRef]
31. Jiang, Y.; King, J.M.; Prinyawiwatkul, W. A review of measurement and relationships between food, eating behaviour and emotion. *Trends Food Sci. Technol.* **2014**, *36*, 15–28. [CrossRef]
32. Santos, M.P.N.; Seixas, S.; Aggio, R.B.M.; Hanazaki, N.; Costa, M.; Schiavetti, A.; Dias, J.A.; Azeiteiro, U.M. Fisheries as a human activity: Artisanal fisheries and sustainability. *J. Integr. Coast Zone Manag.* **2012**, *12*, 405–427.
33. Agh, N.; Jasour, M.S.; Noori, F. Potential development of value-added fishery products in underutilized and commercial fish species: Comparative study of lipid quality indicators. *J. Am. Oil Chem. Soc.* **2014**, *91*, 1171–1177. [CrossRef]

Article

Cricket-Enriched Oat Biscuit: Technological Analysis and Sensory Evaluation

Barbara Biró, Mária Anna Sipos, Anikó Kovács, Katalin Badak-Kerti, Klára Pásztor-Huszár and Attila Gere *

Institute of Food Technology, Faculty of Food Science, Szent István University, Villányi út 29-43, 1118 Budapest, Hungary; barbarabirophd@gmail.com (B.B.); siposa96@gmail.com (M.A.S.); Kovacs.Aniko94@szie.hu (A.K.); Badakne.Dr.Kerti.Katalin@szie.hu (K.B.-K.); Pasztorne.Huszar.Klara@szie.hu (K.P.-H.)
* Correspondence: gereattilaphd@gmail.com or gere.attila@szie.hu; Tel.: +36-20-278-6768

Received: 10 September 2020; Accepted: 21 October 2020; Published: 28 October 2020

Abstract: Insect-containing products are gaining more space in the market. Bakery products are one of the most promising since the added ground insects can enhance not only the nutritional quality of the dough, but technological parameters and sensory properties of the final products. In the present research, different amounts of ground *Acheta domesticus* (house cricket) were used to produce oat biscuits. Colour, hardness, and total titratable acidity (TTA) values were measured as well as a consumer sensory test was completed using the check-all-that-apply (CATA) method. An estimation of nutrient composition of the samples revealed that, according to the European Union's Regulation No. 1924/2006, the products with 10 and 15 g/100 g cricket enrichment (CP10 and CP15, respectively) can be labelled as protein sources. Results of the colour, TTA, and texture measurements showed that even small amounts of the cricket powder darkened the colour of the samples and increased their acidity, but did not influence the texture significantly. Among product-related check all that apply (CATA) attributes, fatty and cheesy flavour showed a significant positive effect on overall liking (OAL). On the other hand, burnt flavour and brown colour significantly decreased OAL. OAL values showed that consumers preferred the control product (CP0) and the product with 5 g/100 g cricket enrichment (CP5) samples over CP10 and rejected CP15.

Keywords: entomophagy; edible insect; novel food; check-all-that-apply method; oat biscuit

1. Introduction

One of the most serious problems of the 21th century is that global food security of the growing population cannot be assured due to the limited freshwater resources and available cultivable land [1,2]. According to the Food and Agricultural Organization of the United Nations' (FAO) 2019 report, the number of people suffering from hunger is about 820 million. Nutritional deficiencies mainly affects Africa, Southern and Western Asia, and Latin America [3]; however, other forms of malnutrition (overweight, obesity, and micronutrient deficiencies) are globally present, not just in the developing countries, but also in the developed countries. In total, about 2 billion people experience some level of food insecurity in the world. Consequently, it is necessary to focus not only on the supply of sufficient quantities of food, but also on the nutritional quality of the diet [4]. Dietary quality scores (DQSs) are used to evaluate the quality of diets. When DQSs are calculated, nutrients are often classified as qualifying or disqualifying, based on the dietary recommendations. According to these classifications, adequate intake of nutrients such as high-quality protein, dietary fibre, vitamin A, vitamin C, vitamin E, folate, calcium, and iron is considered to be qualifying. High saturated fat, added sugar, salt, and cholesterol content, as well as too low or too high energy intake, are classified as disqualifying. These factors are closely linked to undernutrition, obesity, and the development of non-communicable chronic diseases, such as cardiovascular diseases and type 2 diabetes mellitus [5,6].

According to several studies, the food industry is responsible for 26–50% of the greenhouse gas (GHG) emission, while agriculture is responsible for approximately 80% of the anthropological water footprint. High-quality protein production creates the greatest impact on the environment; since livestock breeding requires high amount of feed that is responsible for significant emission of GHGs, among others [7,8]. Based on FAO's statement development on dietary guidelines, a healthy but also sustainable diet contains less food of animal origin (meat, fish, and dairy products) and legumes, wholegrain products, and seeds are instead recommended. The consumption of foods high in fat, salt, and sugar should be very limited; and efforts should also be made to consume minimally processed foods [9].

Several studies from the past decade have focused on the role of edible insects since they can be one of the possible solutions to address the problem of food security. Entomophagy or "eating insects" was common in the prehistoric era and is still a part of the diet of certain areas of Latin America, Africa, and Asia [10,11]. More than 2000 edible species are known, including beetles, caterpillars, wasps, ants, and grasshoppers [12]. The nutritional quality of edible insects is satisfying; however, it depends on the species, the feed, and the stage of the development [13]. Generally, they contain high-quality protein, and their fatty acid composition resembles those of poultry and fish, but with higher amount of polyunsaturated type [14]. Their carbohydrate composition is diverse; due to their exoskeletal chitin content, they are considered as a good source of dietary fibre. Regarding the micronutrients, their vitamin B and E, iron, magnesium, and zinc content are remarkable [12,14–16]. Thus, insects contain most of the nutrients considered to be qualifying, and could be used successfully in malnutrition management [17].

Numerous studies have shown that the farming of edible insects has a lower environmental impact than that of other livestocks'. The land use of insect farming is significantly lower, as insects are suitable for indoor, urban, and vertical farming. Many types of agricultural, industrial, and household wastes can be recycled as insect feed, since they can digest forage higher in dietary fibre. Their GHG and ammonia emissions are minimal compared to beef or pork breeding. The required water for producing a unit amount of insect protein is a fragment of the amount used for the production of other types of animal proteins [11,18].

Using insects as feed or food has been proven to be beneficial from both nutritional and environmental point of view. Commercial farming and insect-based food production require a strict regulatory system, which is still being developed in the European Union. According to the European Food Safety Authority's (EFSA) 2015 scientific opinion, further data generation is required on the field of food safety, especially on microbiological and chemical hazards and allergenicity [19].

The consumer acceptance of insects as food depends on many influencing factors. Western people still have a negative attitude towards them; however, knowledge about edible insects and their nutritional value, previous taste experiences, and sensation seeking seem to increase acceptance. Many studies have proved that using insects in a non-visible form (e.g., insect powders) can increase the willingness to eat them. As a consequence, complementation of our everyday foods (e.g., bread, pasta, and other bakery and snack products), which are able to mask the insect component, can be the first step introducing insect-based foods to the market [20,21]. Several insect-containing food products have been described in the scientific literature recently, such as buckwheat pasta, breads, and energy bars [22–25].

Oat products are very popular nowadays since they are considered to be healthy. Oat *(Avena sativa)* is a cereal belonging to the *Poaceae* family. It has good nutritional qualities because of its relatively high protein, fibre, vitamin B, and mineral content [26]. The most important type of fibre in oats is the soluble β-glucan, which is a heterogeneous group of non-starch polysaccharides. Many studies have proved that this type of fibre has beneficial effects on diseases such as obesity, diabetes mellitus, hypertension, and dyslipidaemia, among others [27]. Shortbread-type biscuits are easy and simple to make, and many recipes are available that includes different type of cereal flours and also specifically oat flour.

The advantageous characteristics of insects and oat products as food ingredients can be mixed and a promising way of introducing insect-containing products would be the enrichment of oat biscuits with insect powders. Based on these, the aims of this study were to:

i. present the usability of different amounts of insect powder in a biscuit product,
ii. examine the technological effect of insect enrichment on the finished products, and
iii. discover how the insect content of the products affects the overall liking (OAL), and which attributes are the drivers of liking.

2. Materials and Methods

2.1. Materials

Four biscuits were created and evaluated, oat and buckwheat flours served as their base. The proportion of buckwheat flour in each flour mixtures was 20 g/100 g, which was determined by carrying out and assessing pre-tests and pre-tastings. The purpose of the addition of the buckwheat flour was to refine the colour differences between the samples without significantly affecting the taste. House cricket (*Acheta domesticus*) powder was added in different amounts as shown in Table 1. Oat flour and buckwheat flour (Első Pesti Malom- és Sütőipari Zrt. Dunaharaszti, Hungary and Bonetta Bt; Lajosmizse, Hungary) were commercially purchased from Hungarian producers, *Acheta domesticus* powder was purchased from JR Unique Foods Ltd. (Udon Thani, Thailand). The other ingredients of the doughs were: unsalted butter, sour cream with 12% fat content (Alföldi Tej Kft; Székesfehérvár, Hungary), baking powder (Dr. Oetker Magyarország Élelmiszer Kft, Budapest, Hungary) and salt. Lactose-free versions of dairy products were used to avoid losing participants due to lactose intolerance. The compositions of each sample are shown in Table 1. Table 2 contains the nutritional composition of the ingredients.

Table 1. The composition of the evaluated biscuits. The amount of ingredients is expressed per 100 g of flour mixtures.

Sample	*Acheta domesticus* * Powder (g)	Oat Flour (g)	Buckwheat Flour (g)	Butter (g)	Sour Cream (g)	Baking Powder (g)	Salt (g)
CP0	0	80	20	33.9	20.3	0.4	0.7
CP5	5	75	20	33.9	20.3	0.4	0.7
CP10	10	70	20	33.9	20.3	0.4	0.7
CP15	15	65	20	33.9	20.3	0.4	0.7

* house cricket, CP0—0 g/100 g house cricket containing flour mixture-based biscuit, CP5—5 g/100 g house cricket containing flour mixture-based biscuit, CP10—10 g/100 g house cricket containing flour mixture-based biscuit, CP15—15 g/100 g house cricket containing flour mixture-based biscuit.

Table 2. The energy value, macronutrient, and fibre composition of the used ingredients. Only the quantitatively relevant ingredients are listed. The nutritional values are given in their density (g or kcal) in 100 g of each ingredient.

Ingredient (100 g)	Energy (kcal)	Protein (g)	Carbohydrates (g)	Fat (g)	Fibre (g)
Cricket powder	457	67.8	5.5	18.2	0.5
Oat flour	370	14.0	55.0	8.0	11.0
Buckwheat flour	334	12.6	70.6	1.0	10.0
Lactose free butter	717	0.8	0.6	81.1	0.0
Lactose free sour cream (12% fat)	135	3.0	3.4	12.0	0.0

Shortbread type biscuit samples were made in laboratory conditions, each in a separate container. For the proper homogenization, batches of 300 g dough were made. After weighing the oat and buckwheat flour and the cricket powder, the four flour mixtures were prepared. Next, the mixtures were mixed and homogenized by kneading and adding the weighed and cut butter, sour cream, baking powder and salt. The homogenization was performed with a Bosch MUM4830 food processor (Robert Bosch Kft; Budapest, Hungary). The prepared dough was rolled out to 3 mm thickness layer, then 5 cm diameter circular pieces were cut out of it. Finally, the biscuits were baked for 10 minutes in a Sveba Dahlen S300 mini batch oven (Sveba Dahlen, Fristad, Sweden) preheated to 180 °C, with the fan function on. The weight of the biscuits was 5.87 ± 1.24 g per piece. The baked biscuits are shown in Figure 1.

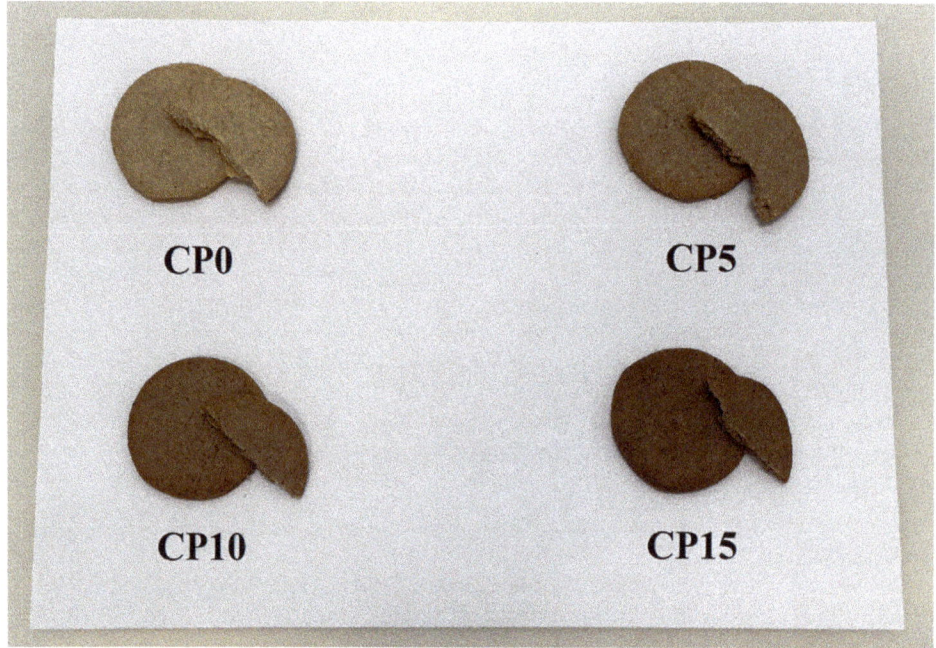

Figure 1. The prepared biscuits. CP0—0 g/100 g house cricket containing flour mixture-based biscuit, CP5—5 g/100 g house cricket containing flour mixture-based biscuit, CP10—10 g/100 g house cricket containing flour mixture-based biscuit, CP15—15 g/100 g house cricket containing flour mixture-based biscuit.

2.2. Methods

2.2.1. Estimated Nutritional Value

Our calculations were based on the nutritional values labeled on the packaging of the used ingredients. The energy values were calculated according to the European Union's Regulation No. 1169/2011 on the provision of food information to consumers. In this present calculation the used ingredients' protein, carbohydrate, and fat content were considered [28]. During the development of the products we worked with a mixture of oat and buckwheat flours. In 100 g flour mixture we replaced 5, 10 and 15 g of the oat flour with cricket powder, while the amount of buckwheat flour in each flour mixture was 20 g/100 g, since our aim was to study the effects of substituting cereal flour with insect powder. Due to the addition of the other ingredients, the amount of cricket powder in 100 *g* of the final products were as follows: CP0: 0 g, CP5: 3.24 g, CP10: 6.49 g, CP15: 9.73 g. Therefore, the products are referred to as "0, 5, 10, and 15 g/100 g house cricket containing flour mixture-based biscuits".

2.2.2. Technological Parameters and Quality Measurements

Colour Measurements

The colour of the baked samples was measured by using a Konica Minolta CR-310 Chroma Meter (Konica Minolta, Chiyoda, Tokyo, Japan). The colour parameters L^*, a^* and b^* are measures of lightness, redness/greenness and yellowness/blueness. The instrument was calibrated against a standard white tile ($L^* = 97.63$, $a^* = 0.78$ and $b^* = 0.25$). Since more biscuits of each batches were tested, the measurements were carried out in fourteen replicates.

Textural Hardness

Textural hardness was measured using a Stable Micro Systems TA.XT2i texture analyser (Stable Micro Systems, Godalming, United Kingdom) calibrated with a 2000 g load cell. For analysis, three-point bend rig was calibrated to a 10 mm height above the sample's surface and programmed to approach at 1 mm·s^{-1}. The samples were placed centrally over the supports. Upon contact with the surface, the probe pressed the samples at 3 mm·s^{-1} for 5 mm distance. The "resistance" of the samples against the moving probe was recorded. The measured maximum force value is referred to as the "hardness" of the sample. Measurements were carried out in fourteen replicates, since more biscuits of each batch were tested.

Total Titratable Acidity (TTA)

Total titratable acidity descirbes the total concentration of undissociated acids and free protons in the sample, that can react with a strong base. In bakery products, TTA is an indicator of microbial activity, since the moist dough is an adequate, nutrient rich media for acid-producing microorganisms [29].

Total titratable acidity was determined as described by Minervini, Lattanzi, De Angelis, Di Cagno, and Gobbetti [30]. 10 g of each biscuit sample was blended with 90 mL distilled water in a porcelain mortar. The suspensions were titrated to the final pH of 8.5 with 0.1 N NaOH. Changes of the pH value were followed by a HANNA Instruments Model pH 209 precision pH meter (HANNA Instruments, Woonsocket, RI, USA). The results are expressed as the amount of NaOH (ml) used. All measurements were carried out in triplicates.

2.2.3. Consumer Sensory Analysis

100 consumers were invited to the test, 67 of whom evaluated the prepared four samples in a one-week session. The assessors were students of Szent István University, Hungary. Gender ratio were 35.82%/64.18% males/females. The age of the participants ranged between 18 and 35 years, 67.16% of them were 18 to 23 years old, 55.22% of them live in the capital, Budapest and 38.81% of them have already tried insects, or insect-based food.

To ensure the reliability of the results, consumers were instructed prior to the evaluation. All participants were informed about the insect content of the samples in both verbal and written forms. A declaration of volunteering was also filled by the assessors, as insects can be allergenic and commercialisation of insect-based products as food is currently not permitted in Hungary.

Recommendations of Kilcast were followed during the sample presentation [31]. According to the international practice, the biscuit samples were labelled with 3-digit random numbers and a balanced block design was also applied for the test [32]. Each assessor was given one piece of biscuit per sample. The average height was 3.30 ± 0.12 mm, the average diameter was 48.12 ± 0.97 mm, while the average weight was 6.13 ± 0.84 g per biscuit. RedJade® software was used to conduct the test (RedJade Sensory Solutions, Martinez, CA, USA).

Check-All-That-Apply (CATA)

Check-all-that-apply questionnaires consist of multiple-choice questions. The participants are presented a list of phrases or words, of which they can select all, which they consider appropriate. Due to its different application possibilities, simplicity, fast execution and effectiveness, CATA experiments has become very popular in the last few years. Recently, the method is used for product sensory analysis with both trained panels and consumers, since the method is quick and easy, the task requires less cognitive effort from the assessor, and consumer-driven sensory characterization could be more useful for product optimization [33,34].

During the experiment, the assessors are asked to identify which sensory attribute from a given list is present in the sample [35]. The list of the attributes is usually built up from the sensory characteristics of the product, but can also contain hedonic terms, emotions, and non-sensory properties. The assessors can select without any constraint on the number of the given terms [33]. According to a 2013 study by Ares & Jaeger, the frequency of usage of the listed attributes can be increased by grouping them (e.g., flavour/taste related terms, odour related terms, etc.) and presenting them in a structural way from the appearance to the flavour [36].

The list of the used terms was compiled with a panel of 10 trained assessors on a consensus basis. The panel consisted of the trained individuals of the Sensory Laboratory, Szent István University, who are continuously trained according to the relevant ISO standard [37]. The panel received product specific training on insect-enriched food products by evaluations of prototype products. The terms are listed in Table 3.

Table 3. Attributes evaluated by consumers using CATA questions for sensory characterization of cricket enriched oat biscuits.

Product Property	CATA Terms
Appearance	too dark, too light, just-about-right colour, brown colour, grainy
Odour	too strong odour, too weak odour, cheesy odour, bitter odour, seedy odour, earthy odour, sunflower-seedy odour, toasty odour, pleasant odour, fishy odour
Texture	friable, hard, soft, crumbly, fatty, crispy, granular, dry, sticky
Flavour	too strong flavour, too weak flavour, cheesy flavour, seedy flavour, spicy flavour, salty taste, sunflower-seedy flavour, toasty flavour, tasty, sweet taste, piquant, fishy flavour, burnt flavour, long lasting taste

The data table of CATA analysis contains the sensory attributes in the columns the assessors and products in the rows. This is a binary table, where 1 means that the attribute is identified by an assessor, while 0 means that the attribute is not perceived by the assessor in the sample.

During the presented study, consumers were also asked to evaluate overall liking (OAL) on a 9-point hedonic scale (1 = "dislike extremely", 9 = "like extremely").

2.2.4. Data Analysis

Analysis of variance (ANOVA) and Tukey HSD *post hoc* tests were used to compare the means obtained during the technological evaluation and quality assessment of the samples, as well as to compare the consumer liking scores. CATA data analysis was done using multiple methods. Cohran's Q test was used to test the independence of products and attributes. Correspondence analysis (CA) was applied to visualize and interpret the products/attributes cross table, where the obtained inertia value indicates the quality of the analysis. Generally, higher inertia means higher quality analysis. Since our analysis contained overall liking assessment of the samples, principal coordinate analysis (PCoA) was run on the CATA and overall liking data. Since CATA variables and overall liking are measured on different scales (CATA—binary scale, overall liking—9-point category scale), they need to be transformed first. PCoA computes the chi-squared distance matrix of the CATA and overall liking variables, then centres the distance matrix by rows and columns, finally decomposes the eigenvalues of the centred distance matrix. This way PCoA enables us to visualize the CATA and overall liking variables on one plot where the distances among the variables express their similarities/dissimilarities [38].

In order to assess the attributes' impact on overall liking, mean drop analysis was also conducted. During mean drop analysis, the mean overall liking is computed for each attribute when the attribute is present (OALpres) and absent (OALabs). OALabs is deducted from OALpres which gives the mean impact of the attribute. If the mean impact is lower than 0, the attribute has a negative impact on overall liking. Similarly, a positive mean impact value indicates positive effect.

ANOVA, Cohran's Q test, CA, and PCoA were calculated using XL-Stat ver. 2019.2.2 (Addinsoft, 2019).

3. Results

3.1. Nutritional Values

Our calculations were based on the nutritional data labeled on the packaging of the used ingredients. Adding house cricket powder to the flour mixture increased the protein content from 9.48 g/100 g to 11.22 g/100 g in the case of 5 g/100 g enrichment (CP5), to 12.97 g/100 g at 10 g/100 g enrichment (CP10), and to 14.71 g/100 g at 15 g/100 g enrichment (CP15). Carbohydrate content was decreased from 38.27 g/100 g to 36.67 g/100 g in the case of CP5, 35.06 g/100 g in the case of CP10, and to 33.46 g/100 g in the case of CP15. The energy value was elevated 8.46 kcal, in the case of 15 g/100 g enrichment. The fat content shows a smaller difference (increased from 23.69 g/100 g to 24.68 in the case of CP15), while the fibre content decreased in proportion to the enrichment (decreased from 7.00 g/100 g to 5.98 g/100 g in the case of CP15) (Table 4).

Similar results were observed in the case of the protein content of wheat flour breads with house cricket and cinereous cockroach powder enrichment [24,25], rice flour cakes with Bombay locust powder enrichment [39], and buckwheat pasta with silkworm enrichment [23].

The reason of the decrease in fibre content is the decreased amount of oat flour, which has higher fibre content, than house cricket powder. The slightly increased fat content of the enriched samples is derived from the higher fat content of the insect powder. Similar results of change in fat content were presented in other studies, however, the fibre content of bread and snacks showed an increasing tendency in proportion to the insect enrichment [40,41]. This may be the consequence of the lower fibre content of the wheat flour, which is used as a control in other studies.

The European Union's Regulation No. 1924/2006 on nutrition and health claims made on foods states: 'a claim that a food is a source of protein, and any claim likely to have the same meaning for the consumer, may only be made where at least 12% of the energy value of the food is provided by protein' [42]. According to this Regulation, the products with higher enrichment can be labelled as protein source, since 12.77% (CP10) and 14.39% (CP15) of their energy value is provided by protein (Table 4).

Table 4. Biscuit compositions and calculated nutritional values. Amounts of each material are presented next to the name of the material. Nutritional values correspond to the quantity each sample contains. The table only lists the ingredients which contain energy-providing nutrients.

Sample	Ingredient	Amount (g/100 g)	Energy (kcal/100 g)	Protein (g/100 g)	Carbohydrates (g/100 g)	Fat (g/100 g)	Fibre (g/100 g)	Protein/Energy Value (%)
CP0	Cricket powder	0.00	0.00	0.00	0.00	0.00	0.00	9.46
	Oat flour	51.88	191.96	7.26	28.53	4.15	5.71	
	Buckwheat flour	12.97	43.32	1.63	9.16	0.13	1.30	
	Butter	21.98	157.63	0.19	0.13	17.83	0.00	
	Sour cream	13.16	17.77	0.39	0.45	1.58	0.00	
	Overall	100.00	410.68	9.48	38.27	23.69	7.00	
CP5	Cricket powder	3.24	14.82	2.20	0.18	0.59	0.02	11.13
	Oat flour	48.64	179.96	6.81	26.75	3.89	5.35	
	Buckwheat flour	12.97	43.32	1.63	9.16	0.13	1.30	
	Butter	21.98	157.63	0.19	0.13	17.83	0.00	
	Sour cream	13.16	17.77	0.39	0.45	1.58	0.00	
	Overall	100.00	413.50	11.22	36.67	24.02	6.66	
CP10	Cricket powder	6.49	29.64	4.40	0.36	1.18	0.03	12.77
	Oat flour	45.40	167.96	6.36	24.97	3.63	4.99	
	Buckwheat flour	12.97	43.32	1.63	9.16	0.13	1.30	
	Butter	21.98	157.63	0.19	0.13	17.83	0.00	
	Sour cream	13.16	17.77	0.39	0.45	1.58	0.00	
	Overall	100.00	416.32	12.97	35.06	24.35	6.32	
CP15	Cricket powder	9.73	44.46	6.60	0.54	1.77	0.05	14.39
	Oat flour	42.15	155.97	5.90	23.18	3.37	4.64	
	Buckwheat flour	12.97	43.32	1.63	9.16	0.13	1.30	
	Butter	21.98	157.63	0.19	0.13	17.83	0.00	
	Sour cream	13.16	17.77	0.39	0.45	1.58	0.00	
	Overall	100.00	419.14	14.71	33.46	24.68	5.98	

CP0—0 g/100 g house cricket containing flour mixture-based biscuit, CP5—5 g/100 g house cricket containing flour mixture-based biscuit, CP10—10 g/100 g house cricket containing flour mixture-based biscuit, CP15—15 g/100 g house cricket containing flour mixture-based biscuit.

3.2. Technological Parameters and Quality Measurements

3.2.1. Colour Measurements

Results of the colour measurement showed that even small amounts of insect enrichment influence the colour of the samples, as all colour measurement values changed proportionally with the amount of cricket powder. The L^* value of the control sample (CP0) was 63.50 ± 1.77, while CP15 showed 50.08 ± 0.73, indicating darkening effect. The a^* value increased from 7.92 ± 0.46 (CP0) to 9.53 ± 0.43 (CP15), which means that more redness appeared in the colour of the samples. Our results are in line with literature data, as the L^* value is also decreased and a^* value is also increased in the case of cockroach enriched bread products and locust enriched rice cakes [24,39]. Similarly to L^*, b^* values were also decreased: CP0 showed 25.47 ± 0.61, while CP15's value was 22.67 ± 0.76, which suggests that the samples became less yellow. According to earlier studies, the b^* value shows an increasing tendency in insect-enriched bakery goods [40]. Our results may differ since the base of the biscuit samples were oat and buckwheat flour. Consequently, the L^*, a^* and b^* values of the control sample were higher than the wheat flour-based products of the cited studies. Significant difference was found among L^* ($F(3,52) = 335.722$, $p < 0.0001$), a^* ($F(3,52) = 42.498$, $p < 0.0001$) and b^* ($F(3,52) = 73.224$, $p < 0.0001$) values of all samples. The results of the colour measurements are listed in Table 5.

Table 5. Means and standard deviations of the colour measurement parameters across samples.

Sample	L^*	a^*	b^*
CP0	63.50 ± 1.77 [a]	7.92 ± 0.46 [a]	25.47 ± 0.61 [a]
CP5	58.24 ± 0.61 [b]	8.66 ± 0.21 [b]	24.88 ± 0.23 [b]
CP10	53.74 ± 1.23 [c]	9.08 ± 0.40 [c]	23.76 ± 0.40 [c]
CP15	50.08 ± 0.73 [d]	9.53 ± 0.43 [d]	22.67 ± 0.76 [d]

L^* value stands for the lightness from black (0) to white (100), a^* value stands for from green (−) to red (+), and b^* value stands for from blue (−) to yellow (+). Superscript letters denote homogenous subgroups defined by Tukey HSD *post hoc* test. OAL denotes overall liking. CP0—0 g/100 g house cricket containing flour mixture-based biscuit, CP5—5 g/100 g house cricket containing flour mixture-based biscuit, CP10—10 g/100 g house cricket containing flour mixture-based biscuit, CP15—15 g/100 g house cricket containing flour mixture-based biscuit.

3.2.2. Hardness

According to the results of hardness, the added amounts of cricket powder do not significantly influence the hardness of the samples ($F(3,52) = 0.887$, $p = 0.454$), however, an increasing tendency of these values is observable in proportion to the cricket powder content. The hardness (maximum force value of the resistance of the pressed probe) of the control sample (CP0) was 160.21 ± 43.90 N. In the case of CP5, CP10 and CP15 the values were 164.62 ± 44.16 N, 185.03 ± 69.48 N, and 183.72 ± 40.79 N, respectively. Other studies have also found that textural hardness of bread products increased with the addition of insect powder [40,43], but showed a decreasing tendency in the case of rice-flour cakes [39].

3.2.3. Total Titratable Acidity (TTA)

TTA showed an increasing tendency in proportion to the amount of house cricket powder in the samples. Significant differences were found among the TTA of the four samples ($F(3,4) = 187.763$, $p < 0.01$). The acidity of the control sample (CP0) was 9.95 ± 0.35, which increased to 13 ± 0.14 in CP5. In the case of CP10 and CP15, TTA were 15 ± 0.42 and 17.65 ± 0.35, respectively. Similar results were presented in other studies on snacks enriched with lesser mealworm and breads enriched with house cricket [25,41].

3.3. Consumer Sensory Analysis

3.3.1. Liking Variables

One-way analysis of variance of the liking variables revealed that significant difference exists among samples for colour, flavour, and overall liking (OAL). The addition of cricket powder significantly changed the colour of the samples, making them darker, redder and less yellow, which was confirmed by the results of colour liking values. Colour of CP5 received the highest liking values, which was not significantly different from that of CP0. Flavour of CP0 and CP5 were rated similarly, indicating that the 5 g/100 g substitution of cricket powder has no effect on the consumer's opinion. On the other hand, CP10 and CP15 received low flavour liking values. Regarding overall liking, consumers preferred CP0 and CP5 and clearly rejected CP15. Odour and texture liking were not affected by the addition of different amounts of cricket powder, the latter is supported by the fact that no significant differences were found among the hardness values of three samples. Other studies showed corresponding results. In the case of biscuits fortified with edible termites, the control sample received the highest overall acceptability, and the panel preferred the 5% insect-containing biscuit among the enriched products. Also, no significant differences were found among the liking variables of the texture. Among the colour of the products, the colour of the sample with 5% insect content received the highest liking values, as well as in our study [44].

10% enrichment was preferred in the case of buckwheat-pasta enriched with silkworm powder, bread enriched with grasshopper powder and bread enriched with cricket powder. Furthermore, bread enriched with 10% flour from cinereous cockroach showed no significant differences from the control [24,25,43]. The results of the consumer sensory analysis are shown in Table 6.

Table 6. Means and standard deviations of the liking variables across samples.

Sample	Colour	Odour	Texture	Flavour	OAL
CP0	7.03 ± 1.78 [ab]	6.46 ± 1.58 [a]	5.93 ± 1.74 [a]	6.55 ± 1.86 [a]	6.57 ± 1.71 [a]
CP5	7.48 ± 1.43 [a]	7.09 ± 1.53 [a]	5.87 ± 1.97 [a]	6.27 ± 2.12 [a]	6.42 ± 1.88 [ab]
CP10	6.42 ± 1.68 [b]	6.55 ± 1.71 [a]	5.93 ± 1.72 [a]	5.21 ± 2.12 [b]	5.49 ± 1.94 [bc]
CP15	5.33 ± 2.00 [c]	6.10 ± 1.93 [a]	5.54 ± 1.92 [a]	4.70 ± 2.40 [b]	4.78 ± 2.04 [c]

Bold indicates significant differences among samples and/or clusters defined by analysis of variance ($p < 0.05$). Superscript letters denote homogenous subgroups defined by Tukey HSD *post hoc* test. OAL denotes overall liking. CP0—0 g/100 g house cricket containing flour mixture-based biscuit, CP5—5 g/100 g house cricket containing flour mixture-based biscuit, CP10—10 g/100 g house cricket containing flour mixture-based biscuit, CP15—15 g/100 g house cricket containing flour mixture-based biscuit.

3.3.2. Check-all-that-apply (CATA)

CATA questionnaire consisted of 38 terms, of which the ten most frequently marked were Crumbly, Friable, Pleasant odour, Long lasting taste, Just-about-right colour, Brown colour, Sticky, Salty taste, Dry and Seedy flavour. The least used five terms were Hard, Spicy flavour, Fishy odour, Sweet taste and Fishy flavour. In order to get more accurate results, the attributes that do not differentiate the samples significantly were filtered out based on the Cochran's Q test. Henceforth, only the remaining 25 significant properties will be used in the analysis. Table 7 shows the frequencies of marking of these attributes in the case of all samples.

Table 7. Frequency of marking of the CATA terms for all four biscuit samples.

Attribute	Marked as Present	Attribute	Marked as Present
Crumbly	196	**Too dark**	67
Friable	182	**Too weak odour**	67
Pleasant odour	136	Sunflower seedy odour	65
Long lasting taste	133	Granular	64
Just-about-right colour	132	Tasty	61
Brown colour	124	**Too strong flavour**	59
Sticky	123	Sunflower seedy flavour	59
Salty taste	115	Crispy	36
Dry	104	**Earthy odour**	33
Seedy flavour	94	**Bitter odour**	27
Cheesy odour	92	**Too light**	26
Grainy	90	**Too weak flavour**	23
Seedy odour	86	Piquant	20
Soft	86	**Too strong odour**	18
Fatty	85	**Fishy flavour**	18
Toasty odour	77	Sweet taste	14
Cheesy flavour	76	**Fishy odour**	13
Burnt flavour	75	Spicy flavour	11
Toasty flavour	68	Hard	8

Attributes that differentiate the samples significantly are highlighted in bold (Cochran's Q test).

According to the correspondence analysis (Figure 2a), the assessors associated different attributes with each sample. CP0 was divisive, as the consumers marked that the flavour and odour of the sample were also not good enough and just-about-right. CP5 was the most liked insect enriched biscuit, which is reflected in this analysis, as *Just-about-right colour*, *Tasty* and *Pleasant odour* attributes are close to the sample, as well as *Soft* and *Fatty* textures. *Grainy appearance* and *Granular texture*, *Brown colour*, *Toasty odour* and *Long lasting taste* appear along with samples CP10 and CP15. Negative properties such as *Burnt odour*, *Too dark colour*, *Too strong taste* and *Too strong odour* were more associated with these samples. Among the animal notes, *Cheese flavour* was mostly marked in the case of CP5, while *Fishy odour* and *Fishy flavour* in the case of CP10. *Hard texture* and *Earthy odour* were chosen by the assessors to describe CP15.

Since principal coordinate analysis (Figure 2b) visualizes the overall liking data (OAL) and the marked CATA terms together, drivers of liking can be easily defined. Hedonic terms with positive meaning *(Tasty, Just-about-right colour, Pleasant odour)* are close to OAL, such as *Toasty odour*, *Friable* texture, and *Cheesy flavour*. Some of the negative meaning hedonic terms *(Too weak odour, Too light, Too weak flavour)* are also close, which means that the assessors gave higher OAL scores when some properties were "not enough", than when they were "too much". *Hard* and *Granular* textures are on the opposite side of OAL, just as *Too dark* colour and *Brown colour*, *Too strong odour*, *Earthy* and *Fishy odour*, *Burnt* and *Fishy flavour*, which means that these attributes were less liked in the products. It can be observed that attributes close to OAL were, according to the correspondence analysis, characteristics of the CP0 and CP5 samples, while the more distant ones were more typical in the case of CP10 and CP15 samples. This confirms the result of the analysis of the liking variables, which showed that CP0 and CP5 were more liked.

Figure 2c presents the results of mean drop analysis of the CATA attributes. Highest mean impact was observed in the case of *Tasty*, however, only 23% of consumers marked as present. *Friable* was rated more than 68% of the consumers, which means strong consensus. Among product-related attributes, *Fatty* and *Cheesy flavour* showed significant positive effect on overall liking. On the other side, *Burnt flavour* and *Brown colour* showed significant negative impact on overall liking.

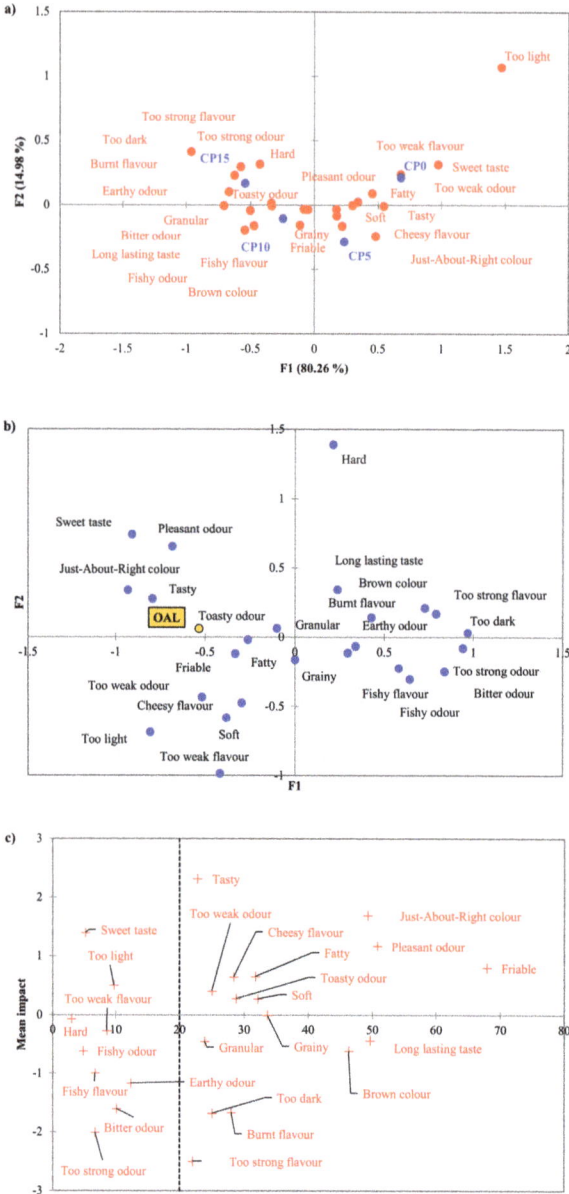

Figure 2. Visualized results of the check-all-that-apply (CATA) analysis of the four biscuit samples. (**a**) Correspondence analysis of the used CATA terms and the four samples. (**b**) Principal Coordinate Analysis of the used CATA terms and Overall Liking (OAL) scores. (**c**) The mean impact of the marked attributes on overall liking, visualized with the percentage of consumer who marked those attributes (dashed line). CP0—0 g/100 g house cricket containing flour mixture-based biscuit, CP5—5 g/100 g house cricket containing flour mixture-based biscuit, CP10—10 g/100 g house cricket containing flour mixture-based biscuit, CP15—15 g/100 g house cricket containing flour mixture-based biscuit. OAL stands for overall liking. Only those 25 CATA terms are shown that significantly differentiate the four samples.

Figure 3 presents separate PCoAs of the four samples. *Too light* attribute is only marked in the case of CP0 (Figure 3a), while *Just-about-right colour* is located near to OAL in the case of CP5 (Figure 3b) and CP10 (Figure 3c). These suggest, that slightly darker colour enhances, while too dark colour decreases OAL. *Tasty* attribute is considered as a driver of liking, since it goes along OAL in every case. However, CP15 (Figure 3d) shows that higher amount of insect enrichment pushes *Tasty* further from OAL. *Sweet taste* is located close to OAL when no enrichment is done, while OAL of CP5 and CP10 is less influenced by the attribute and no participant marked *Sweet taste* while testing CP15. *Too dark* colour and *Burnt flavour* are far away from OAL, meaning these attributes have a decreasing effect on OAL.

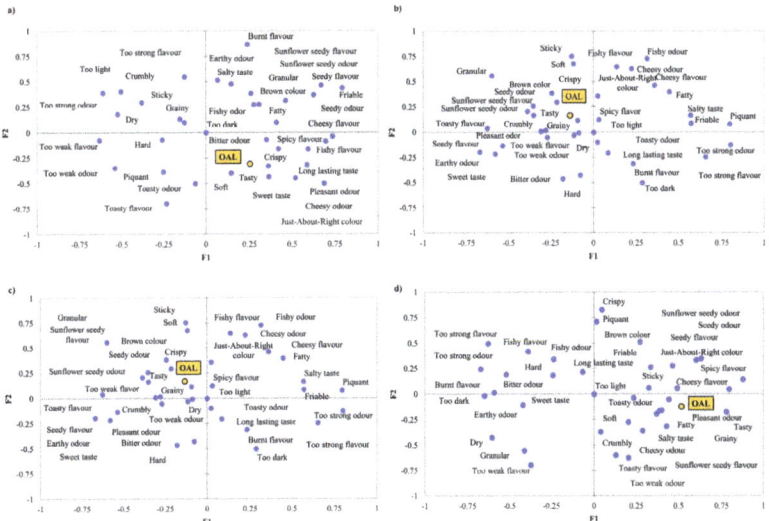

Figure 3. Principal coordinate analysis (PCoA) plots of the four samples. (**a**) CP0—0 g/100 g house cricket containing flour mixture-based biscuit, (**b**) CP5—5 g/100 g house cricket containing flour mixture-based biscuit, (**c**) CP10—10 g/100 g house cricket containing flour mixture-based biscuit, (**d**) CP15—15 g/100 g house cricket containing flour mixture-based biscuit. OAL stands for overall liking.

Literature data is very limited on sensory evaluation of insect-enriched products performed with check-all-that-apply method. Mealworm-containing meatballs and dairy drink products were tested with CATA analysis; however, the attributes of these products are difficult to compare with the biscuits' we developed. Nevertheless, with the addition of ground insects, a few similar properties appeared as in the case of fortified biscuits, e.g., grainy, sticky, and dry [45].

Comparing our study to the international literature, this is the first research which used oat and buckwheat flour as a base of cricket enriched biscuits. These flours have better nutritional characteristics in terms of high protein, higher dietary fibre, vitamin B and mineral content. In the case of functional foods, the acceptance of consumers is higher if the products are considered healthy [46]. However, our products cannot be considered as functional foods, consumer behavior might be similar. Nevertheless, oat and buckwheat can serve as bases of gluten free products. Our results support the evidence that pairing insects with these flours is viable options in order to develop novel gluten free products that could gain the acceptance of consumers [47,48].

From the methodological point of view, there is no existing study, which used check-all-that-apply analysis on ground-insect enriched products. As a result, our study provides a set of CATA descriptors, which can be applied in future studies.

4. Conclusions

The obtained results suggest that samples containing 10 g/100 g flour mixture (CP10) and 15 g/100 g flour mixture (CP15) *Acheta domesticus* powder can be labelled as protein source based on the corresponding EU regulation. However, consumer sensory analysis revealed that CP10 and CP15 were significantly less liked compared with the control and 5 g insect/100 g flour mixture (CP5) based sample. The rejection can be attributed to the changes in appearance and not due to changes of textural attributes, as the technological analysis suggested. The major factors of rejection were *Brown colour* and *Burnt flavour*; hence further product developments should address these issues.

Literature data suggest that consumers prefer insect containing products at different levels, however, there are limited results available about biscuits, since the majority of the publications have focused on other bakery products.

Our results raise the attention of policy makers and producers to the fact that insects enhance the nutritional quality of bakery products even if they are made from gluten free cereals and/or pseudocereals.

Limitations of our study are the lack of representative sampling; however, it is still in line with Næs' recommendation [49]. According to a 2017 study, Hungarian consumers show slight rejection to insects as food, therefore, these results should not be generalized [50].

Further analysis should be carried out to test the effect of different species on the sensory attributes of insect enriched bakery products. Sensory attributes–therefore acceptance–might also be influenced by different base materials (e.g., flour types and mixtures), spices (salted or sweet products) and processing technologies (e.g., drying, frying, cooking, baking).

Author Contributions: Conceptualization, B.B., M.A.S. and A.G; methodology, B.B., K.B.-K. and A.G.; software, B.B.; validation and A.G.; formal analysis, B.B., M.A.S. and K.P.-H.; investigation, B.B. and A.K.; resources, K.P.-H., K.B.-K. and A.G.; data curation, B.B.; writing—original draft preparation, B.B.; writing—review and editing, B.B., K.P.-H., K.B.-K. and A.G.; visualization, B.B.; supervision, K.P.-H., K.B.-K. and A.G.; project administration, A.G.; funding acquisition, A.G. All authors have read and agreed to the published version of the manuscript.

Funding: The Project is supported by the European Union and co-financed by the European Social Fund (grant agreement no. EFOP-3.6.3-VEKOP-16-2017-00005).

Acknowledgments: Barbara Biró and Anikó Kovács thank the support of Doctoral School of Food Sciences, Szent István University. Attila Gere thanks the support of the Premium Postdoctoral Research Program of the Hungarian Academy of Sciences and the support of National Research, Development and Innovation Office of Hungary (OTKA, contracts No. K134260). Barbara Biró was supported by the ÚNKP-20-3-2-SZIE-23 New National Excellence Program of the Ministry for Innovation and Technology. The authors thank Orsolya Tompa for the proofreading.

Conflicts of Interest: The authors declare no conflict of interest.

References

1. Fan, S.; Brzeska, J. Sustainable food security and nutrition: Demystifying conventional beliefs. *Glob. Food Secur.* **2016**, *11*, 11–16. [CrossRef]
2. Meyer-Rochow, V.B. Can Insects Help to Ease the Problem of World Food Shortage? *Search* **1975**, *6*, 261–262.
3. FAO; IFAD; UNICEF; WFP; WHO. *The State of Food Security and Nutrition in the World. Safeguarding Against Economic Slowdowns and Downturns*; Food and Agriculture Organization of the United Nations: Rome, Italy, 2019.
4. Webb, P.; Stordalen, G.A.; Singh, S.; Wijesinha-Bettoni, R.; Shetty, P.; Lartey, A. Hunger and malnutrition in the 21st century. *BMJ* **2018**, *361*, k2238. [CrossRef] [PubMed]
5. Hallström, E.; Davis, J.; Woodhouse, A.; Sonesson, U. Using dietary quality scores to assess sustainability of food products and human diets: A systematic review. *Ecol. Indic.* **2018**, *93*, 219–230. [CrossRef]
6. Khatib, O. Noncommunicable diseases: Risk factors and regional strategies for prevention and care. *East. Mediterr. Health J.* **2004**, *10*, 778–788. [PubMed]

7. Poore, J.; Nemecek, T. Reducing food's environmental impacts through producers and consumers. *Science* **2018**, *360*, 987–992. [CrossRef]
8. Smetana, S.; Palanisamy, M.; Mathys, A.; Heinz, V. Sustainability of insect use for feed and food: Life Cycle Assessment perspective. *J. Clean. Prod.* **2016**, *137*, 741–751. [CrossRef]
9. Gonzalez Fischer, C.; Garnett, T. *Plates, Pyramids and Planets—Developments in National Healthy and Sustainable Dietary Guidelines: A State of Play Assessment*; Food and Agriculture Organization of the United Nations: Rome, Italy, 2016; ISBN 9789251092224.
10. Dobermann, D.; Swift, J.A.; Field, L.M. Opportunities and hurdles of edible insects for food and feed. *Nutr. Bull.* **2017**, *42*, 293–308. [CrossRef]
11. Dossey, A.T.; Morales-Ramos, J.A.; Rojas, M.G. *Insects as Sustainable Food Ingredients*; Elsevier Academic Press: London, UK, 2016; ISBN 9780128028568.
12. Kouřimská, L.; Adámková, A. Nutritional and sensory quality of edible insects. *NFS J.* **2016**, *4*, 22–26. [CrossRef]
13. Gere, A.; Radványi, D.; Héberger, K. Which insect species can best be proposed for human consumption? *Innov. Food Sci. Emerg. Technol.* **2019**, *52*, 358–367. [CrossRef]
14. Chakravorty, J.; Ghosh, S.; Rochow, M.-; Meyer-Rochow, V.B. Chemical Composition of Aspongopus nepalensis Westwood 1837 (Hemiptera; Pentatomidae), a Common Food Insect of Tribal People in Arunachal Pradesh (India). *Int. J. Vitam. Nutr. Res.* **2011**, *81*, 49–56. [CrossRef] [PubMed]
15. Ramos-Elorduy, J.; Manuel, J.; Moreno, P.; Prado, E.E.; Perez, M.A.; Otero, J.L.; Guevara, O.L. Nutritional Value of Edible Insects from the State of Oaxaca, Mexico. *J. Food Compos. Anal.* **1997**, *157*, 142–157. [CrossRef]
16. Rumpold, B.A.; Schlüter, O.K. Nutritional composition and safety aspects of edible insects. *Mol. Nutr. Food Res.* **2013**, *57*, 802–823. [CrossRef]
17. Homann, A.; Ayieko, M.; Konyole, S.; Roos, N. Acceptability of biscuits containing 10% cricket (Acheta domesticus) compared to milk biscuits among 5–10-year-old Kenyan schoolchildren. *J. Insects Food Feed.* **2017**, *3*, 95–103. [CrossRef]
18. Halloran, A.; Flore, R.; Vantomme, P.; Roos, N. *Edible Insects in Sustainable Food Systems*; Springer International Publishing: New York, NY, USA, 2018; ISBN 9783319740119.
19. EFSA. Risk profile related to production and consumption of insects as food and feed. *ESFA J.* **2015**, *13*, 4257.
20. Gere, A.; Zemel, R.; Radványi, D.; Moskowitz, H. Consumer Response to Insect Foods. *Ref. Mod. Food Sci.* **2018**, 1–6. [CrossRef]
21. Looy, H.; Dunkel, F.V.; Wood, J.R. How then shall we eat? Insect-eating attitudes and sustainable foodways. *Agric. Hum. Values* **2013**, *31*, 131–141. [CrossRef]
22. Adámek, M.; Adamkova, A.; Mlcek, J.; Borkovcová, M.; Bednářová, M. Acceptability and sensory evaluation of energy bars and protein bars enriched with edible insect. *Potravin. Slovak J. Food Sci.* **2018**, *12*, 431–437. [CrossRef]
23. Biró, B.; Fodor, R.; Szedljak, I.; Pásztor-Huszár, K.; Gere, A. Buckwheat-pasta enriched with silkworm powder: Technological analysis and sensory evaluation. *LWT* **2019**, *116*, 108542. [CrossRef]
24. De Oliveira, L.M.; Lucas, A.J.D.S.; Cadaval, C.L.; Mellado, M.S. Bread enriched with flour from cinereous cockroach (Nauphoeta cinerea). *Innov. Food Sci. Emerg. Technol.* **2017**, *44*, 30–35. [CrossRef]
25. Osimani, A.; Milanović, V.; Cardinali, F.; Roncolini, A.; Garofalo, C.; Clementi, F.; Pasquini, M.; Mozzon, M.; Foligni, R.; Raffaelli, N.; et al. Bread enriched with cricket powder (Acheta domesticus): A technological, microbiological and nutritional evaluation. *Innov. Food Sci. Emerg. Technol.* **2018**, *48*, 150–163. [CrossRef]
26. Sterna, V.; Zute, S.; Brunava, L. Oat Grain Composition and its Nutrition Benefice. *Agric. Agric. Sci. Procedia* **2016**, *8*, 252–256. [CrossRef]
27. El Khoury, D.; Cuda, C.; Luhovyy, B.L.; Anderson, G.H. Beta Glucan: Health Benefits in Obesity and Metabolic Syndrome. *J. Nutr. Metab.* **2011**, *2012*, 851362. [CrossRef]
28. EC Regulation No 1169/2011 of the European Parliament and the of the Council on the Provision of Food Information to Consumers; 1169/2011/EC. 2011. Available online: http://data.europa.eu/eli/reg/2011/1169/oj (accessed on 10 September 2020).
29. Sadler, G.D.; Murphy, P.A. *pH and Titratable Acidity*; Springer Science and Business Media: Berlin, Germany, 2010; pp. 219–238.

30. Minervini, F.; Lattanzi, A.; De Angelis, M.; Di Cagno, R.; Gobbetti, M. Influence of Artisan Bakery—Or Laboratory-Propagated Sourdoughs on the Diversity of Lactic Acid Bacterium and Yeast Microbiotas. *Appl. Environ. Microbiol.* **2012**, *78*, 5328–5340. [CrossRef] [PubMed]
31. Kilcast, D. *Sensory Analysis for Food and Beverage Quality Control—A Practical Guide*; Woodhead Publishing Limited: Cambridge, UK, 2010; ISBN 978-1-84569-476-0.
32. ISO. *Sensory Analysis—Methodology—General Guidance*; International Organization for Standardization: Geneva, Switzerland, 2017.
33. Ares, G.; Jaeger, S. *Check-All-That-Apply (CATA) Questions with Consumers in Practice: Experimental Considerations and Impact on Outcome*; Elsevier BV: Amsterdam, The Netherlands, 2015; pp. 227–245.
34. Kleij, F.T.; Musters, P.A. Text analysis of open-ended survey responses: A complementary method to preference mapping. *Food Qual. Pref.* **2003**, *14*, 43–52. [CrossRef]
35. Jaeger, S.R.; Beresford, M.K.; Paisley, A.G.; Antúnez, L.; Vidal, L.; Cadena, R.S.; Giménez, A.; Ares, G. Check-all-that-apply (CATA) questions for sensory product characterization by consumers: Investigations into the number of terms used in CATA questions. *Food Qual. Pref.* **2015**, *42*, 154–164. [CrossRef]
36. Ares, G.; Jaeger, S.R. Check-all-that-apply questions: Influence of attribute order on sensory product characterization. *Food Qual. Pref.* **2013**, *28*, 141–153. [CrossRef]
37. ISO. *Sensory Analysis—General Guidelines for the Selection, Training and Monitoring of Selected Assessors and Expert Sensory Assessors*; International Organization for Standardization: Geneva, Switzerland, 2008.
38. Meyners, M.; Castura, J.C.; Carr, B.T. Existing and new approaches for the analysis of CATA data. *Food Qual. Pref.* **2013**, *30*, 309–319. [CrossRef]
39. Indriani, S.; Bin Ab Karim, M.S.; Nalinanon, S.; Karnjanapratum, S. Quality characteristics of protein-enriched brown rice flour and cake affected by Bombay locust (*Patanga succincta* L.) powder fortification. *LWT* **2020**, *119*, 108876. [CrossRef]
40. González, C.M.; Garzón, R.; Rosell, C.M. Insects as ingredients for bakery goods. A comparison study of H. illucens, A. domestica and T. molitor flours. *Innov. Food Sci. Emerg. Technol.* **2019**, *51*, 205–210. [CrossRef]
41. Roncolini, A.; Milanović, V.; Aquilanti, L.; Cardinali, F.; Garofalo, C.; Sabbatini, R.; Clementi, F.; Belleggia, L.; Pasquini, M.; Mozzon, M.; et al. Lesser mealworm (*Alphitobius diaperinus*) powder as a novel baking ingredient for manufacturing high-protein, mineral-dense snacks. *Food Res. Int.* **2020**, *131*, 109031. [CrossRef] [PubMed]
42. EC Regulation No 1924/2006 of the European Parliament and the of the Council on Nutrition and Health Claims Made on Foods; 1924/2006/EC. 2006. Available online: http://data.europa.eu/eli/reg/2006/1924/oj (accessed on 10 September 2020).
43. Haber, M.; Mishyna, M.; Martinez, J.I.; Benjamin, O. The influence of grasshopper (*Schistocerca gregaria*) powder enrichment on bread nutritional and sensorial properties. *LWT* **2019**, *115*, 108395. [CrossRef]
44. Ogunlakin, G.O.; Oni, V.T.; Olaniyan, S.A. Quality Evaluation of Biscuit Fortified with Edible Termite (Macrotermes nigeriensis). *Asian J. Biotechnol. Bioresour. Technol.* **2018**, *4*, 1–7. [CrossRef]
45. Tan, H.S.G.; Verbaan, Y.T.; Stieger, M. How will better products improve the sensory-liking and willingness to buy insect-based foods? *Food Res. Int.* **2017**, *92*, 95–105. [CrossRef]
46. Verbeke, W. Consumer acceptance of functional foods: Socio-demographic, cognitive and attitudinal determinants. *Food Qual. Pref.* **2005**, *16*, 45–57. [CrossRef]
47. Mancini, S.; Fratini, F.; Tuccinardi, T.; Degl'Innocenti, C.; Paci, G. Tenebrio molitor reared on different substrates: Is it gluten free? *Food Control.* **2020**, *110*, 107014. [CrossRef]
48. Nissen, L.; Samaei, S.P.; Babini, E.; Gianotti, A. Gluten free sourdough bread enriched with cricket flour for protein fortification: Antioxidant improvement and Volatilome characterization. *Food Chem.* **2020**, *333*, 127410. [CrossRef]
49. Naes, T.; Brockhoff, P.B.; Tomic, O. *Statistics for Sensory and Consumer Science*; Wiley: Hoboken, NJ, USA, 2010.
50. Gere, A.; Székely, G.; Kovács, S.; Kókai, Z.; Sipos, L. Readiness to adopt insects in Hungary: A case study. *Food Qual. Pref.* **2017**, *59*, 81–86. [CrossRef]

Publisher's Note: MDPI stays neutral with regard to jurisdictional claims in published maps and institutional affiliations.

 © 2020 by the authors. Licensee MDPI, Basel, Switzerland. This article is an open access article distributed under the terms and conditions of the Creative Commons Attribution (CC BY) license (http://creativecommons.org/licenses/by/4.0/).

Article

Functional Tea-Infused Set Yoghurt Development by Evaluation of Sensory Quality and Textural Properties

Katarzyna Świąder [1,*], **Anna Florowska** [2], **Zuzanna Konisiewicz** [1] **and Yen-Po Chen** [3]

1. Department of Functional and Organic Food, Institute of Human Nutrition Sciences, Warsaw University of Life Sciences (SGGW–WULS), 159C Nowoursynowska Street, 02-776 Warsaw, Poland; zuzanna.konisiewicz@gmail.com
2. Department of Food Technology and Assessment, Institute of Food Science, Warsaw University of Life Sciences (SGGW–WULS), 159C Nowoursynowska Street, 02-776 Warsaw, Poland; anna_florowska@sggw.edu.pl
3. Department of Animal Science, The iEGG and Animal Biotechnology Research Center, National Chung Hsing University, Taichung 40227, Taiwan; chenyp@dragon.nchu.edu.tw
* Correspondence: katarzyna_swiader@sggw.edu.pl; Tel.: +48-22-593-70-47

Received: 31 October 2020; Accepted: 8 December 2020; Published: 11 December 2020

Abstract: In the present study, the potential to design natural tea-infused set yoghurt was investigated. Three types of tea (*Camellia sinensis*): black, green and oolong tea as well as lemon balm (*Melissa officinalis* L.) were used to produce set yoghurt. The sensory quality (using Quantitative Descriptive Profile analysis and consumer hedonic test) and texture analysis, yield stress, physical stability and colour analysis were assessed to describe the profile of the yoghurt and influence of quality attributes of the product on the consumer acceptability of infused yoghurts in comparison with plain yoghurt. Among the analyzed plant additives for yoghurt, addition of 2% oolong tea to the yoghurt allows a functional food to be obtained with satisfactory texture and sensory properties, accepted by consumers at the same level as for control yoghurt. Both types of yoghurt were also characterised by high consumer willingness to buy, which confirms the legitimacy of using oolong tea as a natural, functional yoghurt additive that improves the sensory quality of the product. The high overall quality of yoghurt with oolong tea in comparison to other plant extracts was associated with the intensive peach flavour and odour, nectar and sweet odour and flavour, and the highest creaminess and thickness. That was confirmed by principal component analysis (PCA) where the overall sensory quality of yoghurts was mainly positively correlated with peach flavour and odour, sweet odour and yoghurt odour, while it was negatively correlated with herbs flavor and odour, and green tea flavour and odour. The sensory profile confirmed no differences in textural profile between plain yoghurt and the tea-infused one measured in the mouth, which corresponds to the result of textural properties such as firmness and adhesiveness.

Keywords: functional food; yoghurt; tea; oolong tea; sensory quality; texture properties; food design; sensory profile; consumer test

1. Introduction

Functional food is defined as food that is fortified, enhanced, enriched or as a whole food that has a documented health effect beyond that resulting from the presence of nutrients traditionally considered essential [1,2]. Yoghurt, as a nutrient-rich dairy product, can be classified as functional food [3]. The main factors responsible for the beneficial effects of yoghurt are live cultures (*Streptococcus thermophilus* and *Lactobacillus bulgaricus*), proteins (whey and casein), lipids (bioactive fatty acids), vitamins and minerals (calcium and vitamin D) [4,5]. Of all the fermented products, yoghurt is also the most popular in the world [3,6] and best perceived and accepted by consumers [3,7]. Consumers who purchase functional products want the product to be safe, healthy, and natural and to have pleasant taste. They take into consideration such quality-related attributes on the labeling as freshness of the product, healthful properties, and nutritional value [8].

Dairy products belong to the most innovative food sector in Europe. The innovation of these products is based on product improvement, new formulations or new technologies that are used to meet the needs of specific consumers. Based on the research undertaken on French consumers, nutrient fortifications coming from plant sources were the most acceptable for them [9]. Several studies were conducted on yoghurts to know how yoghurt fortification with vitamins such as vitamins C, B9, B12, A and D [10] or minerals such us chromium, iron, magnesium, manganese, molybdenum, selenium and zinc [11] influenced their properties. Apart from yoghurt's fortification with vitamins and minerals, it has become more popular to add plant-based functional ingredients to the yoghurt like pomegranate juice powder (1–5%) [12], dried pomegranate seeds (5–20%) [13], freeze-dried apple pomace powder (1–3%) [14], flaxseed (0–4%) [15], coconut-cake (0–30%) [16], spirulina (0.25–1%) [17], aloe vera gel (1–5%) [18], saffron (0.0125%) [19] or tea [20–25] to improve their technological and sensory quality as well as health-promoting properties.

Tea is the most common functional beverage in the world [1,26] usually prepared by infusing leaves of the plant *Camellia sinensis* (L.) in hot water [27,28]. Tea can be classified according to the degree of fermentation into un-fermented green tea, semi-fermented oolong tea, and fully fermented black tea [1,27,28]. Thanks to the content of polyphenols, especially epigallocatechin-3-gallate, theaflavins and thearubigins, tea from *Camelia sinensis* (L.) provides several health-promoting effects [1,26,29]. Infusions of herbs, fruits, roots and flowers are also referred to as tea, and their health benefits are known and used by people around the world [30]. One of them is *Melissa officinalis*, called lemon balm, known for its many therapeutic properties such as antioxidant, antidepressant, anti-inflammatory, and antimicrobial activities. Lemon balm can be used for both prevention and treatment in medicine as well as in dietary supplements and functional food [31]. The addition of various extracts of black, green and white tea enhanced antioxidant properties of yoghurts [23]. The influence of tea on lactic bacteria during yoghurt fermentation was also verified and it was shown that this addition did not interfere with the fermentation process and did not affect the survival of bacteria. It was also shown that lactic acid bacteria present in yoghurt did not have a negative effect on the content of tea pro-health compounds [22]. In most of the available publications, the purpose of using tea addition to yoghurt was health promotion and enrichment of these products with antioxidants and ingredients that had a positive effect on human health [20,22–24,32]. The sensory quality of yoghurts developed with teas has been evaluated so far by researchers using only consumer tests [15,21,25,33]. However, there is a lack of information on the influence of tea on the sensory quality of yoghurt measured both by expert panels and consumers and supported by yoghurt textural properties that are very valuable during the new product development process.

The aim of developing the new product was to maintain the functional character of nutri-rich yoghurt and, in addition, to introduce a health-promoting plant material into natural yoghurt without added sugar. That is the reason why the aim of the research was to develop functional tea-infused yoghurts by assessing the sensory quality and texture properties of yoghurts.

2. Materials and Methods

2.1. Materials

The material for the actual tests were yoghurts prepared from microfiltered pasteurised cow's milk with 3.2% fat content (Piątnica, Poland), using the thermostatic method. Four types of leafy tea (*Camellia sinensis*) available on the Polish market were used for the production of yoghurt, i.e., green tea (BioFix, Tuszyn, Poland); black tea, Darjeeling FTGFOP1 Blend Lucky Hill (Tea Club Marek Brzezicki, Lubin, Poland) and oolong tea Oolong Milky (Herbaty Szlachetne Sp. Z o. o., Szczecin, Poland) and also the lemon balm (*Melissa officinalis* L., Cesarska Perła, Warszawa, Poland). In order to inoculate the milk, freeze-dried starter culture YO-122 (Serowar, Szczecin, Poland) containing *Streptococcus salivarius* subsp. *thermophilus* and *Lactobacillus delbrueckii* subsp. *bulgaricus* were used.

Yoghurt Processing

The technological process of set yoghurt production was developed based on modifications of two methods [22,23]. All ingredients were weighed on an analytical balance (RADWAG PS 1000/C/2, Radom, Poland). The milk was heated to 85 °C for 30 min. and poured into the beakers with tea leaves (2 g tea/100 mL of milk). It was steeped under a lid for 10 min, from time to time being stirred. Then the solution was manually filtered using gauze filters and cooled to 43 °C. We added 0.1% of starter cultures to milk and stirred thoroughly. Then 100 mL of milk was poured into sterile plastic containers with lids. All samples were thermostated at 43 °C in the incubator set (Memmert INE 500, Schwabach, Germany) until they reached the pH value of 4.5–4.6 (Voltcraft PH-100ATC, Wollerau, Switzerland), which took approximately 4.5 h. The samples were then removed and allowed to cool. The samples were stored at 4 °C for 15 h till the structure was built and then the sensory evaluation and instrumental analysis were undertaken [34–36]. Plain yoghurt was prepared similarly, only tea was not added to it.

2.2. Methods

2.2.1. Sensory Analysis

Expert Test

- **The Method:** The sensory characteristics of the yoghurts were assessed using the Quantitative Descriptive Profile (QDP). A method following the procedure was described in ISO standard 13299:2016 [37]. According to the procedure, the panellists first individually chose the descriptors (attributes) of appearance, odour, consistency and flavour/taste of samples. Then the attributes were discussed, agreed, and defined by the panellists. The final list of 39 attributes with definitions is presented in Supplementary Materials (Table S1). There were 11 characteristics describing the odour of samples (milky, yoghurt, sour, sweet, fat, green tea, black tea, herbs, peach, citrus, nectar), seven describing the appearance of the samples (whey, shine, colour, smoothness visually, adhesiveness, teaspoon filling, consistency uniformity), seven attributes describing the texture felt in the mouth (thickness, melting, firmness, yield stress, fat film, creaminess, smoothness), four describing the taste (sweet, sour, bitter and astringent), eight describing the flavour (milky, yoghurt, quark, green tea, black tea, herbs, peach) and characteristics describing the body and overall quality. The intensity of each attribute was measured by panellist on a linear unstructured 10-point scale (c.u.—contractual units) where 0 means low perception while 10—high perception.
- **The Expert Panel:** Quantitative Descriptive Profile analysis of yoghurt samples was performed by 10 trained panellists (experts), women aged between 35 and 52 with a good knowledge of all of the sensory methods, including profiling and yoghurt analysis. The panellists fulfilled the requirements of ISO standard 8586:2012 [38].
- **Testing Conditions:** Sensory evaluation was performed in the sensory laboratory fulfilling all the requirements of ISO standard 8589:2007 [39]. The assessment was carried out in individual testing booths with controlled lighting, temperature, and humidity. The booths were equipped with a computerised system, ANALSENS, for experiment planning, data acquisition and processing. The assessments were conducted during the morning and early afternoon, with two sessions per day.
- **Sample Preparation and Presentation:** Five types of yoghurt (C—Control, G—Green tea, B—Black tea, O—Oolong tea, M—Lemon balm) were assessed directly from test containers. The samples were prepared in cylindrical containers (ø 50 mm, height 50 mm, volume 100 mL), coded with 3-digit codes, placed randomly on the tray, and served at 7 °C to the evaluators. Still mineral water was used as neutraliser between samples. Each sample was analyzed in two independent

replications, and so the mean values were based on 20 individual results which were used for statistical analysis.

Semi-Consumer Test

The semi-consumer study was conducted at the Institute of Human Nutrition of Warsaw University of Life Sciences (WULS-SGGW) among students from the Faculty of Human Nutrition aged 19–31 recruited on the basis of willingness and interest to participate in the test ($n = 30$), they were randomly recruited. Thirty regular consumers of yoghurt or fermented milk products with no allergy reaction to milk participated in the study. The evaluation was carried out for five types of yoghurt: yoghurt with green tea (G), black tea (B), oolong tea (O), yogurt with lemon balm (M) and plain yoghurt as a control (C). The samples were prepared in cylindrical containers (ø 50 mm, height 50 mm, volume 100 mL), coded with 3-digit codes, placed randomly on the tray, and served at 7 °C to the evaluators. The consumers assessed the acceptability of appearance of yoghurt, their odour, taste, consistency and overall acceptability as well as willingness to buy yoghurt using a structured 9-point hedonic scale, where 1 meant "dislike extremely/will not buy definitely", and 9 "like extremely/will buy definitely" [40,41].

2.2.2. Instrumental Analysis

Textural Properties

Texture analysis of yoghurts was determined using a texture analyzer (TA.XT Plus, Stable Micro Mixtures, Surrey, UK) with a 5 kg load cell at 20 °C. The firmness (N) and adhesiveness (Ns) were analyzed by a 0.5 cm diameter cylindrical flat probe (P/0.5R). The measuring speed was 1.0 mm/s and the trigger force was 1 g. Samples were prepared in cylindrical containers (ø 50 mm, height 50 mm, volume 100 mL) and the penetration depth of the yoghurt was 5 mm. The reported values represented the averages of three replicates. The data were analyzed using the Exponent v6.1.4.0 equipment software [34,42].

Yield Stress

To measure the yield stress of yoghurts a rheometer (DV3T, Brookfield, Middleboro, MA, USA) was used. Measurements were conducted at 20 °C using spindles dedicated to yield stress (Pa) analysis: vane spindle V74 with a torque range HA. Samples were prepared in cylindrical containers (ø 50 mm, height 50 mm, 100 mL volume) and measurement was performed by controlling shear rate from $0.01-100\ \text{s}^{-1}$. The reported values represented the averages of three replicates. The data were analyzed using the software provided with the rheometer [34,43].

Physical Stability—CSA Method

The changes in the yogurt stability were investigated using space- and time-resolved extinction profiles (STEP) technology. This is a new technique employing gravitational fields to accelerate the occurrence of instability phenomena such as sedimentation, flocculation or creaming [43]. The physical stability of yoghurts was determined with an analytical centrifuge LUMiSizer 6120-75 (L.U.M. GmbH, Berlin, Germany) by measuring the intensity of transmitted near-infrared light in suspension, and recording of light intensity profiles as a function of time and position of the sample ("fingerprints") [44,45]. Stability was shown as a space- and time-related transmission profile over the sample length. The parameters used for the analysis were: wavelength 870 nm, volume 1.8 mL of dispersion; light factor: 1; 4000 rpm; experiment time, 50 min; interval time 10 s; temperature 25 °C. The reported values represented the average of six replicates. The data were analyzed by the delivered software (SepView 6.0; LUM, Berlin, Germany) and the instability index was calculated [45].

Colour Parameters

To determine the colour components (L*, a*, and b*) the Minolta CR-200 colorimeter (Minolta, Japan; light source D65, observer 2°, a measuring head hole of 8 mm) was used. Colour parameters were analysed using the CIEL*a*b* system. The measurements were made at the surface of yoghurts. To determine the colour differences between plain yoghurt and infused yoghurts, the parameter of total colour difference ΔE was calculated [46].

$$\Delta E = \sqrt{\left(L_c^* - L_T^*\right)^2 + \left(a_c^* - a_T^*\right)^2 + \left(b_c^* - b_T^*\right)^2}$$

where: L_c^*, a_c^*, b_c^* refers to the colour parameters of plain yoghurts and L_T^*, a_T^*, b_T^* refers to the colour parameters of tea infused yoghurts.

2.2.3. Statistical Analysis

The results of texture, yield stress, stability and colour were statistically analyzed using Statistica 13.3 (TIBICO Software Inc.). To determine the significance of differences between the average values of analyzed parameters of tea-infused yoghurts, one-way analysis of variance (ANOVA) was used. Significant differences between infused yoghurts and plain yoghurt were verified using Tukey's test at significant level $\alpha = 0.05$. The results of sensory analysis were statistically analyzed using Statgraphics Plus 5.1 (Statgraphics Technologies, Inc., Plains, VA, USA). The one-way ANOVA at the significance level ($p \leq 0.05$) was used to check the significance of differences in attributes intensity among analyzed samples. Mean values marked with different indices a, b, c, d differed statistically ($p \leq 0.05$). Principal component analysis (PCA) was used to analyze differences between samples and correlation of selected variables. The ANALSENS NT program was used for the PCA analysis.

3. Results and Discussion

3.1. Sensory Analysis

3.1.1. Preliminary Test

In order to determine the appropriate concentration of the applied tea, preliminary tests were conducted based on the evaluation of the sensory properties of the obtained tea-infused yoghurts. A preliminary study was carried out on green tea yoghurt with the concentration being 1%; 2%; 4%; 6%; 8% for individual samples respectively. After sensory evaluation, it was found that the addition of 2% green tea was sufficient and provided the best sensory characteristics, compared to the others. The higher percentage of tea in yoghurt resulted in deterioration of sensory quality, taste, odour and colour. An improvement in consistency was observed in the 6% and 8% green tea-infused yoghurts, but the intense bitterness, astringency and aroma of these samples resulted in their very low sensory quality. Based on these results, tea with different degree of fermentation (green tea, oolong tea and black tea) and lemon balm (2%) with health-promoting properties was selected for further design studies on the composition of yoghurt.

3.1.2. Quantitative Descriptive Profile Analysis of Tea-Infused Yoghurts in Comparison with Plain Yoghurt (Expert Test)

The sensory quality of developed yoghurts with teas has been evaluated so far by researchers using hedonic methods [15,21,25,33] using a 9 or 7 degree scale, which has enabled them to obtain information on the acceptability of the product, but has not given full information on the sensory profile of the product, which could be obtained by using expert panel research, e.g., Quantitative Descriptive Profile analysis.

In the present study, the sensory profile was assessed for five types of yoghurts: yoghurt with green tea (G), yoghurt with black tea (B), yoghurt with oolong tea (O), yoghurt with lemon balm (M) and plain yoghurt as a control one (C) by using Quantitative Descriptive Profile analysis. The trained panellists defined 39 main attributes with their characteristics (Supplementary Materials Table S1), which described the evaluated yogurt samples. To describe the sensory experience of different food categories, a number of lexicons was developed [47,48]. Coggins and co-authors [49] developed a sensory lexicon for plain yoghurts in the United States based on 12 commercially produced yoghurts, where a trained panel defined 61 sensory descriptors. The plain yoghurt was used in the current study and, based on it, yoghurts with tea were produced. Therefore, our expert panel, in addition to the characteristics of plain yoghurt, also defined additional features describing the characteristics of the plants used in the research for the yoghurt production (Supplementary Materials Table S1).

The results of the profiling analysis of the four tea-infused yoghurts (G—Green tea, B—Black tea, O—Oolong tea, M—Lemon balm) and plain yoghurt (C—Control sample), are presented in Table S2 (Supplementary Materials) and in Figures 1–3. The results showed that the samples were characterised with different sensory profiles (Figures 1–3). Control yoghurt was characterised by milky, yoghurt, sour, quark profile in taste and smell, intense fatty smell, light whey flow and light creamy colour. It was also characterised by dense, creamy, uniform consistency. Green tea infused yoghurt was characterised by intensive green tea, peach and nectar odour and flavour profile and noticeable bitter and astringent taste, more intense flow of whey, darker cream colour and dense, creamy, uniform consistency. All these attributes were related to the quality of the green tea, that is characterised by yellow colour, bitter and astringent taste, as well as floral (we called it nectar), grassy or burn leaf [50]. The profile of oolong tea-infused yoghurt was characterised with the intensive peach and nectar odour and flavour, citrus odour, sweet and astringent taste, more intense flow of whey, cream colour and dense, creamy and uniform consistency. Oolong tea flavour profile depended on the time of fermentation and was described as sweet, floral, green fruity, astringent, bitter and umami [51] and most of these attributes were perceived in oolong tea-infused yoghurt. Black tea-infused yoghurt was characterised by intensive dark cream colour, uniform and dense, and creamy consistency, intense flow of whey, and intensive black tea, and less-intensive peach odour and flavour and bitter and astringent taste. Black tea that was used in the research was Darjeeling tea, Fine Tippy Golden Flowery Orange Pekoe, that was characterised with light and delicate flavour and aroma. The second flush Darjeeling tea produced excellent quality teas that were considered to be better than the first flush as they had a fruitier, less astringent flavour than the earlier teas [52]. The taste and aroma of green, black and oolong tea, depended on the degree of fermentation, and the content of free amino acids, mainly l-teanine and natural amino acids, e.g., glutamic acid and asparagine [53]. The profile of lemon balm-infused yoghurt was characterised with very intensive dark cream colour, uniform, dense and melting consistency, but also very intensive herbs and green tea odour and flavour, as well as bitter and astringent taste and citrus odour. Such a yoghurt profile might result from the addition of lemon balm, which was characterised by lemon taste and odour [31] and light yellow colour [54], but probably their composition might influence changes in the profile during the yoghurt production process.

The yoghurts developed differed significantly from each other in various attributes: odour (milky, yoghurt, sour, sweet, fat, green tea, black tea, herbs, peach, citrus, nectar), appearance (whey and colour intensity), flavours (milky, yoghurt, green tea, black tea, herbs, peach, nectar), body and overall quality. The significance of differences between compared samples in intensity of the attributes is marked in Table S2 and Figures 1–3.

The detailed characteristics of smell, appearance, texture felt in the mouth, taste and flavour as well as body and overall quality of the evaluated yoghurts are described below.

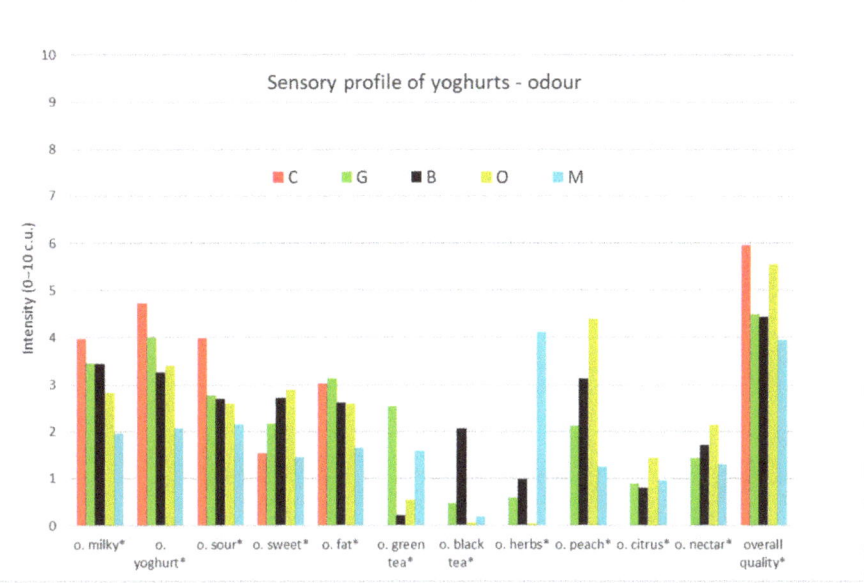

Figure 1. The sensory quality profile of yoghurt samples (odour). C—Control, G—Green tea, B—Black tea, O—Oolong tea, M—Lemon balm (o-odour), (*—differ significantly $p \leq 0.05$).

Figure 2. The sensory quality profile of yoghurt samples (taste and flavour) C—Control, G—Green tea, B—Black tea, O—Oolong tea, M—Lemon balm (t-taste, f-flavour), (*—differ significantly $p \leq 0.05$).

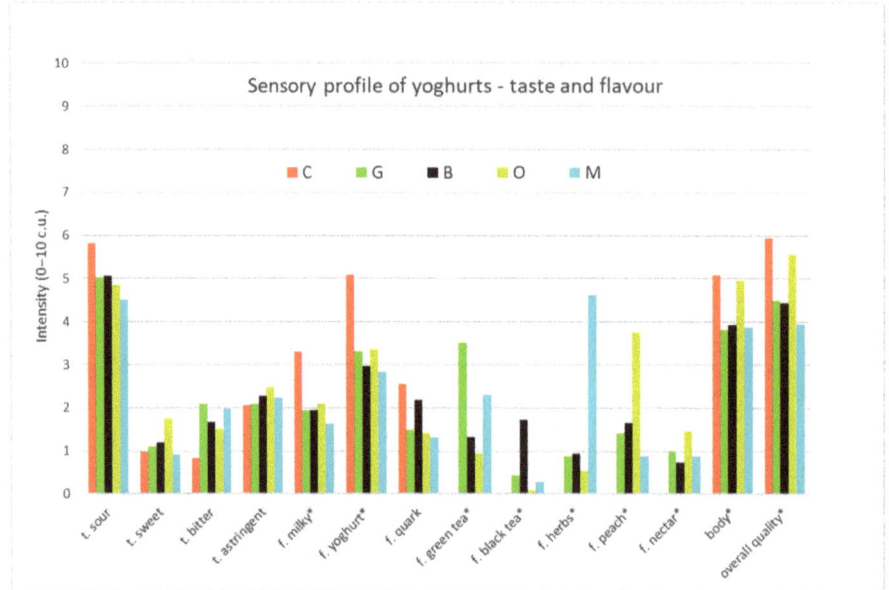

Figure 3. The sensory quality profile of yoghurt samples (appearance and texture). C—Control, G—Green tea, B—Black tea, O—Oolong tea, M—Lemon balm (*—differ significantly $p \leq 0.05$).

Odour

The examined samples of yoghurt differed markedly in all odour attributes (Figure 1). The significantly highest level of milky odour was observed in the control sample (C-4.0 c.u.) (Table S2), while the lowest in yoghurt with lemon balm (M-2.0 c.u.). Also, the milky odour was less intensive in samples G, B and O than in control samples. There were no significant differences in milky odour between samples G, B and O (3.5, 3.4, 2.8 c.u., respectively). Similar dependencies were observed in the yoghurt odour, which was significantly the most intensive in the control sample (C-4.7 c.u.), while the lowest in yoghurt with lemon balm (M-2.1 c.u.). Sour odour was significantly the most noticeable in control sample (C-4.0 c.u.) while significantly least in yoghurt with green tea, black tea, oolong tea and lemon balm (G-2.8, B-2.7, O-2.6, M-2.2 c.u.). Sweet odour was statistically the least noticeable (1.5 c.u.) in the control sample and in yoghurt with lemon balm while significantly the sweetest odour was perceived in yoghurt with oolong tea (2.9 c.u.-twice more intensive than in control one). The significantly most-intensive green tea odour was in yogurt with green tea (G-2.5 c.u.), while black tea in black tea yoghurt (B-2.1 c.u.) and herb odour in yoghurt with lemon balm (M-4.1 c.u.). The oolong tea was characterised with significantly intensive peach (4.4 c.u.) odour. Citrus (0.9-1.4 c.u.) and nectar odour (1.3-2.1 c.u.) were significantly perceived in green tea, black tea, oolong tea and lemon balm. All yoghurts represented a very similar and low level of refreshing odour (2-3.2 c.u.). Tea addition to the plain yoghurt influenced the odour sensation. The smell of milk, yoghurt, sour and fatty smell was lowered, while the sweet smell was emphasised, and a typical smell of infused plants appeared.

Basic Taste and Flavour

The samples of yoghurts examined differed significantly in milky, yoghurt, green tea, black tea, herbs, peach and nectar flavours (Figure 2). All yoghurt samples were described as sour. The intensity of sour taste did not exceed of 5.8 c.u. on the scale in all products (Table S2). The samples showed low intensity of sweet (0.9–1.7) and bitter (0.8–2.1 c.u.) and astringent taste (2–2.5 c.u.). The significantly less bitter was control yoghurt (0.8 c.u.) and the most bitter but still on the lowest level was yoghurt with green tea (2.1 c.u.). The astringent and bitter taste was characteristic for teas due to the polyphenols present in them [26]. The examined samples of yoghurts differed significantly in flavour characteristics. The control sample had significantly more intensive milky flavor (3.3 c.u.), yoghurt flavour (5.1 c.u.) and quark flavour (2.5 c.u.) than other tea-infused yogurts, and there were no plant flavours perceived in plain yoghurt such as green tea, black tea, herbs, peach and nectar flavours. Plain yoghurt should be characterised with a pleasant, clean acid flavour, with no bitter, rancid, oxidised, yeast and unclean flavours, and be firm, with a smooth and homogeneous texture. Yoghurt taken on a spoon should keep its shape without sharp edges and should present a clean, natural white colour, with a smooth, velvety appearance [55]. The plain yoghurt tested met the above requirements, except for the bitter taste, which was perceived in a sample but not very intense. According to the literature [21], to overcome the bitter flavour of the green tea chocolate coupled with honey might be used. The product had good consumer acceptability as indicated by the sensory evaluation [21]. Green tea flavour was significantly perceived more strongly in yoghurt with green tea (3.5 c.u.) than in other plant-infused yoghurts (0–2.3 c.u.). A similar relationship occurred in the case of yoghurt with black tea where black tea flavour was significantly perceived in yoghurt with black tea (1.7 c.u.), while herb flavour was significantly perceived in yogurt with lemon balm (4.6 c.u.) than in other tea-infused yoghurts (0–0.9 c.u.). Peach flavour was significantly perceived in yoghurt with oolong tea (3.8 c.u.), less intensive in other plant yoghurts (0.9–1.7 c.u.) while the nectar flavour was significantly perceived on a similar level (0.7–1.5 c.u.) in tea yoghurts in comparison to the plain one (0 c.u.). In one study, the bitter taste of green tea was effectively masked in the yoghurt by adding chocolate and honey [21]. It was found that three types of tea extract and lemon balm affected the flavour of the plain yoghurt. The control sample was described as having a milky, yoghurt, fatty and quark flavour while in tea-infused yoghurts additional flavour of green tea, black tea, herbs, peach and nectar was perceived.

Visual Attributes/Appearance

While evaluating the visual attributes and appearance of the yoghurt samples, it was observed that they differed significantly only in the presence of whey and the colour of the yoghurt (Figure 3). All yoghurts represented a very similar and high level of shine (6.8–7.5 c.u.) (Table S2). The significantly lower level of whey presence was observed in control sample (1.9 c.u.), then in yoghurts with tea (4.5–4.9 c.u.). In a study analyzing the influence of green tea and moringa on the quality of yoghurt, it was found that moringa showed a greater influence on yoghurt sineresis than green tea [25]. The samples of yoghurts differed significantly in cream colour intensity. The control sample (C) was significantly lighter in cream colour (1.0 c.u.) than the rest of the samples. The yoghurt with lemon balm was characterised by significantly the darkest cream color (5.8 c.u.), as well as yoghurt with black tea (5.2 c.u.). All samples represented similar high visual smoothness (6.1–7.1 c.u.) and thickness measured by spoon resistance (6–6.9 c.u.), as well as filling the teaspoon, and was more conical (6–6.8 c.u.) than flat. All analyzed samples were uniform in consistency (5.2–6.4 c.u.). Yoghurt should have a delicate and smooth texture and a firm body, which is maintained while eating with a spoon [56]. These were the characteristics of the yoghurts developed with the addition of tea and plain yoghurt. The addition of tea resulted in significantly higher whey flow in tea-infused yoghurt than in control yoghurt, as well as its significantly darker colour, especially in yoghurt with lemon balm and black tea.

Texture in the Mouth

The examined samples of yoghurts did not differ significantly in texture attributes felt in the mouth (Figure 3). They were moderately medium thick (5.1–5.6), and melted very well in the mouth (6.4–7.2 c.u.) (Table S2). All yoghurt samples were characterised with medium firmness (5.3–5.8 c.u.) and low yield stress (2.1–2.7 c.u.). Fat film was perceived in all samples on the same level (2.0–2.3 c.u.). All analyzed yoghurts were characterised by moderate creaminess (3.9–5.2 c.u.) and high smoothness in the mouth (6.2–6.8 c.u.). In the case of plain yoghurt's taste and texture, characteristics allowed trained panellists to differentiate, identify and categorise yoghurts, but it was not possible to differentiate or categorise them on the basis of percentage fat content. Natural yoghurts' differentiation was more effective on the basis of taste and texture than aroma and appearance [49]. The addition of tea to yoghurt did not significantly affect the texture felt in the mouth.

All the analyzed samples of yoghurts differed significantly in their bodies, a characteristic that described the harmonisation of all positive attributes. The significantly highest body had control yoghurt (5.1 c.u.) and the oolong tea one (5.0 c.u.). Yoghurts with green tea, black tea and lemon balm had significantly lower body.

Overall quality that depended on all the characteristics perceived in the yoghurt samples was significantly different in all samples. The significantly highest overall quality was the control sample (5.9 c.u.), then yoghurt with oolong tea (5.6 c.u.), while the significantly lowest overall quality was yoghurt with lemon balm (4.0 c.u.). The biggest influence on the lowest overall quality of yoghurt with lemon balm was probably the most intensive herb flavour and odour. The high overall quality of yoghurt with oolong tea was associated with the intensive peach flavour and odour, nectar and sweet odour and flavour, and the highest creaminess and thickness. The highest overall quality of the control sample was associated with the highest yoghurt, milky and quark flavours and sour taste, which were typical for the plain yoghurt.

3.1.3. Principal Component Analysis (PCA) Analysis

For the evaluation of the sensory profile of 5 yoghurts, 39 attributes were defined. Analysing them all in PCA, it was possible to notice that they were all not clearly visible in the graph, so only those attributes with which the samples differed significantly ($p \leq 0.05$) were used for PCA. The principal component analysis of the results of the profile evaluation of the all yoghurt samples showed that the variability of the samples was assigned primarily to the first main component (PC1-53.75% of the total variability) and concerned different colour, intensity of green tea flavour and odour, yoghurt, milky and sour odour, and body of the samples (Figure 4). These were the attributes that differentiated the evaluated samples. The second main component was assigned a smaller percentage of general variability (PC2–28.94%) and concerned a different intensity of peach odour and flavor, and sweet odour. Overall sensory quality of yoghurts was mainly positively correlated with peach flavour and odour, sweet odour and yoghurt odour, while it was negatively correlated with herbs flavour and odour, and green tea flavour and odour. The yoghurt samples differed in sensory quality as evidenced by their location in the space of the PCA system. The samples can be observed to form four distinctive clusters. The first one covered the plain yoghurt, the second yoghurt with lemon balm, the third yoghurt with black tea and green tea, the fourth yoghurt with oolong tea. The yoghurt with oolong tea was relatively close to the overall sensory quality in comparison to other samples and was characterised by intensive peach flavour and odour positively correlated with overall quality of yoghurt. Yogurt with lemon balm was on the opposite site of oolong tea and overall quality because of its very intensive herbal flavour and odour characteristic to that sample. Results of the PCA corresponded to the results obtained in the Quantitative Descriptive Profile analysis.

3.1.4. Acceptability of Yoghurts and Willingness to Buy Evaluated by the Consumers/Semi-Consumer Evaluation of Yoghurts Based on Different Types of Tea

The overall goal when designing a food product is to make it sensorily acceptable and desirable by consumers. In this research, apart from the expert sensory evaluation, it was decided to conduct consumer research. The semi-consumer assessment was carried out among 30 consumers, and their task was to assess the acceptability of the general appearance, odour, consistency, taste/flavour, overall acceptability of the product taking into account all the characteristics and willingness to buy four yoghurt samples differing according to the type of tea added in comparison to the plain yoghurt (Table S3). (Figure 5).

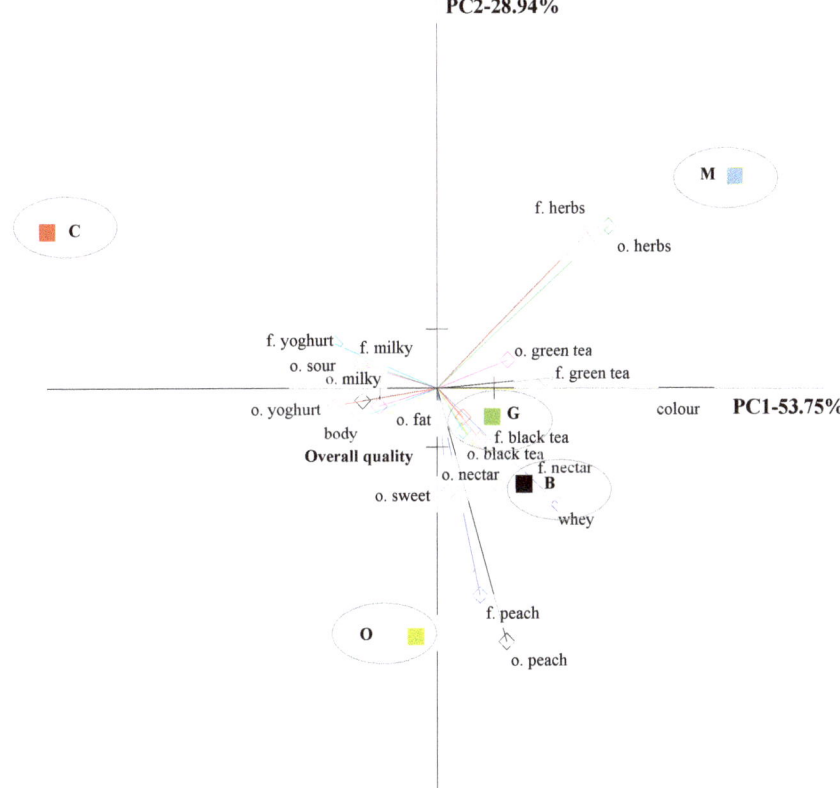

Figure 4. Similarities and differences in the sensory quality of yoghurt samples. C—Control, G—Green tea, B—Black tea, O—Oolong tea, M—Lemon balm (PCA) (f-flavour, o-odour).

On the basis of the results obtained, it was found that the control sample (C) and the sample with the addition of oolong tea (O) (5.7 and 5.9 c.u.) (Table S3) were characterised by the significantly highest and similar overall acceptability, while the significantly least acceptable in terms of overall acceptability were samples with the addition of green tea (G), black tea (B) and lemon balm (3.0, 3.7 and 3.8 c.u. respectively). The significantly highest overall acceptability was presented in the control yoghurt and yoghurt with oolong tea which translated into a willingness to buy those yoghurts. Consumers wanted to buy yoghurt with oolong tea (5.6 c.u.) and control yoghurt (5.3 c.u.) the most, while significantly less the yoghurts with lemon balm, black tea and green tea (3.3, 3.2 and 2.3 c.u. respectively). Overall acceptability was based on the acceptability of general appearance, odour, consistency and taste and flavour. General appearance of control sample and the yoghurt with oolong tea and lemon balm (6.6, 6.5, 5.9 c.u.) were significantly more acceptable for the consumers than yoghurts with green and black tea (3.8 c.u., 4.3 c.u.). The most acceptable odour was that of yoghurt with oolong tea (7.0 c.u.) and control yoghurt (6.2 c.u.) while significantly less acceptable were the odours of yoghurts with green and black tea (4.4. c.u.) and with lemon balm (4.8 c.u.). Samples of yoghurts were also significantly different in taste acceptability which had big influence on the overall acceptability of those yoghurts. Yoghurt with oolong tea had the most acceptable taste (6.1 c.u.), control yoghurt has a significantly less acceptable taste (5.0 c.u.) and the least acceptable taste was that of yoghurt with black tea, yoghurt with lemon balm and with green tea (3.7 c.u., 3.3 c.u. and 2.6 c.u. respectively). Based on the consumer research, yoghurt with oolong tea had significantly the most acceptable consistency (6.3 c.u.), then yoghurt with green tea, black tea and lemon balm (4.4, 4.7 and 5.2 c.u.). It was found that three types of tea extract and lemon balm used in the yoghurt formula affected the acceptability and willingness to buy them by consumers. Of the plant additives used, green tea, oolong, black tea and lemon balm, the addition of oolong tea resulted in a higher acceptability of smell, consistency, but above all of taste, which resulted in a higher overall acceptability of oolong tea yoghurt compared to other plant additives and a comparable acceptability to control yoghurt, which also resulted in a greater willingness of consumers to buy oolong tea-infused yoghurt.

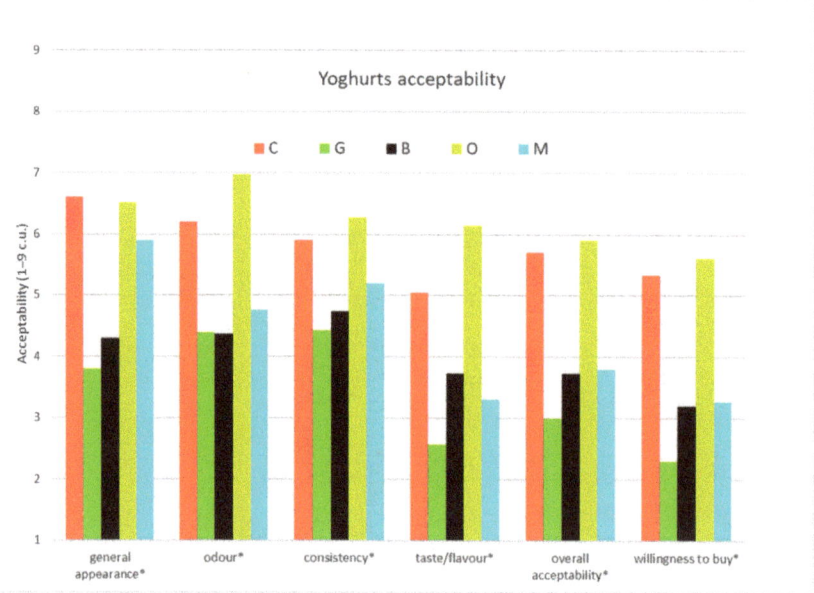

Figure 5. Acceptability of 5 yoghurts with tea (C—Control, G— Green tea, B—Black tea, O—Oolong tea, M—Lemon balm) and willingness to buy them (*—differ significantly $p \leq 0.05$).

Product attributes that affected the level of acceptance were taste, ergonomics, overall form, cost of purchase and functional additives. On the other hand, low nutritional knowledge and lack of knowledge of production technology by the recipients might influence their willingness to try something that sounds foreign negatively, so products with the addition of unknown ingredients or with the use of modern technologies were not accepted by them [57]. The results indicated that yoghurts with innovative functional additives were highly valued by consumers, even better than a control product without additives, and would be willingly bought by them. During the food design process it is very important to use sensory methods to verify the quality of the developed products and their acceptability among consumers for whom the product is dedicated. In this paper the sensory quality of the developed yoghurts was verified by using expert methods such as Quantitative Descriptive Profile analyses and semi-consumer research where hedonic test were used. Both methods confirmed the possibility of using the tea additive, especially oolong teas, as an additive improving the quality of yoghurt by introducing into it a pleasant peach, nectar flavour and odour and at the same time its acceptability and the willingness to purchase it among a selected group of consumers. It should be further emphasised that studies conducted so far have mainly focused on the use of the addition of green tea, while our research indicates that semi-fermented oolong tea has a better effect on the sensory profile of yoghurt and its acceptance by consumers than green tea.

Some authors added different additives to the green tea yoghurt to mask their bitterness and increase their acceptability to consumers. Chatterjee and co-authors [21] developed green tea-infused yoghurts with 9% of chocolate syrup, 1% addition of skim milk powder and 3% of honey. All these ingredients were added to mask the bitterness of the green tea and enrich the taste of the product. Green tea-infused chocolate yoghurt's overall acceptability was 7.35, while taste 6.64. Sugar or honey addition increased the overall acceptability (7.88 and 8.37 respectively) and taste (7.65 and 8.12 respectively). Yoghurt with chocolate syrup had a higher acceptability [21] than that assessed in these studies, which may have resulted from the addition of sugar syrup and honey to the product. In our yoghurt there were only simple ingredients necessary to create the yoghurt and to be in harmony with a clean label. We did not add sugar and any sugar substitutes or additives that increased the sweetness of the product, but we could observe from the sensory profile that the addition of oolong tea increased the perception of sweet taste in comparison with other yoghurt especially the control one. In the next study, it is worth considering the addition of a sweetener to tea yoghurt, but one that will not affect the nutritional value of the product, especially its caloricity, as this is the assumption of this research project. Up to now, studies that have been conducted on tea-infused yoghurt have only been evaluated using consumers [15,21,25,33]. Therefore, the extension of the research to include sensory evaluation by an expert panel allows us to obtain more information about the analyzed product and to verify the factors and attributes influencing its acceptability by consumers, as well as to provide guidelines for further work on the product formula and its possible reformulation.

3.2. Instrumental Analysis

The structure and the rheological properties of yoghurts are important to product quality and shelf life [42]. To evaluate the influence of different tea addition on the yoghurt texture, firmness and adhesiveness were tested. On the basis of statistical analysis, it was found that only the addition of oolong tea had an influence on the firmness of tested yogurts (Table 1). It was observed that oolong tea extract visibly reduces firmness and adhesiveness of yoghurt, which was positively marked in the acceptability of yoghurts and willingness to buy them by consumers in sensory analysis of the yoghurts.

Table 1. Physical properties of tea infused yoghurts. C—Control, G—Green tea, B—Black tea, O—Oolong tea, M—Lemon balm.

Type of Tea	Firmness [N]	Adhesiveness [Ns]	Yield Stress [Pa]	Instability Index
C	$0.73^b \pm 0.01$	$-1.75^a \pm 0.03$	$130.67^{ab} \pm 3.00$	$0.674^a \pm 0.008$
G	$0.84^b \pm 0.03$	$-2.23^a \pm 0.05$	$175.43^b \pm 2.83$	$0.642^a \pm 0.086$
B	$0.69^{ab} \pm 0.07$	$-1.51^{ab} \pm 0.05$	$121.20^{ab} \pm 2.91$	$0.686^a \pm 0.011$
O	$0.43^a \pm 0.02$	$-1.00^b \pm 0.06$	$103.83^a \pm 2.65$	$0.698^a \pm 0.010$
M	$0.68^{ab} \pm 0.03$	$-1.65^a \pm 0.07$	$112.13^a \pm 0.07$	$0.615^a \pm 0.062$

Values are mean ± SD ($n = 3$), $^{a,\ b}$—values followed by the same letter within a column do not differ significantly according to Tukey's test ($p < 0,05$).

Other tea extract did not affect the structure of yoghurts, as their firmness and adhesiveness were very similar to the control sample, which was also confirmed by the sensory analysis. The literature data show that tea extract addition might even improve the texture of yogurts This phenomenon might be explained by the milk protein cross-linking with tea flavanols [35,58,59]. However, data in the literature points to the fact that the texture of tea-infused yogurts depends more on the quantity of added tea that on the kind of tea used [59].

Yield stress is an initial resistance of a probe to flow under stress and it is a measure of the interactions between the components in the product. For tested yoghurts, it was reported that yield stress (Table 1) was not generally influenced by the addition of tea; however it was reported that green tea influenced the yield stress by increasing it. The influence of green tea on the rheological properties of yoghurt such as apparent viscosity was also reported by Amirdivani and Baba [60]. This increase was probably caused by the presence of green tea polyphenolic compounds, which are able to interact with milk proteins [61].

The stability of yoghurts and possible syneresis, serum release from its structure, is regarded as a technological defect in yoghurts. That is why the characteristic of yoghurt stability is very important [62]. The effect of tea addition on yoghurts' stability was examined with the multi-sample analytical centrifuge based on the STEP (space-time resolved extinction profiles) technology. Destabilisation progression of the process is shown in Figure 6. Addition of tea extracts did not affect the instability index (Table 1, Figure 7). All tested yoghurts were not stable during the time of stability determination. The serum release (syneresis), which was not influenced by tea extract addition, was observed. It was also observed in the plain yoghurt. The separation is known to be related to instability of yoghurt's gel network and thus the loss in ability to entrap all the serum phase [63]. It is known from the literature data that polyphenols' secondary plant metabolites presented in the tea extracts have the ability to interact with proteins, resulting in the formation of a protein–polyphenol complex, which are observed as weak interactions, mainly hydrophobic, van der Waals, hydrogen bridge-binding, and ionic interactions and might explain the lack of differences of stability of the tested yoghurts [62].

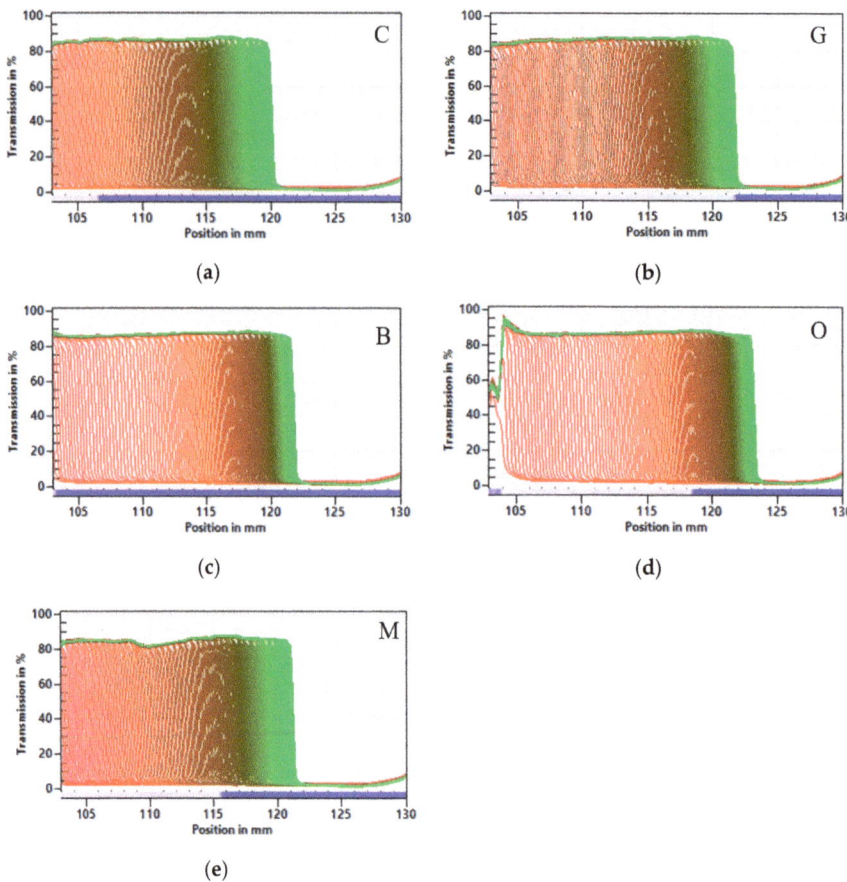

Figure 6. Influence of different tea infusion on the yoghurt transmission profiles presented enabling LUMiSizer® analysis. (**a**) C—Control, (**b**) G—Green tea, (**c**) B—Black tea, (**d**) O—Oolong tea, (**e**) M—Lemon balm.

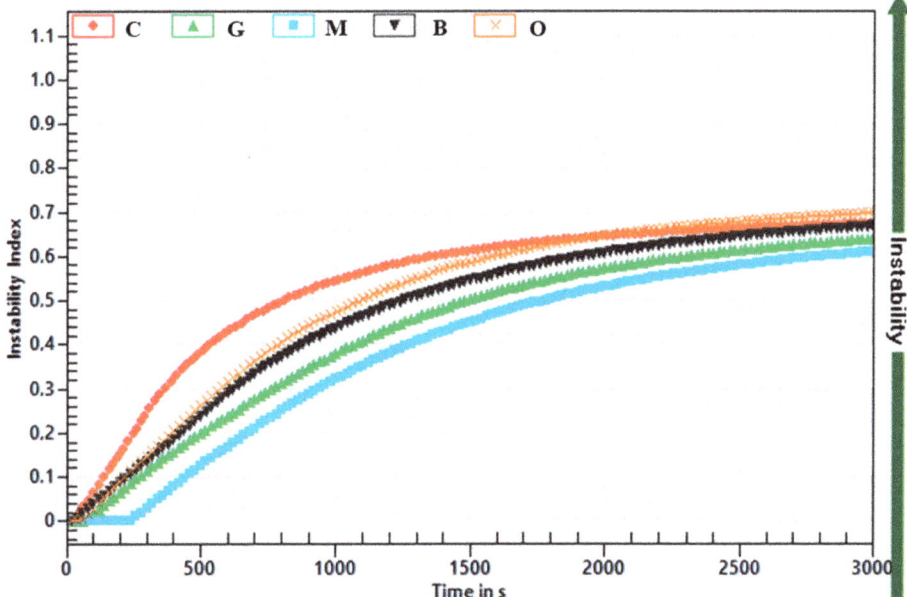

Figure 7. Influence of tea extract addition on stability of yogurt in comparison with plain yoghurt. C—Control, G—Green tea, B—Black tea, O—Oolong tea, M—Lemon balm.

Colour is one of the most important visual attributes in yoghurts. The addition of tea to yoghurts significantly decreased the colour component value L* in comparison with the control sample (Table 2). During tea fermentation, an enzymatic oxidation of polyphenols, especially tea catechins, occurs. This leads to formation of a series of coloured chemical compounds, such as theaflavins (TFs) and thearubigins (TRs), which are responsible for the characteristics of the tea liquor's colour and lightness which, in consequence, might influence the colour of the products obtained with tea [64]. The lowest L* values were noted for the yogurts with the green tea addition, whereas the brightest were yoghurts with lemon balm. The addition of tea also influenced the other colour parameters. Values of the a* component were the lowest for the control yoghurt as well as with green tea addition. The addition of other tested teas resulted in an increment of the a* parameter, which was due to natural colorants that were present in the tea. Also, the values of parameter b* varied between the control yogurts and yogurt with tea addition.

Table 2. Colour parameters and the total colour difference parameter of yogurts obtained without or with the addition of tea. C—Control, G—Green tea, B—Black tea, O—Oolong tea, M—Lemon balm.

Type of Tea	Colour Parameters			ΔE [#]
	L*	a*	b*	
C	90.22 [d] ± 0.22	−0.83 [a] ± 0.04	10.29 [a] ± 0.12	
G	77.98 [a] ± 0.06	−0.88 [a] ± 0.09	12.52 [b] ± 0.28	12.45 ± 0.15
B	81.76 [b] ± 0.22	2.78 [d] ± 0.12	17.52 [d] ± 0.15	11.71 ± 0.43
O	81.85 [b] ± 1.52	1.07 [c] ± 0.18	16.86 [d] ± 0.52	10.84 ± 2.06
M	87.45 [c] ± 0.17	0.66 [b] ± 0.06	14.49 [c] ± 0.21	5.26 ± 0.08

Values are mean ± SD ($n = 3$). [a–d]—values followed by the same letter within a column do not differ significantly according to Tukey's test ($\alpha = 0.05$). [#] total colour difference parameter calculated in relation to the control sample.

Among tested teas, the addition of black and oolong tea had a greater influence on the b* parameter. Those results were in correspondence with the data obtained by Liang et al. [64]. Based on comprehensive analysis of the effect of the tea addition on the colour of the yoghurts, the total colour difference parameter was calculated (ΔE). It was found that the differences in color between yoghurts with and without the addition of tea were noticeable even for the inexperienced observer ($\Delta E > 5$) [46]. That colour analysis was confirmed by the Quantitative Descriptive Profile analysis where plain yoghurt was characterised by a light white colour while tea-infused yoghurts by a darker cream colour.

4. Conclusions

Among the analyzed green tea-, black tea-, oolong tea- and lemon balm-infused yoghurts, the addition of 2% oolong tea to the yoghurt allowed a functional food to be obtained with satisfactory texture, colour and sensory properties, accepted by consumers to the same degree as that of control yoghurt. The research conducted indicated that semi-fermented oolong tea had a better effect on the sensory profile of yoghurt evaluated by the expert panel and its acceptance by consumers than green tea, and these studies can be used to commercialise oolong tea-infused yoghurt. However, further research is planned to increase the acceptability of the product by adding other natural additives which can increase the sensory quality and acceptability of the product, as well as storage research.

Supplementary Materials: The following are available online at http://www.mdpi.com/2304-8158/9/12/1848/s1, Table S1: Sensory attribute for yoghurts profiling, their definitions and intensity range, Table S2: Sensory characteristics of yoghurts evaluated by experts: C—Control, G—Green tea, B—Black tea, O—Oolong tea, M—Lemon balm- the mean values, 2 sessions ($n = 20$), Table S3: Acceptability of yoghurts and willingness to buy evaluated by consumers: C—Control, G—Green tea, B—Black tea, O—Oolong tea, M—Lemon balm- the mean values ($n = 30$).

Author Contributions: Study conception and design, K.Ś.; methodology, K.Ś., A.F.; performed research, K.Ś., A.F., Y.-P.C., Z.K.; analyzed the data, K.Ś., A.F.; interpreted the data, K.Ś., A.F.; writing—original draft preparation K.Ś.; writing—review and editing, K.Ś., A.F., Y.-P.C. All authors have read and agreed to the published version of the manuscript.

Funding: These studies received funding from the Rector of WULS-SGGW as part of the system of financial support for scientists and teams.

Acknowledgments: The authors would like to thank the expert panel members for their support and participation in the samples' evaluation.

Conflicts of Interest: The authors declare no conflict of interest.

References

1. Gaur, S.; Agnihotri, R. Green tea: A novel functional food for the oral health of older adults. *Geriatr. Gerontol. Int.* **2014**, *14*, 238–250. [CrossRef]
2. Sharanya Rani, D.; Penchalaraju, M. A review different types of functional foods and their health benefits. *Int. J. Appl. Nat. Sci.* **2016**, *5*, 19–22.
3. Sarkar, S. Potentiality of probiotic yoghurt as a functional food—A review. *J. Nutr. Food Sci.* **2019**, *2*, 9. [CrossRef]
4. Fernandez, M.A.; Panahi, S.; Daniel, N.; Tremblay, A.; Marette, A. Yogurt and cardiometabolic diseases: A critical review of potential mechanisms. *Adv. Nutr.* **2017**, *8*, 812–829. [CrossRef] [PubMed]
5. Fisberg, M.; Machado, R. History of yogurt and current patterns of consumption. *Nutr. Rev.* **2015**, *73*, 4–7. [CrossRef]
6. Vijaya Kumar, B.; Vijayendra, S.V.N.; Reddy, O.V.S. Trends in dairy and non-dairy probiotic products—A review. *J. Food Sci. Technol.* **2015**, *52*, 6112–6124. [CrossRef] [PubMed]
7. Homayouni Rad, A.; Yari Khosroushahi, A.; Khalili, M.; Jafarzadeh, S. Folate bio-fortification of yoghurt and fermented milk: A review. *Dairy Sci. Technol.* **2016**, *96*, 427–441. [CrossRef]
8. Kraus, A. Development of functional food with the participation of the consumer. Motivators for consumption of functional products. *Int. J. Consum. Stud.* **2015**, *39*, 2–11. [CrossRef]

9. Masson, E.; Debucquet, G.; Fischler, C.; Merdji, M. French consumers' perceptions of nutrition and health claims: A psychosocial-anthropological approach. *Appetite* **2016**, *105*, 618–629. [CrossRef]
10. Keršienė, M.; Jasutienė, I.; Eisinaitė, V.; Pukalskienė, M.; Venskutonis, P.R.; Damulevičienė, G.; Knašienė, J.; Lesauskaitė, V.; Leskauskaitė, D. Development of a high-protein yoghurt-type product enriched with bioactive compounds for the elderly. *LWT* **2020**, *131*, 109820. [CrossRef]
11. Achanta, K.; Aryana, K.J.; Boeneke, C.A. Fat free plain set yogurts fortified with various minerals. *LWT Food Sci. Technol.* **2007**, *40*, 424–429. [CrossRef]
12. Pan, L.H.; Liu, F.; Luo, S.Z.; Luo, J.P. Pomegranate juice powder as sugar replacer enhanced quality and function of set yogurts: Structure, rheological property, antioxidant activity and in vitro bioaccessibility. *LWT* **2019**, *115*, 108479. [CrossRef]
13. Bchir, B.; Bouaziz, M.A.; Blecker, C.; Attia, H. Physico-Chemical, antioxidant activities, textural, and sensory properties of yoghurt fortified with different states and rates of pomegranate seeds (*Punica granatum* L.). *J. Texture Stud.* **2020**, *51*, 475–487. [CrossRef] [PubMed]
14. Wang, X.; Kristo, E.; LaPointe, G. Adding apple pomace as a functional ingredient in stirred-type yogurt and yogurt drinks. *Food Hydrocoll.* **2020**, *100*, 105453. [CrossRef]
15. Mousavi, M.; Heshmati, A.; Daraei Garmakhany, A.; Vahidinia, A.; Taheri, M. Texture and sensory characterization of functional yogurt supplemented with flaxseed during cold storage. *Food Sci. Nutr.* **2019**, *7*, 907–917. [CrossRef]
16. Ndife, J. Production and Quality Assessment of Functional Yoghurt Enriched with Coconut. *Int. J. Nutr. Food Sci.* **2014**, *3*, 545. [CrossRef]
17. Barkallah, M.; Dammak, M.; Louati, I.; Hentati, F.; Hadrich, B.; Mechichi, T.; Ayadi, M.A.; Fendri, I.; Attia, H.; Abdelkafi, S. Effect of Spirulina platensis fortification on physicochemical, textural, antioxidant and sensory properties of yogurt during fermentation and storage. *LWT Food Sci. Technol.* **2017**, *84*, 323–330. [CrossRef]
18. Azari-Anpar, M.; Payeinmahali, H.; Daraei Garmakhany, A.; Sadeghi Mahounak, A. Physicochemical, microbial, antioxidant, and sensory properties of probiotic stirred yoghurt enriched with *Aloe vera* foliar gel. *J. Food Process. Preserv.* **2017**, *41*, e13209. [CrossRef]
19. Gaglio, R.; Gentile, C.; Bonanno, A.; Vintaloro, L.; Perrone, A.; Mazza, F.; Barbaccia, P.; Settanni, L.; Di Grigoli, A. Effect of saffron addition on the microbiological, physicochemical, antioxidant and sensory characteristics of yoghurt. *Int. J. Dairy Technol.* **2019**, *72*, 208–217. [CrossRef]
20. Ünal, G.; Karagözlü, C.; Kinik, Ö.; Akan, E.; Sibel Akalin, A. Effect of Supplementation with Green and Black Tea on Microbiological Characteristics, Antimicrobial and Antioxidant Activities of Drinking Yoghurt. *J. Agric. Sci.* **2018**, *24*, 153–161.
21. Chatterjee, G.; Das, S.; Das, R.S.; Des, A.B. Development of green tea infused chocolate yoghurt and evaluation of its nutritive value and storage stability. *Prog. Nutr.* **2018**, *20*, 237–245. [CrossRef]
22. Jaziri, I.; Ben Slama, M.; Mhadhbi, H.; Urdaci, M.C.; Hamdi, M. Effect of green and black teas (*Camellia sinensis* L.) on the characteristic microflora of yogurt during fermentation and refrigerated storage. *Food Chem.* **2009**, *112*, 614–620. [CrossRef]
23. Muniandy, P.; Shori, A.B.; Baba, A.S. Influence of green, white and black tea addition on the antioxidant activity of probiotic yogurt during refrigerated storage. *Food Packag. Shelf Life* **2016**, *8*, 1–8. [CrossRef]
24. Najgebauer-Lejko, D.; Sady, M.; Grega, T.; Walczycka, M. The impact of tea supplementation on microflora, pH and antioxidant capacity of yoghurt. *Int. Dairy J.* **2011**, *21*, 568–574. [CrossRef]
25. Shokery, E.S.; El-Ziney, M.G.; Yossef, A.H.; Mashaly, R.I. Effect of Green Tea and Moringa Leave Extracts Fortification on the Physicochemical, Rheological, Sensory and Antioxidant Properties of Set-Type Yoghurt. *J. Adv. Dairy Res.* **2017**, *5*, 179. [CrossRef]
26. Adak, M.; Gabar, M.A. Green tea as a functional food for better health: A brief review. *Res. J. Pharm. Biol. Chem. Sci.* **2011**, *2*, 645–664.
27. EFSA Panel on Dietetic Products, Nutrition and Allergies (NDA). Scientific Opinion on the substantiation of health claims related to *Camellia sinensis* (L.) Kuntze (tea), including catechins in green tea, and improvement of endothelium-dependent vasodilation (ID 1106, 1310), maintenance of normal blood pressure (ID 131). *EFSA J.* **2011**, *9*, 2055. [CrossRef]
28. Wang, H.; Provan, G.J.; Helliwell, K. Tea Flavonoids: Their Functions, Utilisation and Analysis. *Trends Food Sci. Tech.* **2000**, *11*, 152–160. [CrossRef]

29. Rasheed, Z. Molecular evidences of health benefits of drinking black tea. *Int. J. Health Sci. (Qassim)* **2019**, *13*, 1.
30. Jin, L.; Li, X.B.; Tian, D.Q.; Fang, X.P.; Yu, Y.M.; Zhu, H.Q.; Ge, Y.Y.; Ma, G.Y.; Wang, W.Y.; Xiao, W.F.; et al. Antioxidant properties and color parameters of herbal teas in China. *Ind. Crops Prod.* **2016**, *87*, 198–209. [CrossRef]
31. Świąder, K.; Startek, K.; Wijaya, C.H. The therapeutic properties of Lemon balm (*Meliss officinalis* L.): Reviewing novel findings and medical indications. *J. Appl. Bot. Food Qual.* **2019**, *92*, 327–335. [CrossRef]
32. Hasni, I.; Bourassa, P.; Hamdani, S.; Samson, G.; Carpentier, R.; Tajmir-Riahi, H.A. Interaction of milk α- And β-caseins with tea polyphenols. *Food Chem.* **2011**, *126*, 630–639. [CrossRef]
33. Karnopp, A.R.; Oliveira, K.G.; de Andrade, E.F.; Postingher, B.M.; Granato, D. Optimization of an organic yogurt based on sensorial, nutritional, and functional perspectives. *Food Chem.* **2017**, *233*, 401–411. [CrossRef] [PubMed]
34. Han, X.; Yang, Z.; Jing, X.; Yu, P.; Zhang, Y.; Yi, H.; Zhang, L. Improvement of the Texture of Yogurt by Use of Exopolysaccharide Producing Lactic Acid Bacteria. *BioMed Res. Int.* **2016**, 7945675. [CrossRef]
35. Abdel-Hamid, M.; Huang, Z.; Suzuki, T.; Enomoto, T.; Hamed, A.M.; Li, L.; Romeih, A. Development of a Multifunction Set Yogurt Using Rubus suavissimus S. Lee (Chinese Sweet Tea) Extract. *Foods* **2020**, *9*, 1163. [CrossRef] [PubMed]
36. Raikos, V.; Grant, S.B.; Hayes, H.; Ranawana, V. Use of β-glucan from spent brewer's yeast as a thickener in skimmed yogurt: Physicochemical, textural, and structural properties related to sensory perception. *J. Dairy Sci.* **2018**, *101*, 5821–5831. [CrossRef]
37. ISO 13299:2016. Sensory Analysis—Methodology—General Guidance for Establishing a Sensory Profile. Available online: https://www.iso.org/standard/58042.html (accessed on 25 October 2020).
38. ISO 8586:2012. Sensory Analysis—General Guidelines for the Selection, Training and Monitoring of Selected Assessors and Expert Sensory Assessors. Available online: https://www.iso.org/standard/45352.html (accessed on 25 October 2020).
39. ISO 8589:2007. Sensory Analysis—General Guidance for the Design of Test Rooms (Amd 1: 2014). Available online: https://www.iso.org/standard/36385.html (accessed on 25 October 2020).
40. Stone, H.; Sidel, J. *Sensory Evaluation Practices*, 3rd ed.; Elsevier Academic Press: San Diego, CA, USA, 2004.
41. ISO 4121:2003. Sensory Analysis—Guidelines for the Use of Quantitative Response Scales. Available online: https://www.iso.org/standard/33817.html (accessed on 25 October 2020).
42. Nguyen, H.T.H.; Ong, L.; Kentish, S.E.; Gras, S.L. Homogenisation improves the microstructure, syneresis and rheological properties of buffalo yoghurt. *Int. Dairy J.* **2015**, *46*, 78–87. [CrossRef]
43. Xu, D.; Wang, X.; Jiang, J.; Yuan, F.; Gao, Y. Impact of whey protein—Beet pectin conjugation on the physicochemical stability of beta-carotene emulsions. *Food Hydrocoll.* **2012**, *28*, 258–266. [CrossRef]
44. Olsen, B. Yogurt quality with fiber addition. In Proceedings of the Cultured Dairy Products Conference, London, UK, 22 May 2007; International Dairy Products Association, Tate & Lyle: London, UK, 2008.
45. Żbikowska, A.; Szymańska, I.; Kowalska, M. Impact of inulin addition on properties of natural yogurt. *Appl. Sci.* **2020**, *10*, 4317. [CrossRef]
46. Mokrzycki, W.S.; Tatol, M. Color difference ΔE—A survey. *Mach. Graph. Vis.* **2011**, *20*, 383–411.
47. Karagül-Yüceer, Y.; Drake, M. Sensory Analysis of Yogurt. In *Manufacturing Yogurt and Fermented Milks*, 2nd ed.; Chandan, R.C., Kilara, A., Eds.; John Wiley & Sons, Inc.: Oxford, UK, 2007; pp. 265–278.
48. Suwonsichon, S. The Importance of Sensory Lexicons for Research and Development of Food Products. *Foods* **2019**, *8*, 27. [CrossRef] [PubMed]
49. Coggins, P.C.; Schilling, M.W.; Kumari, S.; Gerrard, P.D. Conventional Milk Yogurt in the United States. *J. Sens. Stud.* **2008**, *23*, 671–687. [CrossRef]
50. Lee, S.M.; Lee, H.S.; Kim, K.H.; Kim, K.O. Sensory characteristics and consumer acceptability of decaffeinated green teas. *J. Food Sci.* **2009**, *74*. [CrossRef] [PubMed]
51. Liu, P.P.; Yin, J.F.; Chen, G.S.; Wang, F.; Xu, Y.Q. Flavor characteristics and chemical compositions of oolong tea processed using different semi-fermentation times. *J. Food Sci. Technol.* **2018**, *55*, 1185–1195. [CrossRef]
52. Gohain, B.; Borchetia, S.; Bhorali, P.; Agarwal, N.; Bhuyan, L.P.; Ravindranath, R.; Rahman, A.; Gurusubramaniam, G.; Sakata, K.; Mizutani, M.; et al. Understanding Darjeeling tea flavour on a molecular basis. *Plant Mol. Biol.* **2012**, *78*, 577–597. [CrossRef]

53. Weerawatanakorn, M.; Hung, W.L.; Pan, M.H.; Li, S.; Li, D.; Wan, X.; Ho, C.T. Chemistry and health beneficial effects of oolong tea and theasinensins. *Food Sci. Hum. Wellness* **2015**, *4*, 133–146. [CrossRef]
54. Singh Verma, P.P.; Singh, A.; Rahaman, L.; Bahl, J.R. Review Article Lemon Balm (*Melissa officinalis* L.) an Herbal Medicinal Plant With Broad Therapeutic Uses and Cultivation Practices: A Review. *Int. J. Recent Adv. Multidiscip. Res.* **2015**, *2*, 928–933.
55. USDA Specifications for Yogurt, Nonfat Yogurt and Lowfat Yogurt, 2001, USDA. Available online: https://www.ams.usda.gov/sites/default/files/media/yogurtlowfatnonfat.pdf (accessed on 25 October 2020).
56. Aryana, K.J.; Olson, D.W. A 100-Year Review: Yogurt and other cultured dairy products. *J. Dairy Sci.* **2017**, *100*, 9987–10013. [CrossRef]
57. Jeżewska-Zychowicz, M. Uwarunkowania akceptacji konsumenckiej innowacyjnych produktów żywnościowych. *Żywność Nauka Technol. Jakość* **2014**, *6*, 5–17.
58. Avci, E.; Yuksel, Z.; Erdem, Y.K. Green yoghurt revolution. Identification of interactions between green tea polyphenols and milk proteins and resultant functional modifications in yoghurt gel. In Proceedings of the 5th Central European Congress on Food, Bratislava, Slovakia, 22 May 2010; Food Research Institute (VUP): Bratislava, Slovak, 2010.
59. Najgebauer-Lejko, D.; Witek, M.; Żmudziński, D. Changes in the viscosity, textural properties, and water status in yogurt gel upon supplementation with green and Pu-erh teas. *J. Dairy Sci.* **2020**, *103*, 11039–11049. [CrossRef]
60. Amirdivani, S.; Baba, A.S.H. Rheological Properties and Sensory Characteristics of Green Tea Yogurt during Storage. *Life Sci. J.* **2013**, *10*, 378–390. [CrossRef]
61. Ozdal, T.; Capanoglu, E.; Altay, F. A review on protein-phenolic interactions and associated changes. *Food Res. Int.* **2013**, *51*, 954–970. [CrossRef]
62. Dönmez, Ö.; Mogol, B.A.; Gökmen, V. Syneresis and rheological behaviors of set yogurt containing green tea and green coffee powders. *J. Dairy Sci.* **2017**, *100*, 901–907. [CrossRef] [PubMed]
63. Zare, F.; Boye, J.I.; Orsat, V.; Champagne, C.; Simpson, B.K. Microbial, physical and sensory properties of yogurt supplemented with lentil flour. *Food Res. Int.* **2011**, *44*, 2482–2488. [CrossRef]
64. Liang, Y.; Lu, J.; Zhang, L.; Wu, S.; Wu, Y. Estimation of black tea quality by analysis of chemical composition and colour difference of tea infusions. *Food Chem.* **2003**, *80*, 283–290. [CrossRef]

Publisher's Note: MDPI stays neutral with regard to jurisdictional claims in published maps and institutional affiliations.

© 2020 by the authors. Licensee MDPI, Basel, Switzerland. This article is an open access article distributed under the terms and conditions of the Creative Commons Attribution (CC BY) license (http://creativecommons.org/licenses/by/4.0/).

Article

Consumer-Based Sensory Characterization of Steviol Glycosides (Rebaudioside A, D, and M)

Ran Tao [1] and Sungeun Cho [2],*

[1] Department of Food Science & Human Nutrition, Michigan State University, East Lansing, MI 48824, USA; taoran3@msu.edu
[2] Department of Poultry Science, Auburn University, Auburn, AL 36849, USA
* Correspondence: sungeun.cho@auburn.edu

Received: 3 July 2020; Accepted: 29 July 2020; Published: 31 July 2020

Abstract: Rebaudioside (Reb) D and M are the recent focus of the food industry to address the bitter taste challenge of Reb A, which is the most commonly used steviol glycoside in natural sweetener stevia. This study evaluated the sensory characteristics of Reb A, D, and M, compared to 14% (w/v) sucrose, using a consumer panel and explored the relationship between 6-n-Propylthiouracil (PROP) taster status (i.e., non-tasters, medium tasters, supertasters) and the perceived intensity of sweet and bitter tastes of the three steviol glycosides. A total of 126 participants evaluated the intensities of in-mouth, immediate (5 s after expectorating), and lingering (1 min after expectorating) sweetness and bitterness of 0.1% Reb A, D, M, and 14% sucrose and described the aftertaste of the sweeteners by using a check-all-that-apply (CATA) question. The results showed that in-mouth sweetness and bitterness of Reb D and M were not significantly different from sucrose, unlike Reb A which showed significant bitterness. However, Reb D and M showed more intense lingering sweetness than sucrose. The CATA analysis resulted that Reb D and M were closer to positive attribute terms and also to sucrose than Reb A, but Reb D and M were still considered artificial, which may cause them to be perceived negatively. When comparing among PROP taster groups, no significant differences in the perceived sweetness and bitterness of the three steviol glycosides were found. This study generates important information about Reb A, D, and M for the food industry, especially working with products formulated to deliver reductions in sugar using a natural high-intensity sweetener, stevia.

Keywords: stevia; steviol glycosides; rebaudiosides; Reb A; Reb D; Reb M; aftertaste; CATA; PROP

1. Introduction

Artificial sweeteners are widely used in a variety of foods and beverages as a sugar substitute that mimics the effect of sugar on taste without adding calories. However, consumers have a negative perception of artificial sweeteners not only due to aversive sensations such as bitter off-taste [1,2] but also due to potential health risks and demand more natural options [3]. To respond to the consumers' demand for natural sugar substitutes with low/zero calories, the food industry has focused on stevia, which is a natural high-intensity non-nutritive sweetener. Stevia (*Stevia Rebaudiana* Bertoni) is a shrub native to Paraguay, and the leaves of stevia have been used to sweeten teas for hundreds of years in Paraguay and Brazil [4,5]. Stevia is the source of many different types of steviol glycosides, which are the sweetening compounds in stevia leaves [6]. Stevioside and rebaudioside (Reb) A are the major sweet compounds among the steviol glycosides [6] and are the most widely used steviol glycosides on the market according to a Mintel Global New Products Database (GNPD) product search [7]. However, stevioside and Reb A exhibit bitter and licorice off-taste [8–13], which pose challenges to product formulation.

To overcome the taste challenges of stevioside and Reb A, the researchers and food industry have looked into other minor steviol glycosides in the stevia leaves to provide better sugar reduction solutions. Several studies have reported that the two minor steviol glycosides, Reb D and M, elicit significantly less bitterness with better sweetness than Reb A and also work well in products without sacrificing the taste [14–19]. Prakash et al. [16] reported that Reb M had less bitterness and astringency than Reb A. Most of the studies investigating sensory characteristics of steviol glycosides were conducted within a specific range of 5–10% sweetness equivalency related to sucrose (SE) [8,10,11,13,16]. Little research was done at high concentrations for high-sugar applications such as frozen desserts, which generally contain 13–22% sucrose w/v [20], although sweetness potency of stevia heavily depends on the SE [13].

For sensory characterization of food products, sensory descriptive analysis using trained assessors is the most widely used method, but it is time-consuming to train a panel. Less time consuming and more flexible methodologies such as check-all-that-apply (CATA) or intensity scales using consumers have been discussed in the last two decades [21]. It has been reported that consumers were capable of evaluating sensory attributes of various products, showing good agreement between consumers and trained assessors in terms of discrimination, reproducibility, and consensus [22–25]. Although Worch et al. [24] found that the trained panel showed greater consensus among each other, the larger sample size of consumers compensated for the higher variability. Moskowitz [26] suggested that a minimum of 40–50 people was needed to get stable averages, and the averages would not be affected by the base size much once the participant number exceeded 80. Ares et al. [27] also indicated that 80 consumers would be sufficient to get stable results when samples had large differences, but caution would be needed if samples had smaller differences or more complex attributes.

CATA is also often used to determine the characteristics of a product from a consumer perspective, which allows the consumers to describe a product by selecting terms from a given list that would match the product [28]. CATA questions have been used for a variety of foods and beverages [28–32], and these studies showed that CATA was a simple way to understand consumer perception on the sensory profile of a product.

Phenylthiocarbamide (PTC) and 6-n-Propylthiouracil (PROP) are bitter-tasting compounds that have been used to test people's sensitivity to bitter taste. Supertasters are a group of people who perceived intense bitter taste from PTC and PROP, while non-tasters barely detect the bitterness of them [33]. It has been reported that individuals have different sensitivity to the aftertaste of high-intensity sweeteners [34], and thus researchers have long been interested in understanding the relationship between PROP status (e.g., non-tasters vs. supertasters) and perceived taste intensities of high-intensity sweeteners. Bartoshuk [35] and Drewnowski et al. [36] found a significant difference between non-tasters and supertasters in the bitterness of saccharin at low concentrations. Zhao and Tepper [37] also suggested that supertasters perceived more bitterness and sweetness than non-tasters in carbonated soft drinks with artificial sweeteners, including sucralose, aspartame, acesulfame-K. However, Horne et al. [38] did not find a relationship between PROP taster status and the sweetness and bitterness of saccharin and acesulfame-K. Rankin et al. [39] failed to find any significant difference in bitterness between supertasters and non-tasters in cola drink sweetened with artificial sweeteners either. Risso et al. [40] looked into the effect of genetic variations on stevioside and found that the bitter taste receptor for PROP did not predict the bitterness perception of stevioside. However, little research was done to investigate the influence of PROP status on the perceived sweet and bitter taste intensities of novel steviol glycosides such as Reb D and M.

The primary objective of this study was to determine sensory characteristics of Reb A, D, and M, compared to 14% (w/v) sucrose, using a consumer panel. A secondary objective was to determine if there is a relationship between PROP taster status and the perceived intensities of the three steviol glycosides.

2. Materials and Methods

2.1. Materials

Sweeteners used in the study were 95% Reb A (ENLITEN® 30000015 High Intensity Sweetener, Ingredion, Westchester, IL), 95% Reb D (BESTEVIA® Reb D stevia leaf sweetener, Ingredion, Westchester, IL), 95% Reb M (BESTEVIA® Reb M stevia leaf sweetener, Ingredion, Westchester, IL, USA), and sucrose (Smidge & Spoon™, Kroger, Cincinnati, OH, USA). PROP (6-Propyl-2-thiouracil, #P3755, Sigma-Aldrich, St. Louis, MO), NaCl (Sigma-Aldrich, St. Louis, MO, USA), and filter papers (1.5 dia. cm, VWR Scientific Products, West Chester, PA, USA) were used to make paper disks for supertaster screening.

2.2. PROP Status Determination

The paper disks for PROP status determination were prepared following the method described by Zhao et al. [41]. Blank, NaCl, and PROP disks were prepared. Blank disks were used as the control. NaCl disks were made by placing filter papers in 1.0 mol/L NaCl solution for 30 s at room temperature and oven-dried for 1 h at 121 °C (250 °F). 50-mmol/L PROP solution at boiling temperature was used for PROP disks.

PROP testing and classification were based on Zhao et al. [41] and Zhao and Tepper [37]. Michigan State University SONA Paid Research Pool (https://msucas-paid.sona-systems.com) was used to recruit participants with age between 18 and 55. Participants were instructed to rinse their mouth with water, taste the paper disk for 15 s or until the disk is wet, discard the paper disk, and then rate the perceived intensity of the taste on the labeled magnitude scale (LMS). The participants would taste a blank, a NaCl, and a PROP disk in order with a 30 s break in between samples to minimize fatigue and carryover. The set was repeated after a 5 min break.

The LMS is a 100 mm quasi-logarithmic spacing vertical scale with verbal labels from "barely detectable" to "strongest imaginable" [42]. The scale set up was "no sensation" = 0, "barely detectable" = 1.5, "weak" = 6, "moderate" = 17, "strong" = 35, "very strong" = 52, and "strongest imaginable" = 100 [43,44]. The PROP score of participants was calculated based on the mean of the two replicates. Because the LMS is not equal in spacing, the difference between two scores when both ratings are at the higher end is less than when ratings are at the lower end. If the difference between two ratings was bigger than 30 mm, or bigger than 40 mm when both ratings were higher than "very strong", the participant would be considered having bad reproducibility and would not be invited to the following water solution testing. Out of 224 participants, 27 were excluded.

Initially, "moderate" or below (≤17 mm on the LMS) and "very strong" or above (≥52 mm on the LMS) of PROP score were used to group participants into non-tasters and supertasters. The group means and 95% confidence interval were then calculated to set new cut-off scores. The new cut-off score for non-tasters was 10.3 and for supertasters was 70.7. Participants with scores in between were classified as medium tasters. When the PROP score of a participant was borderline, the NaCl score was used to help classify the person [41]. A participant would be classified as a non-taster if the person gave a non-taster borderline score and rated NaCl much higher than PROP (~30 mm difference on the LMS). When a participant was at the supertaster borderline and gave a much lower NaCl score than PROP, the person would be classified as a supertaster. Out of 197 remaining participants, 25 were identified as non-tasters and 55 were supertasters.

2.3. Subjects Demographics

Following the PROP test, participants were asked to provide some basic demographic information, including age, gender, ethnicity, educational level, weight, height, health condition, consumption frequency of low/zero sugar added products, consumption of sweeteners on a regular basis (at least once a month), and familiarity with stevia.

2.4. Consumer Testing

2.4.1. Samples and Sample Preparation

All solutions were prepared using deionized water and the concentration of the sample is expressed in g/L (*w/v*). Sucrose at 14% was chosen as the control. Reb A, D, and M at 0.09% were used in a preliminary test ($n = 31$) to determine the relative sweetness to 14% sucrose. The result showed that 0.09% Reb M were not statistically different from 14% sucrose in sweetness intensity ($p = 0.16$), but there was still a 1.1 difference on a marked 15-cm line scale with descriptors of "not at all" and "extremely" as endpoint anchors. Another preliminary test ($n = 65$) was then conducted to prove the sweetness equivalency of Reb M to sucrose, using 0.09% and 0.12% Reb M and 10% and 14% sucrose. The result indicated that both 0.09% and 0.12% Reb M were not significantly different from 14% sucrose ($p = 0.34$ and $p = 0.11$, respectively), with 0.09% Reb M closer to 14% sucrose at 0.5 difference on a 15-cm line scale, comparing to a 1.0 difference between 0.12% Reb M and 14% sucrose. Since 0.09% Reb M was again lower in intensity on the scale, Reb M at 0.10% was chosen for the consumer testing. Reb A and D at 0.10% were used to compare the sensory characteristics of the three steviol glycosides at the same concentration. Thus, samples used for the testing were Reb A, D, and M at 0.10%, and 14% sucrose. The consumer test lasted four days and fresh samples were made 1 day before testing each day. Ten milliliters of each solution was measured into a 1 oz soufflé cup and stored in the refrigerator (4 °C) prior to serving.

2.4.2. Testing Procedure

This study was approved by the University Institutional Review Board of the Michigan State University (East Lansing, MI) (Study ID: STUDY00004019). SIMS 2000 software (SIMS Sensory Software, Morristown, NJ, USA) was used to create and administer the questionnaire.

Consumers were instructed to rate the sweetness and bitterness intensities of the solutions on a 15-cm line scale three times, which were while the solution was in the mouth, 5 s after expectorating it, and 1 min after expectorating it. Consumers were asked to pinch their nose while holding the solution in the mouth to focus on the taste. The sweet and bitter tastes perceived at this time would be called in-mouth sweetness and bitterness throughout this paper. The perceived intensities of sweet and bitter tastes 5 s after expectorating would be referred to as immediate sweetness and bitterness. A check-all-that-apply (CATA) question on the taste was followed after evaluating the immediate tastes, including terms collected from an open-ended question in the two preliminary tests ($n_1 = 31$ and $n_2 = 65$), asking if the consumers noticed any aftertaste. The term *pleasant* was added to the list as a positive word, *and spicy* was added as an attention check to identify careless respondents and would be removed from the correspondence analysis. The final list of CATA consisted of 15 terms, which were *artificial, bitter, chemical, honey, licorice, metallic, minty, pleasant, pungent, spicy, sweet, tangy, tart, tingling,* and *vanilla,* and the terms were listed in alphabetical order. A 45 s break was enforced after the CATA question, which was before evaluating the sweet and bitter tastes 1 min after expectorating. The perceived intensities would be considered as lingering sweetness and bitterness. Water and crackers were provided as palate cleansers in between samples.

2.5. Statistical Analyses

Data analysis was performed using XLSTAT (AddinSoft, New York, NY, USA). Intensity data were analyzed using a one-factor ANOVA model. For CATA analysis, the frequencies of each attribute were counted. Cochran's Q test was performed for each attribute to compare the difference among samples. Multiple pairwise comparisons using critical difference (Sheskin) were performed when the attribute was significant ($p < 0.05$). Correspondence analysis (CA) was generated to visually show the relationship between sensory attributes and samples. A two-way ANOVA model was used to determine the effect of PROP taster status, sweetener, and their interaction. Fisher's least significant difference (LSD) post hoc test was performed when $p < 0.05$. Agglomerative hierarchical clustering

(AHC) was used as a second way to classify PROP groups. Pearson correlation test was performed and correlation coefficients were calculated between PROP bitterness and sweet and bitter tastes of Reb A, D, and M combined over time (in-mouth, immediate, lingering sweetness and bitterness).

3. Results

3.1. Participant Characteristics

A total of 126 naïve consumers completed the study, with an average age of 23 ± 1.7 years and an average BMI of 24.7 ± 4.6 kg/m^2 based on self-reported height and weight. None of the participants had heart disease, cancer, or diabetes. The socio-demographics of participants are shown in Table 1. The majority were female (72.2%) and 60.3% of the participants identified themselves as white. Table 2 listed out the responses of sweetener consumption behavior questions. Sucrose (81.0%) and honey (69.8%) were the most commonly consumed sweetener on a regular basis (at least once a month), followed by stevia (19.8%), sucralose (19.0%), and aspartame (19.0%), which were high-intensity sweeteners. Other sweeteners consumed (8.7%) included maple syrup, brown sugar, xylitol, high fructose corn syrup, and acesulfame K. Sixty-seven percent of participants consumed low or zero sugar added products at least once a month. More than half of the participants (54.8%) said they were somewhat or very familiar with stevia.

Table 1. Socio-demographic characteristics of participants ($n = 126$).

Variable	Definition	Frequency	%
Gender			
	Male	35	27.8%
	Female	91	72.2%
Ethnicity			
	White	76	60.3%
	Hispanic or Latino	5	4.0%
	Asian or Pacific Islander	33	26.2%
	Black or African American	7	5.6%
	Native American or American Indian	0	0.0%
	Other	3	2.4%
	Prefer not to respond	2	1.6%
Education			
	Less than high school	0	0.0%
	High school diploma or GED	29	23.0%
	2-year college degree	4	3.2%
	4-year college degree	50	39.7%
	Graduate degree (Master's, Doctorate, etc.)	43	34.1%

Table 2. Participants' sweetener consumption behavior ($n = 126$).

Characteristic	Definition	Frequency	%
Sweetener consumption [1]			
	Agave nectar	16	12.7%
	Aspartame	24	19.0%
	Erythritol	10	7.9%
	Honey	88	69.8%
	Monk fruit extract	9	7.1%
	Saccharin	12	9.5%
	Stevia	25	19.8%
	Sucralose	24	19.0%
	Sucrose	102	81.0%
	Others	11	8.7%

Table 2. Cont.

Characteristic	Definition	Frequency	%
Low/zero sugar added product consumption frequency			
	More than 3 times a week	16	12.7%
	1–2 times a week	29	23.0%
	2–3 times a month	27	21.4%
	Once a month	12	9.5%
	Every other month	6	4.8%
	1–2 times per 6 months	15	11.9%
	Less than once a year	7	5.6%
	Almost never	14	11.1%
Familiarity with stevia			
	Very unfamiliar	25	19.8%
	Somewhat unfamiliar	21	16.7%
	Neutral	11	8.7%
	Somewhat familiar	55	43.7%
	Very familiar	14	11.1%

[1] This is a check-all-that-apply question.

3.2. Sensory Characteristics

3.2.1. Intensities of Sweet and Bitter Tastes

Table 3 summarizes the mean intensity ratings (± SEM) for four sweetener solutions evaluated by all participants. The decreasing trend in sweetness and bitterness intensities from in-mouth to immediate (5 s after expectorating the samples) to lingering (1 min after expectorating the samples) indicated that consumers followed the directions and evaluated the samples correctly, since a fading in intensity over time was expected.

Table 3. Mean intensity scores (± SEM) of sweetener solutions by participants ($n = 126$).

Sweetener	Sweetness [1]			Bitterness [2]		
	In-Mouth	Immediate	Lingering	In-Mouth	Immediate	Lingering
Sucrose	8.3 ± 0.3 [3] a [4]	7.1 ± 0.3 b	3.6 ± 0.3 b	0.8 ± 0.1 b	0.6 ± 0.1 c	0.4 ± 0.1 b
Reb A	7.2 ± 0.3 b	6.5 ± 0.3 b	4.3 ± 0.3 b	3.5 ± 0.3 a	3.5 ± 0.3 a	1.6 ± 0.2 a
Reb D	7.8 ± 0.3 ab	7.2 ± 0.3 b	4.5 ± 0.3 ab	1.1 ± 0.2 b	1.3 ± 0.2 b	0.6 ± 0.1 b
Reb M	8.6 ± 0.3 a	8.2 ± 0.3 a	5.3 ± 0.3 a	1.0 ± 0.2 b	0.9 ± 0.1 bc	0.6 ± 0.1 b

[1,2] In-mouth tastes (sweetness and bitterness) were evaluated when the solution was in the mouth; Immediate tastes were evaluated 5 s after expectorating the sample; Lingering tastes were evaluated 1 min after expectorating the sample. [3] Intensities were evaluated on a marked 15-cm line scale anchored with "not at all" to "extremely". [4] Different letters in the same column show the significant differences between sample means at $p < 0.05$ by Fisher's LSD.

The in-mouth sweetness of 14% sucrose and 0.1% Reb M were not significantly different ($p = 0.55$). The in-mouth sweetness of Reb D was slightly lower than sucrose but was still considered to be not different from sucrose ($p = 0.19$). Reb A showed significantly less in-mouth sweetness than Reb M and sucrose ($p < 0.01$ and $p < 0.05$, respectively). Reb M had the highest immediate sweetness among the samples and was significantly different from others. The sweetness of Reb M remained the highest after one minute. The lingering sweetness of Reb M (intensity = 5.3) was higher than Reb D (intensity = 4.5), but the difference was not significant ($p = 0.05$). Reb D was higher in lingering sweetness than sucrose (intensity = 3.6), but it was not significantly different ($p = 0.05$). The participants rated the in-mouth bitterness of sucrose, Reb D, and Reb M around 1, while the rating of Reb A was at 3.5 on a 15-cm line scale. The bitterness of Reb A persisted after 5 s (intensity = 3.5). Reb D was perceived to have more immediate bitterness than sucrose ($p < 0.05$), and there was no significant difference in the immediate bitterness between Reb M and sucrose ($p = 0.27$). While the lingering bitterness of sucrose, Reb D, and Reb M was at a minimum, Reb A still had detectable bitterness remaining (intensity = 1.6).

3.2.2. CATA

Table 4 summarizes the total counts of CATA attributes selected by the consumer panel ($n = 126$) to describe the aftertaste of each sweetener solution. The term *sweet* was the most frequently used term, and *spicy* was the least, which were as expected. Significant differences among samples were found in 10 out of 15 attributes ($p < 0.05$). Reb A, D, and M were described as *artificial* more frequent than sucrose. The *bitter* and *chemical* tastes of Reb A were significantly higher than other sweeteners, and fewer participants considered Reb A as *sweet* and *pleasant*. *Honey* and *vanilla* were checked the most for sucrose, followed by Reb D and M, while Reb A was rarely associated with these two terms. *Licorice, metallic, minty, pungent, spicy, tangy, tart*, and *tingling* were rarely selected by participants, with no more than 15 counts for each sample. Among those 8 less-checked terms, *licorice, pungent, spicy, tangy*, and *tingling* were not significantly different among samples.

Table 4. Total counts of check-all-that-apply attributes for sweetener solutions.

Attribute	Sucrose	Reb A	Reb D	Reb M
Artificial ***	38 b	83 a	64 a	69 a
Bitter ***	8 b	66 a	17 b	12 b
Chemical ***	9 b	42 a	17 b	18 b
Honey ***	41 a	8 c	25 b	24 b
Licorice ns	5	8	4	6
Metallic *	6 a	15 a	6 a	6 a
Minty **	7 ab	0 b	9 a	3 ab
Pleasant ***	65 a	25 b	49 a	57 a
Pungent ns	2	7	2	6
Spicy ns	0	0	2	1
Sweet***	110 a	83 b	110 a	114 a
Tangy ns	1	5	4	8
Tart *	2 ab	6 a	1 ab	0 b
Tingling ns	4	7	6	7
Vanilla ***	27 a	7 b	15 ab	15 ab

* indicates $p < 0.05$, ** indicates $p < 0.01$, *** indicates $p < 0.001$, and ns indicates no significant differences among samples. Different letters in the same row indicate the significant differences between sample means at $p < 0.05$ by Critical Difference (Sheskin).

The sensory attributes of sweeteners were summarized visually in Figure 1. The first two dimensions explained 96% of the variation. *Honey* and *vanilla* were associated with sucrose. Reb A was close to *metallic, bitter*, and *chemical*. Reb D and M were similar to each other and were closer to sucrose as compared to Reb A. Reb D and M were mostly associated with the positive words, but *artificial* was between Reb A and Reb D and M.

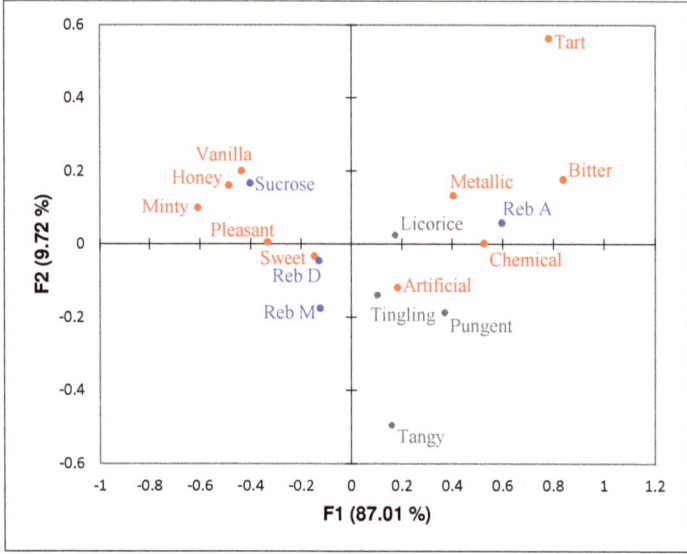

Figure 1. Correspondence analysis (CA) of sweeteners. Gray color indicates non-significant attributes; Red color indicates significant attributes; Samples are in blue.

3.3. PROP Bitterness

3.3.1. PROP Taster Groups

Out of 126 participants who completed the consumer test, there were 15 non-tasters, 81 medium tasters, and 30 supertasters. The interaction between taster groups and samples was not significant for all taste evaluations of each Reb A, D, and M solutions. There was also no significant difference when examining the main effect of taster groups on perceived intensity scores of sweet and bitter tastes of the three sweeteners combined over time (in-mouth, immediate, and lingering) (Figure 2).

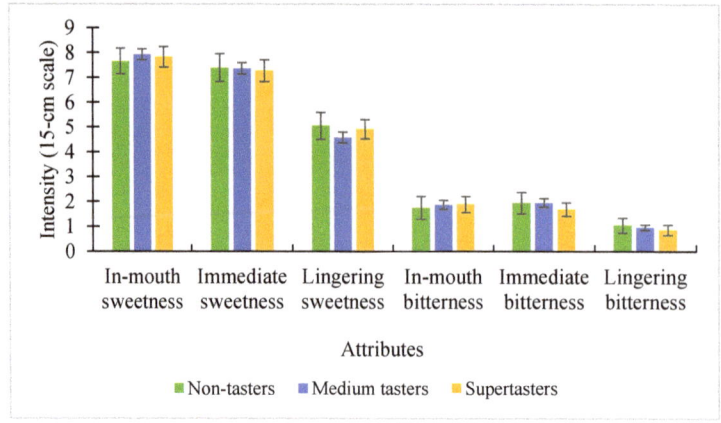

Figure 2. The influence of 6-n-Propylthiouracil (PROP) taster groups on the perceived intensities (± SEM), with sweeteners combined (Reb A, D, and M).

Due to the disproportional ratio of people in each taster groups, agglomerative hierarchical clustering (AHC) was used to group people based on their dissimilarity on the PROP rating (data not shown). Three groups were generated with 55, 44, and 30 people, corresponding to low, medium, and high-sensitive clusters, respectively. However, no significant difference in the main effect of clusters was found.

3.3.2. Relationships with Perceived Intensities of Reb A, D, and M

Pearson correlation tests were conducted to determine the association between PROP bitterness and perceived intensities of sweet and bitter tastes of Reb A, D, and M over time (in-mouth, immediate, lingering). No significant relationships existed between PROP bitterness and the rated intensities of the three steviol glycosides ($p > 0.05$).

4. Discussion

The present study investigated the sweetness and bitterness of Reb A, D, and M compared to sucrose at a high sucrose equivalent level (14% w/v) using consumers. To compare with 14% (w/v) sucrose solution, the solution concentration of the three steviol glycosides was determined by two small scale consumer tests as a preliminary test (see 2.4.1 for details). Briefly, the sweetness of 0.1% (w/v) Reb M was proved to be not significantly different from a 14% sucrose solution, and the same concentration was used for Reb A and D to compare the sensory characteristics of the three steviol glycosides at the same concentration. Prakash et al. [16] estimated that Reb M was about 200–350 times sweeter than sucrose, and the sweet potency at 10% SE was calculated to be 159. In the present study, the sweet potency of Reb M at 14% SE was calculated as 140. This is in line with the model from Prakash et al. [16], sweet potency of high-intensity sweeteners tended to decrease as the sucrose sweetness equivalent level increased [45].

The three steviol glycosides showed significant differences in sweetness and bitterness at the same concentration (0.1% w/v). The in-mouth sweetness of Reb D and M at 0.1% were not statistically different from sucrose at 14%, while 0.1% Reb A was less sweet than sucrose. Reb A was significantly less sweet than Reb M as well ($p < 0.01$) but was not significantly different from the in-mouth sweetness of Reb D ($p = 0.24$) with a tendency of being less sweet. Reb D and M were not significantly different in in-mouth sweetness ($p = 0.06$), but there was also a clear tendency of Reb D to be less sweet than Reb M. These results were consistent with the previous studies investigating the sweetness of Reb A, D, and/or M at different concentrations showing that Reb M was the sweetest sweetener and Reb A was the least sweet sweetener among Reb A, D, and M at the same concentration [15,16,46].

The sweetness temporal profile of Reb M was studied by Prakash et al. [16], who compared the sweetness appearance time and extinction time to examine the change in perception over 3 min. The sweetness of Reb M elicited later and persisted longer than sucrose at 10% SE in water. The descriptive panel rated the lingering sweetness of Reb M higher than that of sucrose as well [16]. In the present study, Reb M had a similar in-mouth sweetness to sucrose, but the lingering sweetness was significantly higher than that of sucrose, which corresponded with Prakash's finding. Reb A was also found to have a longer extinction time than sucrose [13] and exhibited persistent flavor duration in the mouth [8]. When at a similar sweetness level (i.e., at 8% SE), the lingering sweetness of Reb A and Reb M were not different [16]. Even though, in this study, there was no significant difference in lingering sweetness between sucrose and Reb A ($p = 0.12$), the lingering sweetness of Reb A became higher than sucrose after being rated less sweet in-mouth, which suggested that if the sweetness of Reb A was at the same level as sucrose, the lingering sweetness might be significantly higher than sucrose. Reb D, like Reb M, also had a similar in-mouth sweetness to sucrose, but had marginally higher lingering sweetness than sucrose ($p = 0.05$). When comparing Reb D to Reb M, the lingering sweetness of Reb D was marginally less than that of Reb M ($p = 0.05$) similar to the in-mouth sweetness of Reb D that was almost significantly less than Reb M ($p = 0.06$). Although the lingering sweetness of Reb M seemed to be stronger than Reb D, it may due to its higher initial sweetness than Reb D.

The bitterness of Reb A stood out among the samples when consumers first tasted the sample and the bitterness continued to be significantly different from others even after one minute. Many other researchers have reported the bitterness of Reb A [13,15]. On the other hand, Reb D and M did not show much in-mouth bitterness and had a similar intensity to sucrose. Even though Reb D exhibited a significantly higher immediate bitterness than sucrose, it was still considered low. A trained panel did not detect any significant bitter taste of Reb M when comparing to sucrose at 10% SE [16]. Hellfritsch et al. [15] and Ko et al. [47] indicated that Reb D elicited a lot less bitterness than Reb A. Our results confirmed that naïve consumers like trained assessors did not detect much bitterness from Reb D and M.

Based on the total counts of CATA and the CA, Reb A was associated with some negative perception terms, such as *bitter*, *chemical*, and *artificial*. The *bitter* and *chemical* attributes were significantly more selected for Reb A than Reb D, Reb M, and sucrose. The *bitter* attribute was in agreement with the bitterness intensity rating. The bitterness and chemical sensation of Reb A was reported by Fujimaru et al. [48] as well. Significantly less *sweet* was checked for Reb A than the other sweeteners, suggesting that the bitterness and chemical sensation might overshadow the sweetness of it. Reb D and M appeared to have good taste profiles because they were close to positive terms and sucrose. However, even though Reb D and M had low citations for *bitter* and higher citations for *pleasant* than Reb A, many participants still checked *artificial* significantly more frequent than sucrose. Waldrop and Ross [49] reported that consumers did not like stevia because of its association with *artificial* flavor. Thus, the *artificial* attribute may cause negative consumer perception of Reb D and M even though they are natural sweeteners without the aversive bitter aftertaste. Interestingly, the *artificial* attribute was also selected for sucrose by 38 participants. It is not common to drink pure sugar water in daily life, so the participants may not be familiar with the taste of sucrose solutions, and thus might select *artificial* for sucrose solution.

Licorice, *pungent*, *spicy*, *tangy*, and *tingling* were rarely selected by the participants and were not significant to discriminate the samples. Thus, these five terms may not be appropriate terms for consumers to describe the three steviol glycosides, even though *licorice* has been commonly used to describe the aftertaste of Reb A by researchers [10,13,50] and media [18]. The *licorice* taste of Reb A did not exhibit at low SE levels, but was elicited at higher SE levels [13], and this was further proved by Reyes et al. [50] that Reb A at 0.1% had more notable *licorice* taste than at 0.012%. In this study, we did not find the correlation between *licorice* and Reb A at 0.1%, since only 8 people out of 126 selected it, which suggested that *licorice* may not be an appropriate term for consumers to describe the aftertaste of Reb A or to discriminate Reb A, D, and M.

The CATA analysis also found that *vanilla* and *honey* were associated with sucrose. A consumer survey showed that honey was the most popular sugar alternative, which was natural, and natural sweeteners were perceived better than artificial sweeteners in general [3]. In this study, those who checked *honey* for steviol glycosides might imply that the sample gave them a sense of natural. As for *vanilla*, Lavin and Lawless [51] showed that an added vanilla flavor enhanced the perception of sweetness in milk, and Wang et al. [52] also indicated a taste-aroma interaction between perceived sweetness and vanilla flavor in skim milk. Vanilla was congruent with sweetness, so participants might choose the term even though the attribute was not presented in the solution.

A secondary objective of this study was to investigate the influence of consumers' PROP taster status on the sweetness and bitterness of Reb A, D, and M. We found that there were no significant differences in the perceived sweetness and bitterness of Reb A, D, and M (in-mouth, immediate, and lingering) among PROP taster groups. Risso et al. [40] reported that there was no correlation between PROP bitterness and stevioside bitterness. Humans have about 25 bitter taste receptors from the taste 2 receptors (hTAS2Rs) gene family [53]. Each receptor responds to different compounds but may have overlapped molecular range [54]. The sensitivity to the bitterness of PROP/PTC is mainly associated with TAS2R38 bitter taste receptor [55,56]. TAS2R4 and TAS2R14 responded to the bitterness of stevioside and Reb A in vitro, while TAS2R38 did not react [15]. Meyerhof et al. [54] sorted receptors into 4 groups, and both TAR2S4 and TAS2R14 were not in the same group as TAS2R38. The different responses in bitter taste receptors might explain why no relationship was found between PROP taster status and perceived bitterness intensity of Reb A, D, and M.

Some studies suggested that PROP bitterness sensitivity influenced other oral sensations, such as sweetness [36,37,57–61]. Drewnowski et al. [36] found a weak and marginal significant difference in sweetness perception of sucrose and saccharin between PROP tasters and non-tasters, and the difference was more significant at lower concentrations. Allen et al. [59] reported that the sweetness of acesulfame potassium was positively associated with PROP bitterness. A large sample size study ($n > 1500$) found a weak association between sweetness and PROP bitterness, suggesting that a bigger size sample is required to detect weak association with PROP [60]. A recent study confirmed that PROP bitterness was positively correlated with sweetness of sucrose [61]. However, some of the previous studies also indicated that there was no relationship between PROP sensitivity and sweet taste responsiveness [62–64]. Here, we found no significant differences in perceived sweetness intensity among PROP groups and further, no correlation between PROP bitterness and sweetness of the steviol glycosides. In this study, the test stimuli were singles (i.e., each sweetener solution), but the aftertaste of the three steviol glycosides, especially the sweet-bitter Reb A at a high concentration, might cause difficulties for participants to evaluate intensities of sweet and bitter tastes. Horne et al. [38] reported that sweet-bitter stimuli might be more difficult to evaluate than single taste stimuli due to taste–taste interactions. Expansive, linear, and compressive phases of psychophysical functions could be used to predict how taste stimuli would behave when mixed at low, medium, and high intensity/concentration [65]. For example, perceptual enhancement and suppression has been extensively reported at low and high intensity/concentration mixtures, corresponding to the expansive phase and compressive phase of the psychophysical function, respectively [65]. Ly and Drewnowski [63] showed a reduced difference in bitterness between PROP taster groups was found when the caffeine solution was sweetened, even though PROP tasters rated caffeine solution without sweetener as more bitter than non-tasters [63]. The perceptual suppression as a result of sweet-bitter interaction at a high intensity/concentration may explain no differences in perceived sweetness intensity among PROP groups and no correlations between PROP bitterness and perceived sweetness of Reb A, D, and M.

One possible limitation of the study was that consumers did not swallow the solution, which limited the number of taste buds utilized for the evaluation. Taste buds are distributed not only in the mouth but also in the throat [66]. Consumers were asked to expectorate the sample to reduce fatigue, however, the swallowing sensation could be different and might impact the perceived intensities. No hedonic question was asked in this study because it might be difficult for naïve consumers to rate the likings of pure sweetener solutions when the solutions were not regularly consumed in daily life. However, no association could be drawn between the negative CATA attributes and the likings of the sweeteners. Another limitation was the disproportional size of PROP groups, which only had 15 non-tasters. The data from non-tasters might be less variable if more non-tasters were recruited.

5. Conclusions

The present study investigated the sensory profile of Reb A, D, and M at 14% SE using a consumer panel, and the influence of PROP taster groups on the perceived sweet and bitter tastes of the three glycosides. Reb D and M had sensory profiles that were closer to sucrose, compared to Reb A, but were still associated with negative sensation, such as *artificial*, which may cause negative perception toward Reb D and M. The lingering sweetness of Reb D and M was also a concern. The sensory characteristics of Reb A, D, and M in this study can be used as a reference for the food industries working with steviol glycosides in high-sugar applications, such as frozen desserts. Furthermore, there were no significant differences among non-tasters, medium tasters, and supertasters on the perceived sweetness and bitterness of Reb A, D, and M as well as no significant correlations between PROP bitterness and perceived sweet and bitter tastes, suggesting that supertasters who experience more intense taste sensations may not report aversive sensations from stevia. Further studies on the consumer acceptance of Reb A, D, and M are needed to determine if these characteristics would affect the likings of these sweeteners. Since the sweeteners may perform differently in a food matrix than in aqueous solution, more research using steviol glycosides in final food products are needed to determine the sensory profile and acceptance of them.

Author Contributions: R.T. contributed to study design, data collection, data analysis, and manuscript writing. S.C. contributed to study design, data collection, data analysis, and critical revision of the manuscript. S.C. supervised the project. All authors have read and agreed to the published version of the manuscript.

Funding: This work was supported by the United States Department of Agriculture Specialty Crop Research Initiative (grant number 2017-5181-26828).

Acknowledgments: The authors thank Ingredion Incorporated for providing Reb A, D, and M.

Conflicts of Interest: The authors declare no conflict of interest.

References

1. Chattopadhyay, S.; Raychaudhuri, U.; Chakraborty, R. Artificial sweeteners—A review. *J. Food Sci. Technol.* **2014**, *51*, 611–621. [CrossRef] [PubMed]
2. Ott, D.B.; Edwards, C.L.; Palmer, S.J. Perceived taste intensity and duration of nutritive and non-nutritive sweeteners in water using time-intensity (T-I) evaluations. *J. Food Sci.* **1991**, *56*, 535–542. [CrossRef]
3. Mintel. Sugar and Alternative Sweeteners—US–December 2018—Market Research Report. Available online: https://reports-mintel-com.proxy2.cl.msu.edu/display/860879/ (accessed on 25 June 2020).
4. Brandle, J.E.; Starratt, A.N.; Gijzen, M. *Stevia rebaudiana*: Its agricultural, biological, and chemical properties. *Can. J. Plant. Sci.* **1998**, *78*, 527–536. [CrossRef]
5. Geuns, J.M.C. Stevioside. *Phytochemistry* **2003**, *64*, 913–921. [CrossRef]
6. Kinghorn, A.D. *Stevia: The Genus Stevia*; Taylor & Francis: London, UK, 2002; ISBN 978-0-203-16594-2.
7. Mintel Global New Products Database (GNPD). Available online: https://www.mintel.com/global-new-products-database (accessed on 25 June 2020).
8. Gwak, M.-J.; Chung, S.-J.; Kim, Y.J.; Lim, C.S. Relative sweetness and sensory characteristics of bulk and intense sweeteners. *Food Sci. Biotechnol.* **2012**, *21*, 889–894. [CrossRef]
9. Jenner, M.R.; Grenby, T.H. *Progress in Sweeteners*; Elsevier: London, UK, 1989; ISBN 978-1-85166-364-4.
10. Kim, M.-J.; Yoo, S.-H.; Jung, S.; Park, M.-K.; Hong, J.-H. Relative sweetness, sweetness quality, and temporal profile of xylooligosaccharides and luo han guo (Siraitia grosvenorii) extract. *Food Sci. Biotechnol.* **2015**, *24*, 965–973. [CrossRef]
11. Li, X.E.; Lopetcharat, K.; Drake, M.A. Parents' and children's acceptance of skim chocolate milks sweetened by monk fruit and stevia leaf extracts. *J. Food Sci.* **2015**, *80*, S1083–S1092. [CrossRef]
12. Medeiros, A.C.; Filho, E.R.T.; Bolini, H.M.A. Impact of natural and artificial sweeteners compounds in the sensory profile and preference drivers applied to traditional, lactose-free, and vegan frozen desserts of chocolate flavor. *J. Food Sci.* **2019**, *84*, 2973–2982. [CrossRef]
13. Prakash, I.; DuBois, G.E.; Clos, J.F.; Wilkens, K.L.; Fosdick, L.E. Development of rebiana, a natural, non-caloric sweetener. *Food Chem. Toxicol.* **2008**, *46*, S75–S82. [CrossRef]
14. Gray, N. Evolva and Cargill Publish Patent Application for "Next Generation" Fermented Sweeteners. Available online: https://www.foodnavigator.com/Article/2014/08/18/Evolva-and-Cargill-publish-patent-application-for-next-generation-fermented-sweeteners (accessed on 25 June 2020).
15. Hellfritsch, C.; Brockhoff, A.; Stähler, F.; Meyerhof, W.; Hofmann, T. Human psychometric and taste receptor responses to steviol glycosides. *J. Agric. Food Chem.* **2012**, *60*, 6782–6793. [CrossRef]
16. Prakash, I.; Markosyan, A.; Bunders, C. Development of next generation stevia sweetener: Rebaudioside, M. *Foods* **2014**, *3*, 162–175. [CrossRef] [PubMed]
17. PureCircle. Stevia Use in Food and Beverages Accelerated Significantly in 2018. Available online: https://purecircle.com/news/stevia-use-in-food-and-beverages-accelerated-significantly-in-2018/ (accessed on 25 June 2020).

18. Watson, E. GLG: We'll be Able to Breed Stevia Plants with More. Available online: https://www.foodnavigator.com/Article/2015/09/16/GLG-We-ll-be-able-to-breed-stevia-plants-with-more-Reb-D-Reb-M (accessed on 25 June 2020).
19. Watson, E. Amyris Bids for 30% Slice of Stevia Sweetener Market by 2022. Available online: https://www.foodnavigator.com/Article/2018/12/03/Amyris-bids-for-30-slice-of-stevia-sweetener-market-by-2022-with-Reb-M (accessed on 25 June 2020).
20. Goff, H.D. Ice cream and frozen desserts. In *Ullmann's Encyclopedia of Industrial Chemistry*; American Cancer Society: Atlanta, GA, USA, 2015; pp. 1–15; ISBN 978-3-527-30673-2.
21. Varela, P.; Ares, G. Sensory profiling, the blurred line between sensory and consumer science. A review of novel methods for product characterization. *Food Res. Int.* **2012**, *48*, 893–908. [CrossRef]
22. Moskowitz, H.R. Experts versus consumers: A comparison. *J. Sens. Stud.* **1996**, *11*, 19–37. [CrossRef]
23. Husson, F.; Le Dien, S.; Pagès, J. Which value can be granted to sensory profiles given by consumers? Methodology and results. *Food Qual. Prefer.* **2001**, *12*, 291–296. [CrossRef]
24. Worch, T.; Lê, S.; Punter, P. How reliable are the consumers? Comparison of sensory profiles from consumers and experts. *Food Qual. Prefer.* **2010**, *21*, 309–318. [CrossRef]
25. Ares, G.; Bruzzone, F.; Giménez, A. Is a consumer panel able to reliably evaluate the texture of dairy desserts using unstructured intensity scales? Evaluation of global and individual performance. *J. Sens. Stud.* **2011**, *26*, 363–370. [CrossRef]
26. Moskowitz, H.R. Base size in product testing: A psychophysical viewpoint and analysis. *Food Qual. Prefer.* **1997**, *8*, 247–255. [CrossRef]
27. Ares, G.; Tárrega, A.; Izquierdo, L.; Jaeger, S.R. Investigation of the number of consumers necessary to obtain stable sample and descriptor configurations from check-all-that-apply (CATA) questions. *Food Qual. Prefer.* **2014**, *31*, 135–141. [CrossRef]
28. Dooley, L.; Lee, Y.; Meullenet, J.-F. The application of check-all-that-apply (CATA) consumer profiling to preference mapping of vanilla ice cream and its comparison to classical external preference mapping. *Food Qual. Prefer.* **2010**, *21*, 394–401. [CrossRef]
29. Adams, J.; Williams, A.; Lancaster, B.; Foley, M. Advantages and uses of check-all-that-apply response compared to traditional scaling of attributes for salty snacks. *Sens. Sci. Symp.* **2007**, *16*.
30. Ares, G.; Barreiro, C.; Deliza, R.; Giménez, A.; Gámbaro, A. Application of a check-all-that-apply question to the development of chocolate milk desserts. *J. Sens. Stud.* **2010**, *25*, 67–86. [CrossRef]
31. Ares, G.; Varela, P.; Rado, G.; Giménez, A. Are consumer profiling techniques equivalent for some product categories? The case of orange-flavoured powdered drinks. *Int. J. Food Sci. Technol.* **2011**, *46*, 1600–1608. [CrossRef]
32. Plaehn, D. CATA penalty/reward. *Food Qual. Prefer.* **2012**, *24*, 141–152. [CrossRef]
33. Bartoshuk, L.M.; Duffy, V.B.; Miller, I.J. PTC/PROP tasting: Anatomy, psychophysics, and sex effects. *Physiol. Behav.* **1994**, *56*, 1165–1171. [CrossRef]
34. Simons, C.T.; Adam, C.; LeCourt, G.; Crawford, C.; Ward, C.; Meyerhof, W.; Slack, J.P. The "bitter-sweet" truth of artificial sweeteners. In *Sweetness and Sweeteners*; Weerasinghe, D.K., DuBois, G.E., Eds.; ACS Symposium Series; American Chemical Society: Washington, DC, USA, 2008; pp. 335–354, ISBN 978-0-8412-7432-7.
35. Bartoshuk, L.M. Bitter taste of saccharin related to the genetic ability to taste the bitter substance 6-n-Propylthiouracil. *Science* **1979**, *205*, 934–935. [CrossRef]
36. Drewnowski, A.; Henderson, S.A.; Shore, A.B. Genetic sensitivity to 6-n-propylthiouracil (PROP) and hedonic responses to bitter and sweet tastes. *Chem. Senses* **1997**, *22*, 27–37. [CrossRef]
37. Zhao, L.; Tepper, B.J. Perception and acceptance of selected high-intensity sweeteners and blends in model soft drinks by propylthiouracil (PROP) non-tasters and super-tasters. *Food Qual. Prefer.* **2007**, *18*, 531–540. [CrossRef]
38. Horne, J.; Lawless, H.T.; Speirs, W.; Sposato, D. Bitter taste of saccharin and acesulfame-K. *Chem. Senses* **2002**, *27*, 31–38. [CrossRef]
39. Rankin, K.M.; Godinot, N.; Tepper, B.J.; Kirkmeyer, S.V.; Christensen, C.M. Assessment of different methods for PROP status classification. In *Genetic Variation in Taste Sensitivity*; Prescott, J., Tepper, B.J., Eds.; Marcel Dekker: New York, NY, USA, 2004; pp. 63–88, ISBN 0-8247-4087-4.

40. Risso, D.; Morini, G.; Pagani, L.; Quagliariello, A.; Giuliani, C.; De Fanti, S.; Sazzini, M.; Luiselli, D.; Tofanelli, S. Genetic signature of differential sensitivity to stevioside in the Italian population. *Genes Nutr.* **2014**, *9*, 401. [CrossRef]
41. Zhao, L.; Kirkmeyer, S.V.; Tepper, B.J. A paper screening test to assess genetic taste sensitivity to 6-n-propylthiouracil. *Physiol. Behav.* **2003**, *78*, 625–633. [CrossRef]
42. Green, B.G.; Dalton, P.; Cowart, B.; Shaffer, G.; Rankin, K.; Higgins, J. Evaluating the 'labeled magnitude scale' for measuring sensations of taste and smell. *Chem. Senses* **1996**, *21*, 323–334. [CrossRef]
43. Bartoshuk, L.M. Comparing sensory experiences across individuals: Recent psychophysical advances illuminate genetic variation in taste perception. *Chem. Senses* **2000**, *25*, 447–460. [CrossRef] [PubMed]
44. Low, J.Y.Q.; McBride, R.L.; Lacy, K.E.; Keast, R.S.J. Psychophysical evaluation of sweetness functions across multiple sweeteners. *Chem. Senses* **2017**, *42*, 111–120. [CrossRef] [PubMed]
45. Wee, M.; Tan, V.; Forde, C. A comparison of psychophysical dose-response behaviour across 16 sweeteners. *Nutrients* **2018**, *10*, 1632. [CrossRef] [PubMed]
46. Kinghorn, A.D.; Kim, N.C.; Kim, D.H.L. Terpenoid glycoside sweeteners. In *Naturally Occurring Glycosides*; Ikan, R., Ed.; John Wiley & Sons: Hoboken, NJ, USA, 1999; pp. 399–429; ISBN 978-0-471-98602-7.
47. Ko, W.-W.; Kim, S.-B.; Chung, S.-J. Effect of concentration range on the accuracy of measuring sweetness potencies of sweeteners. *Food Qual. Prefer.* **2020**, *79*, 103753. [CrossRef]
48. Fujimaru, T.; Park, J.-H.; Lim, J. Sensory characteristics and relative sweetness of tagatose and other sweeteners. *J. Food Sci.* **2012**, *77*, S323–S328. [CrossRef]
49. Waldrop, M.E.; Ross, C.F. Sweetener blend optimization by using mixture design methodology and the electronic tongue. *J. Food Sci.* **2014**, *79*, S1782–S1794. [CrossRef]
50. Reyes, M.M.; Castura, J.C.; Hayes, J.E. Characterizing dynamic sensory properties of nutritive and nonnutritive sweeteners with temporal check-all-that-apply. *J. Sens. Stud.* **2017**, *32*, e12270. [CrossRef]
51. Lavin, J.G.; Lawless, H.T. Effects of color and odor on judgments of sweetness among children and adults. *Food Qual. Prefer.* **1998**, *9*, 283–289. [CrossRef]
52. Wang, G.; Hayes, J.E.; Ziegler, G.R.; Roberts, R.F.; Hopfer, H. Dose-response relationships for vanilla flavor and sucrose in skim milk: Evidence of synergy. *Beverages* **2018**, *4*, 73. [CrossRef]
53. Chandrashekar, J.; Mueller, K.L.; Hoon, M.A.; Adler, E.; Feng, L.; Guo, W.; Zuker, C.S.; Ryba, N.J.P. T2Rs function as bitter taste receptors. *Cell* **2000**, *100*, 703–711. [CrossRef]
54. Meyerhof, W.; Batram, C.; Kuhn, C.; Brockhoff, A.; Chudoba, E.; Bufe, B.; Appendino, G.; Behrens, M. The molecular receptive ranges of human TAS2R bitter taste receptors. *Chem. Senses* **2010**, *35*, 157–170. [CrossRef]
55. Duffy, V.B.; Davidson, A.C.; Kidd, J.R.; Kidd, K.K.; Speed, W.C.; Pakstis, A.J.; Reed, D.R.; Snyder, D.J.; Bartoshuk, L.M. Bitter receptor gene (TAS2R38), 6-n-Propylthiouracil (PROP) bitterness and alcohol intake. *Alcohol. Clin. Exp. Res.* **2004**, *28*, 1629–1637. [CrossRef]
56. Kim, U.; Jorgenson, E.; Coon, H.; Leppert, M.; Risch, N.; Drayna, D. Positional cloning of the human quantitative trait locus underlying taste sensitivity to phenylthiocarbamide. *Science* **2003**, *299*, 1221–1225. [CrossRef] [PubMed]
57. Hayes, J.E.; Bartoshuk, L.M.; Kidd, J.R.; Duffy, V.B. Supertasting and PROP bitterness depends on more than the TAS2R38 gene. *Chem. Senses* **2008**, *33*, 255–265. [CrossRef] [PubMed]
58. Tepper, B.J. Nutritional implications of genetic taste variation: The role of PROP sensitivity and other taste phenotypes. *Annu. Rev. Nutr.* **2008**, *28*, 367–388. [CrossRef] [PubMed]
59. Allen, A.L.; McGeary, J.E.; Knopik, V.S.; Hayes, J.E. Bitterness of the non-nutritive sweetener acesulfame potassium varies with polymorphisms in TAS2R9 and TAS2R31. *Chem. Senses* **2013**, *38*, 379–389. [CrossRef] [PubMed]
60. Hwang, L.-D.; Breslin, P.A.S.; Reed, D.R.; Zhu, G.; Martin, N.G.; Wright, M.J. Is the Association Between Sweet and Bitter Perception due to Genetic Variation? *Chem. Senses* **2016**, *41*, 737–744. [CrossRef] [PubMed]
61. Nolden, A.A.; McGeary, J.E.; Hayes, J.E. Predominant Qualities Evoked by Quinine, Sucrose, and Capsaicin Associate with PROP Bitterness, but not TAS2R38 Genotype. *Chem. Senses* **2020**, *45*, 383–390. [CrossRef] [PubMed]
62. Drewnowski, A.; Henderson, S.A.; Shore, A.B.; Barratt-Fornell, A. Nontasters, tasters, and supertasters of 6-n-Propylthiouracil (PROP) and hedonic response to sweet. *Physiol. Behav.* **1997**, *62*, 649–655. [CrossRef]

63. Ly, A.; Drewnowski, A. PROP (6-n-Propylthiouracil) Tasting and sensory responses to caffeine, sucrose, neohesperidin dihydrochalcone and chocolate. *Chem. Senses* **2001**, *26*, 41–47. [CrossRef] [PubMed]
64. Lim, J.; Urban, L.; Green, B.G. Measures of individual differences in taste and creaminess perception. *Chem. Senses* **2008**, *33*, 493–501. [CrossRef] [PubMed]
65. Keast, R.S.J.; Breslin, P.A.S. An overview of binary taste–taste interactions. *Food Qual. Prefer.* **2003**, *14*, 111–124. [CrossRef]
66. Kikut-Ligaj, D.; Trzcielińska-Lorych, J. How taste works: Cells, receptors and gustatory perception. *Cell. Mol. Biol. Lett.* **2015**, *20*, 699–716. [CrossRef]

© 2020 by the authors. Licensee MDPI, Basel, Switzerland. This article is an open access article distributed under the terms and conditions of the Creative Commons Attribution (CC BY) license (http://creativecommons.org/licenses/by/4.0/).

Article

Hydrothermally Treated Soybeans Can Enrich Maize Stiff Porridge (Africa's Main Staple) without Negating Sensory Acceptability

Martin Kalumbi [1], **Limbikani Matumba** [2,*], **Beatrice Mtimuni** [1], **Agnes Mwangwela** [3] **and Aggrey P. Gama** [3]

1. Department of Human Nutrition and Health, Lilongwe University of Agriculture and Natural Resources (LUANAR), Box 219 Lilongwe, Malawi; martinkalumbi@yahoo.com (M.K.); bmtimuni@gmail.com (B.M.)
2. Food Technology and Nutrition Research Group—NRC, Lilongwe University of Agriculture and Natural Resources (LUANAR), Box 143 Lilongwe, Malawi
3. Department Food Science, Lilongwe University of Agriculture and Natural Resources (LUANAR), Box 219 Lilongwe, Malawi; amwangwela@luanar.ac.mw (A.M.); aggreyg@yahoo.co.uk (A.P.G.)
* Correspondence: lmatumba@luanar.ac.mw or alimbikani@gmail.com; Tel.: +265-999682549

Received: 28 October 2019; Accepted: 3 December 2019; Published: 6 December 2019

Abstract: Maize-based stiff porridge, a starchy protein-deficient staple food, dominates among the populations in sub-Saharan Africa (SSA). Unfortunately, this is often consumed along with leafy vegetables since the majority of the population in this region lack resources for the purchase of high protein animal source foods, a situation that exacerbates protein-energy malnutrition. Considering this, the current study evaluated the effect of enriching maize-based stiff porridge with flour made from hydrothermally treated soybeans on consumer acceptability. A total of nine experimental flours were prepared from maize and maize-soybean mixtures following a 3^2 factorial design involving two factors, namely maize flour type (whole maize, non-soaked dehulled maize, and soaked dehulled maize) and soybean flour proportion (0%, 20%, and 30%). A total of 125 adult consumers from a rural setting in Malawi evaluated maize-based stiff porridges made thereof using a 7-point hedonic scale. Subsequently, the participants were asked to guess an ingredient that was added to some of the test samples. The 10% and 20% soybean-enriched maize-based stiff porridges scored 5/7 and above, with some being statistically similar to plain maize-based stiff porridges. No participant recognized that soybeans were incorporated into the maize-based stiff porridges. The study has clearly demonstrated the potential of enriching maize-based stiff porridge with hydrothermally treated soybeans without compromising consumer acceptability. This innovation could significantly contribute towards reducing the burden of energy-protein under-nutrition in SSA.

Keywords: maize-based stiff porridge; enrichment; hydrothermally treated soybeans; acceptability

1. Introduction

Protein-energy malnutrition is a major public health concern in sub-Saharan Africa (SSA) [1,2]. Beside food insufficiency, it is because diets in SSA are predominantly starchy with low levels of proteins. Maize (*Zea mays* L.), a cereal with low protein content (10%), is the key staple crop in SSA which is largely consumed as a main dish during lunch and dinner in the form of a stiff porridge. The stiff porridge is usually prepared using flour made from whole maize, non-fermented degermed-dehulled maize or fermented degermed-dehulled maize and is called *nsima* in Malawi, *nshima* in Zambia, *sadza* in Zimbabwe, *ugali* in Kenya and Tanzania, *mielie pap* in South Africa, and many other names across SSA. Unfortunately, the majority of the population in SSA lack resources for the purchase of high protein animal source foods such as meat, milk, eggs, and fish protein [3]. Consequently, they consume this protein-deficient staple along with vegetable relish, thus exacerbating protein malnutrition in the region [4].

Legumes are nutritionally valuable protein sources (20–45%) that offer a cheap option for upgrading the nutritive value of cereal-based foods. Owing to the high protein content, soybeans are the most popular legume used to enrich porridge served as breakfast for the whole family, and as a complementary food for infants and young children in SSA [5,6]. The soybeans are usually roasted or are extruded together with the maize to take care of anti-nutritional factors, off-beany flavor, and reduce the porridge preparation time. However, the roasted and extruded soybean products cannot be used to enrich stiff porridge due to the formation of flavor compounds. Consumers in sub-Saharan Africa are very accustomed to a maize-based stiff porridge that is essentially flat and flavorless, and any deviations from this norm are likely to be objectionable [7–10]. Likewise, it is also impractical to use raw maize-soybean flour mixture as it takes so long to cook and besides, the stiff porridge prepared in this manner retains objectionable off-beany flavors.

Off-beany flavors in soybeans described as "painty, bean, green, and unpleasant" are an activity of endogenous lipoxygenase enzymes which catalyze the insertion of oxygen into polyunsaturated fatty acids producing hydroperoxides, which subsequently participate in autoxidation process leading to the production of off-flavor compounds—particularly hexanal and heptanal [11]. Lipoxygenase works almost instantaneously during the grinding of soybeans in an ambient environment [12]. However, soymilk research has proven that it is possible to hydrothermally deactivate lipoxygenase before it catalyzes the oxidation of polyunsaturated fatty acids at 80 °C or higher and produce acceptable milk [12–14].

Despite the feasibility and widespread application of hydrothermal deactivation of lipoxygenase in soymilk industry, no report is available about the evaluation of this technique in the preparation of odorless products for the enrichment of protein-deficient maize stiff porridge. Therefore, the objective of this study was to investigate the effect enriching maize-based stiff porridge with flour made from hydrothermally treated soybeans on consumer acceptability. In this study, the effect of soybean proportion and the maize flour type are investigated, and the nutrition implication of soybean enrichment is discussed.

2. Materials and Methods

2.1. Experimental Flours

A total of nine experimental flours were prepared from maize and maize-soybean mixtures following a 3^2 factorial design involving two factors, namely maize flour type (whole maize, non-soaked dehulled maize, and soaked dehulled maize) and soybean proportion (0%, 20%, and 30%) as summarized in Figure 1.

The soybeans (Tikolore variety, obtained from Chitedze Research Station, Lilongwe, Malawi) were first hydrothermally treated to take care of the beany-grassy flavor in the following manner: Clean soybeans were gradually introduced into boiling water without causing the water to stop boiling (one part soybean to three parts water) for an hour, cooled in water, and manually dehulled by means of scrubbing the soybeans between two hands to force the hulls from the cotyledons. The dehulled soybeans were then washed and sundried for two days. To ensure homogeneity of the maize-soybean flour mixtures, the dried thermally processed soybean beans were thoroughly mixed with the maize portion (dry maize or dry dehulled maize grits) wherever applicable before the milling process (see Figure 1). In this study the authors assumed general nutrient composition data available in literature for soybean and a maize. In general, soybeans contain approximately 40% proteins and 20 % lipids [15], while maize contains only about 10% proteins and 4.5% lipids [16].

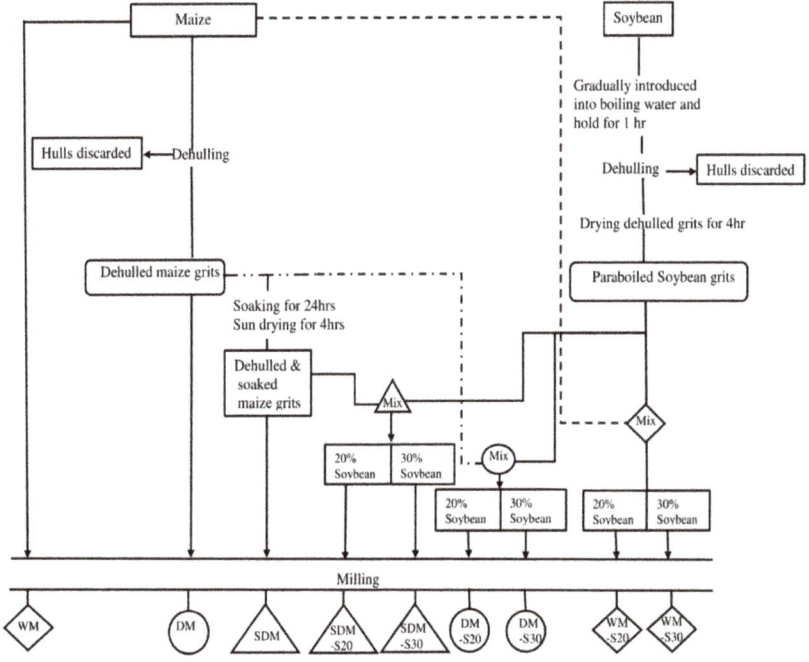

Figure 1. Scheme of flour preparation. DM, dehulled maize; DM-S20, dehulled maize with 20% soybean; DM-S30, dehulled maize with 30% soybean; WM, whole maize; WM-S20, whole maize with 20% soybean; WM-S320, whole maize with 30% soybean; SDM, soaked dehulled maize; SDM, soaked dehulled maize; SDM-S20, soaked dehulled maize with 20% soybean; SDM-S30, soaked dehulled maize with 30% soybean.

2.2. Preparation of Maize Stiff Porridges and Warming

Five experienced women were engaged to jointly prepare nine maize stiff porridges, using each of the nine flours described in Figure 1, 2–3 h before the sensory evaluation test. The maize stiff porridge recipe involved cooking a mixture of 1500 mL of water and 600 g of grain flour for 40 min on an electric cooker, which was then immediately transferred into in tightly fitted food warmer ready for sensory evaluation.

2.3. Consumer Acceptability

A total of 125 adult consumers (aged 18 years and above), without soy allergy, who regularly consume maize-based stiff porridge, were recruited from Lilongwe, rural Malawi, for this study after giving their informed consent. The recruits (participants) were predominantly female (63.2%), and more information about the participants is given in Supplementary Table S1.

Considering that consumer opinions are influenced by how much information has been presented to them about the products before tasting [17,18], the panelist were not informed that some of the test samples (maize-based stiff porridges) had been enriched with soybean flour. However, to screen out soy allergic consumers without compromising the blinding of participants (regarding the presence of soybean in some test maize-based stiff porridges), the participants were individually asked to indicate if they have problems with common allergens including fish, pork, meat, egg, peanut, tree nut, wheat, and soy.

Each participant independently evaluated all of the nine maize-based stiff porridge samples (~8 g of each sample). The samples were presented one after the other, coded with random 4-digit numbers.

The samples were served on white plates using a completely randomized balanced block design. In additional to asking the participants to rinse their palates with distilled water, five-minute breaks were also included between sample evaluations to minimize carry-over effects. A 7-point facial hedonic scale (1 = super bad, 7 = super good) was used to score the participant's overall acceptability of each sample to accommodate those who did not attend any form of education or had low-level education attainment (Supplementary Table S1). A score of 5 (good) was taken as the lower limit of acceptability.

Upon evaluating the nine test samples, the participants were asked to guess an ingredient that was added to some of the test samples. Finally, the participants were informed about the inclusion of soybean and the processes involved.

2.4. Data Analysis

A two-way ANOVA was used to test the main effects and interactions among the independent variables (maize flour type and soybean proportion). Post-hoc mean separations were carried out using Tukey's honestly significant difference test at a significance level of 0.05 ($p < 0.05$). All statistical analyses were done in XLSTAT 2017 (version 19.01, Addinsoft, New York, NY, USA).

3. Results

Results of overall hedonic ratings of the maize-based stiff porridge samples are presented in Figure 2. Irrespective of maize flour type, all blends incorporating 20% soybean (DM-S20, SDM-S20, and WM-S20) were regarded as acceptable with average hedonic scores of at least 5, the lower limit for acceptability on the 7-point hedonic scale. Except for the maize-based stiff porridge prepared from flour blend of 30% soybean and dehulled maize flour (DM-S30), incorporation of 30% soybean had a negative effect on acceptability.

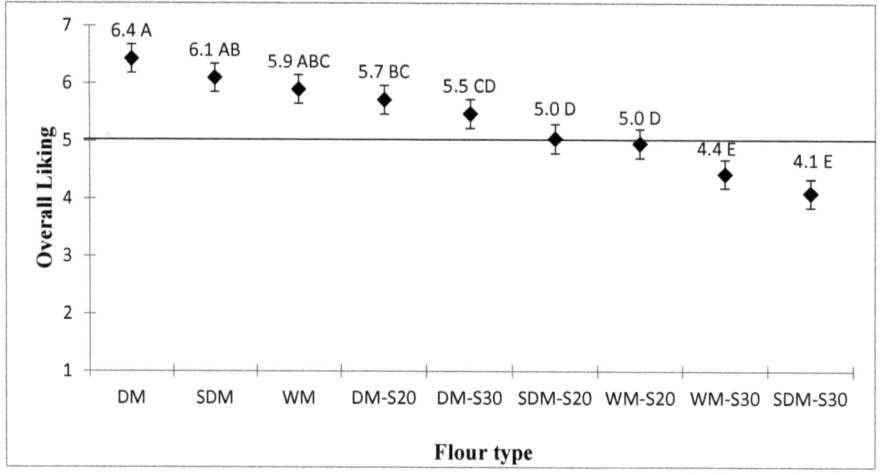

Figure 2. Overall liking mean score of maize-based stiff porridges. DM, dehulled maize; DM-S20, dehulled maize with 20% soybean; DM-S30, dehulled maize with 30% soybean; WM, whole maize; WM-S20, whole maize with 20% soybean; WM-S20, whole maize with 30% soybean; SDM, soaked dehulled maize; SDM-S20, soaked dehulled maize with 20% soybean; SDM-S30, soaked dehulled maize with 30% soybean. The different letters indicate significant differences among the flour types.

However, stiff porridges made from blended flours of non-soaked dehulled maize (DM), irrespective of soybean proportion, had relatively higher acceptability scores that were not significantly different from soaked dehulled flour (SDM) and whole maize (WM) plain flours, respectively ($p > 0.05$). It is therefore not surprising that maize flour type, soybean proportion, and the interaction between maize flour type and soybean proportion had a significant effect on the acceptability of the resultant stiff porridges as shown in Table 1 as well as Figures 2 and 3.

Table 1. Statistical results for effect of maize flour type and soybean proportion.

Variable	DF	Sum of Squares	Mean Squares	F	Pr > F
Maize flour type [1]	2	154.483	77.241	38.496	<0.0001
Soybean proportion [2]	2	418.332	209.166	104.246	<0.0001
Maize flour type * Soybean proportion	4	35.593	8.898	4.435	0.001

[1] Maize flour types [dehulled maize (DM); whole maize (WM); soaked dehulled maize (SDM)]; [2] Soybean proportions (0% soybean; 20% soybean; 30% soybean); DF, degrees of freedom; F, Fisher-Snedecor distribution; Pr, probability. * stands for interactions between variables.

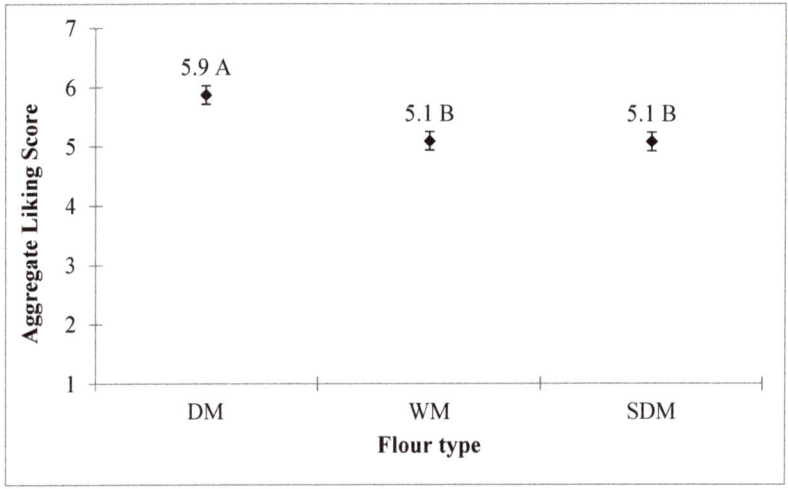

Figure 3. Effect of flour type on acceptance of the resultant stiff porridges. The aggregated liking score is the mean score computed based on flour type, irrespective of the proportion of soybean in the flour mixture. DM, dehulled maize; WM, whole maize; SDM, soaked dehulled maize. The different letters indicate significant differences among the flour types.

When asked to guess what was added to the stiff porridge, none of the participants successfully recognized that soybean flour was incorporated in the maize-based stiff porridges. Nonetheless, participants reported that they suspected that something might have been added to some of the maize-based stiff porridges and distorted their texture.

4. Discussion

Owing to the high prevalence rates of malnutrition in SSA, a search for practical ways of improving nutrient densities of staple diets in the region is merited. This is the first study to report on the effect of incorporating of hydrothermally treated soybeans into maize-based stiff porridge on consumer acceptability. The results have shown that it is possible to enrich maize-based stiff porridge with a substantial proportion of hydrothermally treated soybeans without compromising consumer acceptability. Consumers in sub-Saharan Africa are very accustomed to a maize-based stiff porridge that is essentially flavorless, and any deviations from this norm is objectionable [7–10]. Therefore, the findings of this study suggest that unlike roasting, the hydrothermal treatment of soybeans did not result in the formation of other extraneous flavors, likely because of the unfavorable conditions for browning reactions such as Maillard reaction.

Although most participants reported that they consume stiff porridge made from whole maize flour (WM) more often than from dehulled maize flour (DM) (Supplementary Table S1), in this study the participants highly liked stiff porridges made from DM. Likewise, among the blended flours, all blends of DM and soybean were highly ranked. Strange as it may seem, it is worth noting that sometimes socioeconomic restrictions override preferences, such that food is eaten just for survival in some situations [19,20]. The dehulling process results in approximately 20% loss of the maize flour quantity [21]. With the usual scarcity of food in SSA, people are likely to opt for WM even if they do not like it much to avoid the processing loss. It is, therefore, encouraging that WM blended with 20% soybean (WM-S20) was still acceptable as it could be a viable nutritious option even when food is scarce.

The novel maize-soybean stiff porridge provides a viable nutritious alternative staple for people living in the SSA. The stiff porridge made from maize-soybean blended flour not only contains higher amounts of protein, but also a higher content of lipids and micronutrients than the common maize-based stiff porridge. Since soybeans contain approximately 40% proteins and 20% lipids [15], while maize contains only about 10% proteins and 4.5% lipids [16], the novel 20% soybean stiff porridge typically represents a boost of 60% in protein and 69% in lipid composition (flour blend compositions: 16.0% protein and 7.6% lipid). This study preceded our published research, where we evaluated the efficacy of utilizing soybean-enriched stiff porridge as a strategy for managing human immunodeficiency virus (HIV)-related wasting among resource-poor people [22]. The 20% soybean-enriched stiff porridge induced a mean 2.9 kg cumulative mean weight gain among HIV-positive rural women within 3-month study period corresponding to body mass index (BMI) improvement from 19.3 to 21.1 kg/m^2.

It is worth highlighting here that there have been exploratory initiatives aimed at enriching staple foods with legumes before. Notably, Bressani et al. [23] evaluated the effect of enriching tortillas with soybeans. Later, Nyotu et al. [24] evaluated the effect of incorporating commercial defatted soy flour into stiff maize porridge using an untrained panel consisting of 12 Kenyan students, and recorded a positive outcome. Likewise, several researchers explored the enrichment of cassava-based stiff porridges (fufu) with cowpea and Bambara flours at a laboratory scale with success [25,26]. However, there is no record of the scaling up such innovations, possibly due to some socioeconomic hindrances. It is therefore imperative that future initiatives focus on understanding potential deterrents of the current innovation in order to make our innovation successful beyond the pilot scale.

The study had certain limitations in that it did not systematically optimize the maize stiff porridges as it only investigated products with soybean proportions of 0, 20, and 30%. While stiff porridges containing 30% of soybean were less liked by the participants, it is possible that higher proportions of soybean between 20% and 30% would give a favorable outcome. Further, the study did not investigate the effect of various processes involved in the preparation of the soybean on nutrient retention and bioavailability. Such information would help to optimize nutrient losses and improve nutrient use efficiency. Nonetheless, given the positive outcome recorded in our research with HIV-positive women [22], there is clear evidence that the new soybean-enriched stiff porridge is superior to the traditional stiff porridge.

5. Conclusions

The present study has demonstrated that it is possible to enrich maize-based stiff porridge with hydrothermally treated soybeans without compromising acceptability. Although the study involved participants from a single locality (Lilongwe, Malawi), the results herein may have wide application across SSA, particularly among the resource-limited poor who have minimal access to animal source protein foods, considering that soybeans potentially grow almost everywhere across the region. Unlike soybean-based ordinary porridges that are served as breakfast for the whole family and as complementary food for infants and young children, soybean-enriched maize stiff porridges have the potential of significantly contributing to the fight of energy-protein under-nutrition as stiff porridges are consumed by almost everyone in optimal quantities (dry matter basis) and regularly during lunch and supper. Future efforts should assess the feasibility of this innovation based on cost.

Supplementary Materials: The following are available online at http://www.mdpi.com/2304-8158/8/12/650/s1, Table S1: Demographic and socioeconomic characteristics of study participants and the type of the maize which they traditionally consume at their homes.

Author Contributions: M.K. was involved in designing the study, as well as acquisition, analysis, and interpretation of data, and drafting and revision of the manuscript. L.M. and B.M. were involved in the conception and design of the study, as well as analysis, and interpretation of data, and revision of the manuscript. A.M. and A.P.G. were involved in the analysis and interpretation of data, and revision of the manuscript.

Funding: The study was funded by USAID World Learning through the Malawi Scholarship Program.

Conflicts of Interest: The authors declare that they have no conflict of interest.

Ethical approval: All procedures performed in studies involving human participants were in accordance with the ethical standards of the National Health Science Research Committee under the Ministry of Health and Population is registered with the USA Office for Human Research Protections (OHRP) as an international IRB number IRB00003905FWA00005976 and registered as study number NHSRC#16/6/1614 and with the 1964 Helsinki declaration and its later amendments or comparable ethical standards.

Data availability: The data that support the findings of this study are available from the corresponding author upon reasonable request.

References

1. Schönfeldt, H.C.; Hall, N.G. Dietary protein quality and malnutrition in Africa. *Br. J. Nutr.* **2012**, *108*, S69–S76. [CrossRef] [PubMed]
2. Akombi, B.J.; Agho, K.E.; Merom, D.; Renzaho, A.M.; Hall, J.J. Child malnutrition in sub-Saharan Africa: A meta-analysis of demographic and health surveys (2006-2016). *PLoS ONE* **2017**, *12*, e0177338. [CrossRef] [PubMed]
3. Hetherington, J.B.; Wiethoelter, A.K.; Negin, J.; Mor, S.M. Livestock ownership, animal source foods and child nutritional outcomes in seven rural village clusters in Sub-Saharan Africa. *Agric. Food Secur.* **2017**, *6*, 9. [CrossRef]
4. Uusiku, N.P.; Oelofse, A.; Duodu, K.G.; Bester, M.J.; Faber, M. Nutritional value of leafy vegetables of sub-Saharan Africa and their potential contribution to human health: A review. *J. Food Compos. Anal.* **2010**, *23*, 499–509. [CrossRef]
5. Flax, V.L.; Phuka, J.; Cheung, Y.B.; Ashorn, U.; Maleta, K.; Ashorn, P. Feeding patterns and behaviors during home supplementation of underweight Malawian children with lipid-based nutrient supplements or corn-soy blend. *Appetite* **2010**, *54*, 504–511. [CrossRef] [PubMed]
6. Mangani, C.; Maleta, K.; Phuka, J.; Cheung, Y.B.; Thakwalakwa, C.; Dewey, K.; Ashorn, P. Effect of complementary feeding with lipid-based nutrient supplements and corn–soy blend on the incidence of stunting and linear growth among 6- to 18-month-old infants and children in rural Malawi. *Matern. Child Nutr.* **2015**, *11*, 132–143. [CrossRef]
7. De Groote, H.; Kimenju, S.C. Consumer preferences for maize products in urban Kenya. *Food Nutr. Bull* **2012**, *33*, 99–110. [CrossRef]
8. Khumalo, T.P.; Schönfeldt, H.C.; Vermeulen, H. Consumer acceptability and perceptions of maize meal in Giyani, South Africa. *Dev. S. Afr.* **2011**, *28*, 271–281. [CrossRef]
9. Pillay, K.; Derera, J.; Siwela, M.; Veldman, F.J. Consumer acceptance of yellow, provitamin A-biofortified maize in KwaZulu-Natal. *S. Afr. J. Clin. Nutr.* **2011**, *24*, 186–191. [CrossRef]
10. Stevens, R.; Winter-Nelson, A. Consumer acceptance of provitamin A-biofortified maize in Maputo, Mozambique. *Food Policy* **2008**, *33*, 341–351. [CrossRef]
11. Chedea, V.S.; VICAS, S.; Socaciu, C. Kinetics of soybean lipoxygenases are related to pH, substrate availability and extraction procedures. *J. Food Biochem.* **2008**, *32*, 153–172. [CrossRef]
12. Zhang, Y.; Guo, S.; Liu, Z.; Chang, S.K. Off-flavor related volatiles in soymilk as affected by soybean variety, grinding, and heat-processing methods. *J. Agric. Food Chem.* **2012**, *60*, 7457–7462. [CrossRef]
13. Lv, Y.C.; Song, H.L.; Li, X.; Wu, L.; Guo, S.T. Influence of Blanching and Grinding Process with Hot Water on Beany and Non-Beany Flavor in Soymilk. *J. Food Sci.* **2011**, *76*, S20–S25. [CrossRef]

14. Sun, C.X.; Cadwallader, K.R.; Kim, H. Comparison of key aroma components between soymilks prepared by cold and hot grinding methods. In *Chemistry, Texture and Flavor of Soy*; Cadwallader, K., Chang, S.K.C., Eds.; ACS Symposium Series 1059; American Chemical Society: Washington, DC, USA, 2010; pp. 361–373.
15. Nishinari, K.; Fang, Y.; Guo, S.; Phillips, G.O. Soy proteins: A review on composition, aggregation and emulsification. *Food Hydrocoll.* **2014**, *39*, 301–318. [CrossRef]
16. Nuss, E.T.; Tanumihardjo, S.A. Maize: A paramount staple crop in the context of global nutrition. *Compr. Rev. Food Sci. Food Saf.* **2010**, *9*, 417–436. [CrossRef]
17. Bowers, J.A.; Saadat, M.A.; Whitten, C. Effect of liking, information and consumer characteristics on purchase intention and willingness to pay more for a fat spread with a proven health benefit. *Food Qual. Prefer.* **2003**, *14*, 65–74. [CrossRef]
18. Cho, H.Y.; Chung, S.J.; Kim, H.S.; Kim, K.O. Effect of sensory characteristics and non-sensory factors on consumer liking of various canned tea products. *J. Food Sci.* **2005**, *70*, s532–s538. [CrossRef]
19. Steenhuis, I.H.; Waterlander, W.E.; De Mul, A. Consumer food choices: The role of price and pricing strategies. *Public Health Nutr.* **2011**, *14*, 2220–2226. [CrossRef]
20. Gama, A.; Adhikari, K.; Hoisington, D. Peanut Consumption in Malawi: An Opportunity for Innovation. *Foods* **2018**, *7*, 112. [CrossRef]
21. Ranum, P.; Peña-Rosas, J.P.; Garcia-Casal, M.N. Global maize production, utilization, and consumption. *Ann. N. Y. Acad. Sci.* **2014**, *1312*, 105–112. [CrossRef]
22. Bhima, K.; Mtimuni, B.; Matumba, L. Tackling protein–energy under-nutrition among resource-limited people living with HIV/AIDS in Malawi using soybean-enriched maize-based stiff porridge (nsima): A pilot study. *Nutr. Diet.* **2019**, *76*, 257–262. [CrossRef]
23. Bressani, R.; Elías, L.G.; Braham, J.E. Improvement of the protein quality of corn with soybean protein. *Adv. Exp. Med. Biol.* **1978**, *105*, 29–65.
24. Nyotu, H.G.; Alli, I.; Paquette, G. Soy supplementation of a maize based Kenyan food (ugali). *J. Food Sci.* **1986**, *51*, 1204–1207. [CrossRef]
25. Agbon, C.A.; Ngozi, E.O.; Onabanjo, O.O. Production and nutrient composition of fufu made from a mixture of cassava and cowpea flours. *J. Culin. Sci. Technol.* **2010**, *8*, 147–157. [CrossRef]
26. Olapade, A.A.; Babalola, Y.O.; Aworh, O.C. Quality attributes of fufu (fermented cassava) flour supplemented with bambara flour. *Int. Food Res. J.* **2014**, *21*, 2025.

© 2019 by the authors. Licensee MDPI, Basel, Switzerland. This article is an open access article distributed under the terms and conditions of the Creative Commons Attribution (CC BY) license (http://creativecommons.org/licenses/by/4.0/).

MDPI
St. Alban-Anlage 66
4052 Basel
Switzerland
Tel. +41 61 683 77 34
Fax +41 61 302 89 18
www.mdpi.com

Foods Editorial Office
E-mail: foods@mdpi.com
www.mdpi.com/journal/foods

www.ingramcontent.com/pod-product-compliance
Lightning Source LLC
LaVergne TN
LVHW070405100526
838202LV00014B/1391